Energy — Present and Future Options

Volume 2

Edited by

David Merrick
National Coal Board
Cheltenham

A Wiley—Interscience Publication

JOHN WILEY & SONS
Chichester · New York · Brisbane · Toronto · Singapore

Copyright © 1984 by John Wiley & Sons Ltd.

Library of Congress Cataloging in Publication Data
(Revised for volume 2)
Main entry under title:

Energy — present and future options.
 'A Wiley–Interscience publication.'
 Includes bibliographical references and indexes.
 1. Power resources — Addresses, essays, lectures.
I. Merrick, David. II. Marshall, Richard, Ph.D.
TJ163.24.E55   1981   333.79   80–41416

ISBN 0 471 27922 6 (v.1)
ISBN 0 471 90416 3 (v.2)

British Library Cataloguing in Publication Data:

Energy — present and future options.
 Vol. 2
 1. Power resources
 I. Merrick, David
 333.79'12   TJ163.2

ISBN 0 471 90416 3

Typeset by Mathematical Composition Setters Ltd,
7 Ivy St, Salisbury, Wilts
Printed by Page Bros (Norwich) Ltd., Norwich.

List of Contributors

T. D. BEYNON — Department of Physics, University of Birmingham, PO Box 363, Birmingham B15 2TT, UK

P. N. COOPER — Department of Physics, University of Aston in Birmingham, Gosta Green, Birmingham B4 7ET, UK

C. A. MCAULIFFE — Department of Chemistry, University of Manchester Institute of Science and Technology, PO Box 88, Manchester M60 1QD, UK

J. T. MCMULLAN — Energy Study Group, School of Physical Sciences, The New University of Ulster, Coleraine BT52 1SA, Northern Ireland

A. F. POSTLETHWAITE — Central Electricity Generating Board, Generation Development and Construction Division, Barnett Way, Barnwood, Gloucester GL4 7RS, UK

R. P. SHAH — General Electric Company, Research & Development Center, PO Box 8, Schenectady, NY 12301, USA

K. R. SHAW — Arthur Andersen and Co., 1 Surrey Street, London WC2.

F. C. TREBLE — 43 Pierresfondes Avenue, Farnborough, Hants GU14 8PA, UK

B. WESTON — National Coal Board, Coal House, Lyon Road, Harrow, Middlesex, UK

Contents

Preface

Since the first volume of *Energy: Present and Future Options* was published, several changes to the energy scene have taken place. Possibly the most striking is how the combined effects of the world economic recession and conservation measures have caused world oil demand to moderate, so that the 'energy crisis' that appeared to be facing the world has receded.

Although energy shortages are no longer regarded as an immediate threat to the economies of the industrialized countries, energy continues to command public attention. In particular, there is a continuing concern about the environmental impact of energy production and use, and an awareness that high energy prices have been a contributory factor to the recession.

The purpose of *Energy: Present and Future Options* is to provide technical background to the main options, together with an assessment of their potential contribution, feasibility, and economics. The first volume covered conventional hydro power, wave energy, tidal power, sea thermal power, microbial fuel production, petroleum source bed formation, coal conversion, energy conservation in buildings, the economics of alternative energy sources, and energy analysis. This second volume is intended to be complementary to the first, and covers the following topics: nuclear fission, nuclear fusion, photovoltaic solar energy conversion, the heat pump, hydrogen and energy, advanced fossil fuel power generation systems, combined heat and power, and energy systems.

I would like to express my thanks to the authors who have contributed to the present volume, and to friends and colleagues who have provided valuable assistance.

19 July 1983   D. MERRICK

ix

Energy—Present and Future Options, Volume 2
Edited by D. Merrick
© 1984 John Wiley & Sons Ltd

P. N. COOPER
Department of Physics
University of Aston in
Birmingham, UK

1

Fission

1.1 THE NUCLEUS

Nuclei are assemblies of positively charged protons and zero charged neutrons. By convention the number of protons in an atom is indicated by a subscript to the symbol for the element and the total number of neutrons and protons (i.e. the mass number) by a superscript. For example, uranium with 92 protons and 143 neutrons is written as $^{235}_{92}U$. The protons and neutrons each weigh about 1.67×10^{-27} kg and contain nearly all the mass of the atom. A cloud of electrons, equal in number to the number of protons, surrounds the nucleus and the overall diameter of the atom is about 3×10^{-10} m, whereas the diameter of the nucleus is of the order of 10^{-14} m.

Since the electron structure of an atom determines its chemical properties, all atoms containing the same number of protons are called isotopes of an element irrespective of the number of neutrons. Different isotopes are chemically identical but have widely varying characteristics. If the neutron to proton ratio of an isotope is not close to the optimum value for the element then it may be radioactive. Also the nuclear reactions obtained when the nucleus is struck by a fundamental particle (such as a neutron) depend on the composition of the nucleus.

Neutrons are useful particles for inducing nuclear reactions since they have no charge and hence are not repelled from the nucleus. For low or moderate energy neutrons the main reactions are elastic scattering in which the neutron rebounds from the struck nucleus with a loss in energy, and neutron capture, which produces a new isotope of the same element, often radioactive, together with gamma-radiation which removes excess energy resulting from the reaction. Only one of the naturally occurring isotopes, $^{235}_{92}U$, can undergo a third type of reaction called fission when struck by low or moderate energy neutrons. When $^{235}_{92}U$ captures a neutron it can split into two roughly equal halves releasing two or three neutrons and a large amount of energy. This energy arises from the conversion of mass into energy since about 0.1% of the mass is lost in the fission process. A single fission releases only about 3×10^{-11} J (190 MeV) but if all of the ^{235}U atoms in 1 kg are fissioned the total release is about 8.6×10^{13} J. To obtain the same energy from combustion of oil about 2200 tonnes would have to be consumed, or over two million times the weight of ^{235}U consumed. In addition this would produce about 6700 tonnes of CO_2 and 3000 tonnes of water as a result of the combustion, compared with 1 kg of waste products from the fission process. These fission products are, however, highly radioactive and therefore constitute a serious waste disposal problem. Since the mass numbers of the fission products range from about 80 to 160, due to the different ways in which fission can occur, the number of radioactive isotopes produced is large. Figure 1.1 shows the yield of fission products as a function of mass number. Each primary fission product will undergo three or four stages of beta decay before becoming stable and will also emit gamma-radiation in the decays. The half lives of these fission products range from fractions of a second to thousands of years but the majority of the energy released in radioactive decay is emitted within a few days of fission. Of the energy released in the fission process, about 80% is given to the fission

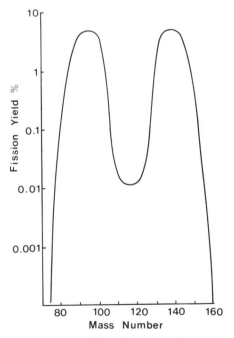

Fig. 1.1. Yield of fission products

fragments as kinetic energy. Since these fission fragments travel only about 10 μm before coming to rest, this kinetic energy is rapidly converted to heat in the uranium. About 3% of the energy is given to the neutrons emitted in fission and the remainder appears later as the energy emitted during radioactive decay.

Certain heavy atoms, notably $^{238}_{92}$U and $^{232}_{90}$Th can be made to fission with fast neutrons of velocity greater than about 1.6×10^7 m/s or energy about 2.2×10^{-13} J (1.4 MeV), but the competing neutron capture reaction is more probable. For ^{235}U typically about 20% of the incident neutrons will lead to neutron capture instead of fission so that about 1.3 kg of ^{235}U are consumed for each 1.0 kg fissioned, the alternative neutron capture reaction leading to production of ^{236}U.

1.2 CONTROLLED FISSION

Naturally occurring uranium consists of two main isotopes, ^{235}U present as 1 part in 139 and ^{238}U, present as 138 parts in 139. Only ^{235}U can be made to fission by neutrons of any energy whereas ^{238}U needs neutrons of at least 1.4 MeV to cause fission. Neutrons emitted in fission have a range of energies as shown in Fig. 1.2 with an average of 2 MeV and a most probable energy of 0.72 MeV. Since the main reaction for ^{238}U is neutron capture and the proportion of ^{235}U is small it is impossible to sustain a chain reaction in pure natural uranium. If the proportion of ^{235}U can be increased sufficiently then a chain reaction in uranium alone becomes possible, but enrichment is an expensive and energy consuming process and only low degrees of enrichment to three or four times the natural content are generally economic.

In order to use natural or low enrichment uranium the neutron energy must

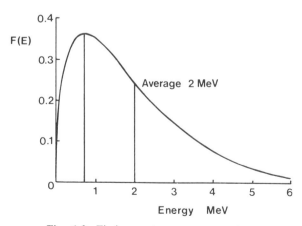

Fig. 1.2. Fission neutron energy spectrum

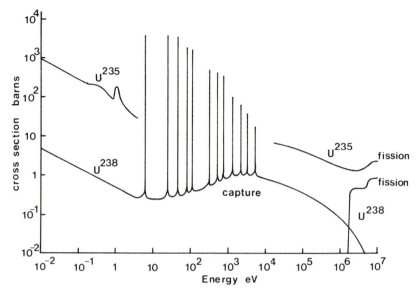

Fig. 1.3. Uranium cross-sections and neutron energy

be reduced, because the probability of fission of a ^{235}U nucleus increases about
400 times as the neutron energy decreases from 2 MeV (velocity 2×10^7 m/s)
to 0.025 eV (velocity 2200 m/s), where 2200 m/s is the most probable velocity
of a neutron at equilibrium with a scattering material at $20\,^\circ$C(293 K). Over
the same range the neutron capture probability for ^{238}U only increases about
ten times. Instead of probabilities it is more usual to speak of the cross-section
of a reaction, where cross-section means the effective target size of a nucleus
in barns (1 barn $= 10^{-28}$ m^2). The cross-section for a reaction is not related
to the geometrical size of the nucleus. Figure 1.3 shows the variation of cross-
section for the uranium isotopes. Note the large peaks or 'resonances' over the
energy range 1000 eV to 5 eV for ^{238}U. These have important consequences for
reactor design. If neutrons can be reduced in energy to the order of 0.025 eV
then under suitable conditions a self-sustaining chain reaction becomes
possible.

1.2.1 Moderators

Neutrons can be slowed down by a series of elastic collisions with light nuclei.
The relative masses of a neutron and a nucleus of mass number A are about
$1:A$. Consider a neutron of energy E_0 colliding with a nucleus in a perfectly
elastic collision. If it is scattered through an angle θ then conservation of

Table 1.1 Slowing down in moderators and reactor materials

Moderator	Mass number	Collisions from fission to thermal
^{1}H	1	18
^{2}H	2	25
C	12	114
O	16	150
Na	23	215
U	238	2000

energy and momentum show that

$$\frac{E_\theta}{E_0} = \frac{A^2 + 1 - 2\sin^2\theta + 2\cos\theta\,\sqrt{A^2 - \sin^2\theta}}{(A + 1)^2}$$

Note that if $A = 1(^1\text{H})$, θ (maximum) $= 90\,^\circ$, and $E_{90} = 0$.

The maximum energy loss is $E_0\,4A/(A + 1)^2$ which shows that low values of A are most effective for slowing down neutrons. Table 1.1 gives the number of collisions needed to slow a neutron from 2 MeV to 0.025 eV, which is the energy corresponding to the most probable velocity at a temperature of 20 °C. Figure 1.4 shows the Maxwellian distribution of neutron velocities at 20 °C. Materials used for slowing down are called moderators. It can be seen from Table 1.1 that uranium is useless as a moderator and hydrogen the best, but it has significant cross-section for capture of slow or thermal neutrons. The only suitable form of hydrogen for a moderator is as water and for heavy hydrogen or deuterium (^{2}H) as heavy water. It can be seen that the oxygen does not play a great part in the slowing down process. Carbon may be used in its pure form as graphite, and sodium is included for completeness since it is used

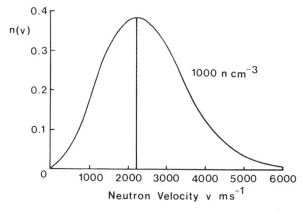

Fig. 1.4. Maxwellian thermal neutron spectrum (20 °C)

as coolant in fast reactors, in which the neutron energy is kept as high as possible.

Of the remaining light nuclei not so far considered, helium is gaseous and so has too low a density, lithium and boron are strong neutron absorbers and beryllium is scarce, difficult to work, and a health hazard.

It is possible to make a neutron chain reactor with natural uranium and graphite or heavy water as the moderator, but if the ordinary (light) water is the moderator it is necessary to use uranium enriched to three or four times the natural content in order to overcome the thermal neutron absorption in hydrogen.

1.3 THERMAL NEUTRON REACTORS

In order to design a thermal neutron reactor it is necessary to consider neutron production from fission, neutron losses during slowing down, competing reactions for the slowed down neutrons, and neutron leakage. The coolant medium to be used, the fuel canning, and heat transfer from can to coolant are also important.

Reaction rates depend on the number of nuclei (N per m^3), their cross-sections σ, and the rate at which they are being bombarded with neutrons. Since neutron directions are nearly random in a reactor it is usual to describe this by neutron flux (ϕ) which is a product of neutron density (n) and neutron velocity (v), where the neutron velocity distribution closely follows a Maxwellian distribution in a weakly absorbing moderator. Reaction rates can therefore be expressed in integral form

$$\text{Reaction rate per unit volume} = N \int_0^\infty \sigma(E)\phi(E)\,dE$$

$$= N\bar{\sigma}\bar{\phi} = N\bar{\sigma}n\bar{v}$$

The normal convention is to take $\bar{v}$ as 2200 m/s and this has the great advantage that for the many absorbers which show a $1/v$ variation of cross-section the average cross-section $\bar{\sigma}$ is given by σ at 2200 m/s in a Maxwellian flux. The cross-section at 2200 m/s is the normal tabulated value.

1.3.1 Production Rate of Fast Neutrons

The production rate of fast neutrons from fission induced by thermal neutrons is

$$\nu N \sigma_{\text{fission}} \bar{\phi} \ m^{-3}$$

where ν is the average number of fast neutrons released per fission, typically 2.5. From now on $N\sigma$ will be written as Σ so that

$$N\sigma_{fission} = \Sigma_f$$

The production rate is therefore

$$\nu \Sigma_f \bar{\phi} \ \mathrm{m}^{-3}$$

We have already noted that neutrons with energies above about 1.4 MeV can cause ^{238}U to fission and a small addition to the number of fast neutrons is caused by this effect. The enhancement of fast neutrons by fast fission of ^{238}U is called the fast fission factor (ϵ) which, in a graphite moderated reactor, typically has a value of 1.03. The total production rate of fast neutrons is therefore

$$\epsilon \nu \Sigma_f \bar{\phi} \ \mathrm{m}^{-3}$$

1.3.2 Losses during Slowing Down

The predominant uranium isotope is ^{238}U and this has many large peaks (or resonances) in the neutron absorption cross-section for the range 1000 eV to 5 eV (see Fig. 1.3). Neutrons which are slowing down therefore will have a finite chance of being captured by ^{238}U as they cross this energy range. The fraction of neutrons that are slowed down and escape capture by ^{238}U in the resonance region is therefore known as the resonance escape probability (p). This factor depends strongly on the geometrical arrangement of fuel and moderator and is lowest for a uniform mixture of fuel and moderator. By concentrating the fuel into rods distributed uniformly through the moderator (a heterogeneous reactor) many of the neutrons can slow down without re-entering a fuel rod and p is increased considerably. Heterogeneous cores are also better suited to fuel production and the use of canning to prevent corrosion of the fuel in the coolant and leakage of the highly radioactive fission products. They also make heat transfer easier to achieve and simplify fuel replacement. From the definition of p the rate at which neutrons reach thermal energies is

$$p \epsilon \nu \Sigma_f \bar{\phi} \ \mathrm{m}^{-3}$$

There are many possible fates for neutrons reaching thermal energies. They can be absorbed by the fissile material either to cause fission or neutron capture and they can be absorbed in the ^{238}U, canning, coolant moderator, and structural materials. This can be expressed as:

$$\text{Absorption rate} = (\Sigma_f + \Sigma_{fcap} + \Sigma_{238} + \Sigma_{can} + \Sigma_{mod} + \Sigma_{coolant})\bar{\phi}$$
$$= \Sigma\bar{\phi}$$

Of the slowed down neutrons, a fraction Σ_f/Σ will lead to fissions in the next generation. Hence the fission rate in the next generation will be $p\epsilon_n(\Sigma_f\bar{\phi})\,\Sigma_f/\Sigma$ fissions m^{-3}. The neutron multiplication can therefore be defined as the ratio of fission rates in successive generations. Since leakage has so far implicitly been neglected this defines the infinite multiplication factor (k_∞).

$$k_\infty = p\epsilon\nu(\Sigma_f\bar{\phi})\,\Sigma_f/\Sigma\;(1/\Sigma_f\bar{\phi}) = p\epsilon\nu\Sigma_f/\Sigma$$

Similar arguments apply to fast reactors, but these are designed with as little low atomic weight material as possible to minimize slowing down and so keep the average neutron energy as high as possible. The neutron energy distribution is a somewhat degraded form of Fig. 1.2 rather than Maxwellian and the average neutron energy is in the region of 0.1 MeV to 1 MeV. Hence resonance capture is much lower than in a thermal reactor.

1.3.3 Neutron Leakage

This depends on the reactor shape and size, the slowing down distance and the thermal neutron scattering and absorption mean free paths. A spherical reactor has minimum leakage but is not a practical shape. All power reactors are cylindrical in shape with the fuel rods inserted axially. Provided that the reactor has uniform composition, the neutron flux distributions radially and axially are as shown in Fig. 1.5. The flux at any point (r, z) is given by $\phi(r, z) = \phi(r) \times \phi(z)$ (where $\phi(r)$ is a J_0 Bessel function and $\phi(z)$ a cosine distribution) and to a first approximation the fraction of neutrons leaking from the core is proportional to the reciprocal of the square of the dimensions.

One major effect of neutron leakage is that since the power production rate

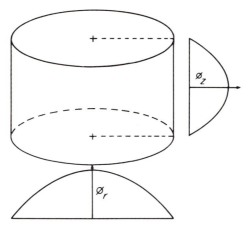

Fig. 1.5. Flux distributions in bare cylindrical reactor

at any point is $\Sigma_f \phi$ this is a maximum at the centre falling to zero at the outer boundaries. The average power production rate is only 0.275 of the maximum value at the core centre. For maximum utilization of the fuel, ways must therefore be found of obtaining a much more uniform power distribution.

1.3.4 Neutron Reflectors

If the reactor is surrounded with a layer of pure moderator this will scatter back into the core some of the escaping thermal neutrons and will also slow down and return some of the escaping faster neutrons. The effect of this is to raise the neutron flux at the edges of the core above zero and so to increase the average power input. Figure 1.6(a) shows the effect on the radial flux distribution of adding a reflector. The axial flux distribution is generally similar in shape to the radial distribution.

About 1 m of graphite or about 20 cm of water are sufficient as a reflector. Any increase beyond this in thickness has negligible additional effect.

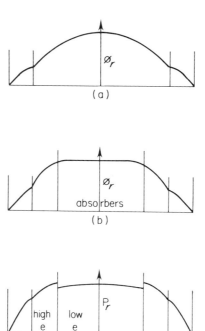

Fig. 1.6. (a) Power distribution in reflected core (b) Power distribution in reflected and flattened core (c) Power distribution of two enrichment zones

1.3.5 Flux Flattening

In natural uranium fuelled reactors there is generally excess multiplication so that neutron absorbers can be added to the central region in order to give a more uniform or flattened flux. Boron is a suitable neutron absorber for this purpose, generally in the form of boron steel rods in the moderator running parallel to the fuel rods. Figure 1.6(b) shows the effect of flux flattening on the radial flux, but the axial flux is not altered in shape.

1.3.6 Varying Enrichment

In reactors using enriched fuel, such as water moderated reactors, a more uniform power distribution can be obtained by having the central radial zone of the reactor at a lower enrichment than the outer zone. Figure 1.6(c) shows the effect on radial power distribution of two enrichment zones. In the previous cases of reflector and flux flattening, radial power follows the same profile as radial flux since the enrichment is uniform. Some reactors use up to four radial enrichment zones with enrichment increasing from the centre to the outside in order to obtain a uniform radial power distribution.

1.3.7 Excess Multiplication and Reactor Control

Any power reactor must have excess multiplication. If such a reactor were designed to have an effective multiplication of unity (after allowing for neutron losses by leakage), it would soon stop working since burn-up of fissile material and growth of fission products, which are neutron absorbers, would upset the balance of neutron production, absorption, and leakage. In fact it would prove impossible to start a reactor with an effective multiplication of exactly one since the power level could not increase. In addition temperature effects tend to reduce multiplication as the bulk temperature of the reactor rises. Sufficient excess multiplication must be designed into the reactor to allow for all these effects and in a graphite moderated natural uranium fuelled reactor the maximum excess effective multiplication obtainable is about 1.04 to 1.05. Some method of changing the multiplication while the reactor is in operation must therefore be provided. This is achieved by means of control rods containing a neutron absorber such as boron which can be inserted into the core sufficiently to absorb the excess multiplication. Additionally a separate set of absorbers which are normally held out of the core is also provided. In an emergency these safety rods can be rapidly inserted to shut down the reactor by reducing the effective multiplication well below unity.

The ability to control a reactor with the control rods depends on the rate of rise of power caused by a change in multiplication. This in turn depends on the neutron lifetime or the time between successive generations. Lifetime is the sum of the delay between neutron absorption in the fissile material and fast neutron emission, the time taken to slow down and the time spent as a thermal neutron before absorption or leakage. The sum of the two last effects, known as the prompt lifetime, is only about 10^{-3} s in a graphite moderated reactor and about 10^{-4} s in a water moderated reactor. Most of the neutrons emitted in fission are instantaneous or prompt but in ^{235}U about 0.65% of the neutrons are considerably delayed. These neutrons do not arise directly in the fission process but are emitted by certain fission products formed by the decay of primary fission products. The daughter nucleus formed by a beta decay can sometimes instantaneously emit a neutron which therefore appears to be emitted with the half-life of the preceding decay. About fifty fission products are known to give rise to such a decay but an adequate mathematical description for modelling purposes is given by assuming that there are six groups of delayed neutrons, each group having a fraction β_i and half-life τ_i. The energy of delayed neutrons is about one-tenth of the mean energy of fission neutrons but this does not have any significant effect except in small reactors. Table 1.2 summarizes the main characteristics of the six delayed groups for ^{235}U.

For slow changes in power the six groups can be regarded as one effective group of fraction 0.0065 and half-life 9 s mean life 12.9 s. (Half-life is the time taken for the number of radioactive nuclei to decay to half of its initial value and mean life is the time taken for the number of radioactive nuclei to decay to $1/e$ of its initial value.) Therefore the effective lifetime is the sum of prompt lifetime and the mean life of delayed neutrons weighted by the delayed neutron fraction. This is 0.084 s in a graphite moderated reactor and 0.083 s in a water moderated reactor. Thus it can be seen that the delayed neutrons are responsible for the majority of the effective neutron lifetime. Provided that the excess multiplication is kept well below $1 + \beta$ (1.0065 for a ^{235}U fuelled reactor) power changes are slow. For example, an excess multiplication of 0.001 (effective multiplication 1.001) leads to a doubling of reactor power every 50 s. Therefore if multiplication changes are made slowly reactor control is easily carried out manually, although computer control is now common. This is

Table 1.2. Delayed neutron data for ^{235}U

Group i	Fraction β_i (%)	Half-life τ_i (s)
1	0.025	54.5
2	0.138	21.8
3	0.122	6.00
4	0.265	2.23
5	0.083	0.496
6	0.017	0.179
Total	0.650	

more necessary in large power reactors which are many times the minimum size for criticality and where different sectors of the reactor may be controlled individually to achieve uniform power distribution.

1.3.8 Breeding of Fissile Material

From the point of view of criticality the direct effect of neutron capture in ^{238}U is a lowering of neutron multiplication. Indirectly, however, it does lead to a new fissile material which is an alternative to ^{235}U by the following chain of events:

$$^{238}_{92}U + n \longrightarrow {}^{239}_{92}U \xrightarrow[23.5 \text{ min}]{\beta-} {}^{239}_{93}Np \xrightarrow[2.35 \text{ days}]{\beta-} {}^{239}_{94}Pu$$

The end product, ^{239}Pu can be made to fission by neutrons of any energy. It is radioactive with a half-life of 24 000 years, decaying by alpha particle emission to ^{235}U. In a reactor environment, its fission cross-section is about twice that of ^{235}U, but its neutron capture cross-section for the reaction leading to production of ^{240}Pu is high, almost half the fission cross-section. In addition, ^{240}Pu has a high neutron capture cross-section and can only be fissioned by neutrons above 1.4 MeV in energy. Apart from being significantly radioactive and a consequent health hazard if inhaled, since it then tends to be deposited in the skeleton, plutonium as a chemical is hightly poisonous, so great care must be taken in handling. It is not at present fed back into fuel for thermal reactors but is the best fissile material for fast reactors, which can be designed to produce more plutonium than is consumed. Conversion ratio is defined as the number of fissile atoms per fissile atom destroyed. It depends both on resonance capture and on thermal capture of neutrons in ^{238}U relative to the neutron absorption in fissile material. This second factor decreases with increasing degree of enrichment. Natural uranium reactors are therefore expected to have the highest conversion ratio. For example, in the natural uranium Magnox reactors the conversion ratio is about 0.8 whereas in the enriched AGR plants the conversion ratio is about 0.45. None of the reactors containing ^{238}U as fertile material has a conversion ratio approaching or exceeding unity, so excess production is not possible. Build-up of plutonium in the fuel does, however, partially compensate for burn-up of ^{235}U and so prolongs fuel life before loss of multiplication. One disadvantage resulting from the build-up of plutonium is its small delayed neutron fraction of only 0.21% which leads to a lowering of the effective neutron lifetime. For a reactor completely fuelled with plutonium the effective neutron lifetime would only be 0.026 s, less than one-third of the value for ^{235}U.

An alternative source of fissile material is the only naturally occurring isotope of thorium, $^{232}_{90}$Th. Workable thorium deposits exist so this could be a future source of fuel for reactors.

Table 1.3. Excess plutonium production

Reactor type	Excess Pu kg/GWa(e)	Excess Pu kg/GWa(th)
U_{nat}/graphite (Magnox)	617	154
UO_2/graphite (AGR)	173	69
UO_2/H_2O (PWR)	270	90
UO_2/D_2O (CANDU)	493	125

$$^{232}_{90}Th + n \longrightarrow {}^{233}_{90}Th \xrightarrow[22 \text{ min}]{\beta-} {}^{233}_{91}Pa \xrightarrow[27 \text{ days}]{\beta-} {}^{233}_{92}U$$

^{233}U has a neutron capture cross-section of only about 10% of the fission cross-section and offers the possibility of breeding excess ^{233}U in a thermal reactor fuelled with a $^{233}U/^{232}Th$ mixture. Large quantities of ^{233}U would have to be created by loading thorium into conventional uranium fuelled reactors before power reactors could be fuelled with ^{233}U. Rather like ^{239}Pu, ^{233}U also has a small delayed neutron fraction of 0.26% and so an effective neutron lifetime of 0.033 s.

Excess plutonium production by thermal reactors depends on the length of time the fuel remains in the reactor as well as on the enrichment, fuel design, and moderator used, but approximate figures can be given for various major reactor types. Table 1.3 gives the excess plutonium production per year per GW(e) at 100% load factor and it will be seen that the enriched reactors (AGR, PWR, BWR) produce less plutonium than those fed with natural uranium.

1.3.9 Fuel Lifetime and Burn-up

The main factors affecting fuel life are:

(1) Loss of criticality due to burn-up of fissile material.
(2) Gain in criticality due to plutonium production.
(3) Loss of criticality due to build-up of fission products.
(4) Loss of criticality due to build-up of higher plutonium isotopes.
(5) Radiation damage to fuel and can.

Burn-up is commonly expressed in units of megawatt days (heat) per tonne of fuel, which includes ^{238}U as well as ^{235}U. A burn-up of 1000 MWd/t corresponds to about 0.1% of the atoms fissioned, either directly as ^{235}U or indirectly via plutonium build-up from ^{238}U. This corresponds to about 0.14% of the atoms being consumed due to the competing reaction of non-fissile neutron capture.

In a natural uranium/graphite moderated system a single fuel charge would have a maximum burn-up of 2000 to 2500 MWd/t before loss of criticality.

If, however, the fuel is replaced on a continuous cycle involving refuelling a few channels each day, then burn-ups of around 5000 MWd/t may be achieved. This occurs because the rapid build-up of plutonium in young fuel compensates for the slower increase in fission products and higher plutonium isotopes in the older fuel and so enables a fuel element to stay longer in the reactor. At these higher burn-ups, however, radiation damage is beginning to become significant. Since the plutonium is not recycled in such a reactor, about 0.7% of the atoms will have been consumed at discharge.

For reactors using enriched fuel, higher burn-ups are possible. The oxide form of the fuel is more resistant to radiation damage and the degree of enrichment can be chosen to allow for loss of multiplication. Light water moderated reactors have an enrichment of about 3% and burn-ups of 30 000 MWd/t are possible, implying the consumption of about 4.2% of the atoms in the fuel. This figure is somewhat misleading in that the total uranium feed to the enrichment plant must be considered. Figure 1.7 shows the input and output flows for an enrichment plant. Balancing the input and outputs gives

$$n = fe + (1 - f)d$$

From this the fraction f of the original fuel in the enriched stream is

$$f = (n - d)/(e - d)$$

from which it will be seen that the enrichment d in the depleted stream should be as low as possible. A low value of d would mean a large and consequently energy expensive enrichment plant and so d is usually limited to about 0.25% compared with n which is 0.7%. For an enrichment, e, of 3%, 83.6% of the original uranium appears in the depleted stream and only 16.4% in the enriched stream, which contains 70% of the original ^{235}U. Therefore a burn-up of 30 000 MWd/t, although equivalent to 4.2% of the atoms consumed in the fuel fed to the reactor, represents a consumption of only 0.7% of the original uranium atoms fed to the enrichment plant. Enrichment does not, therefore,

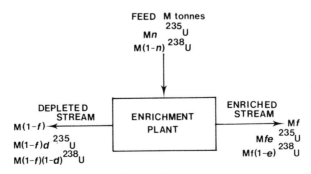

Fig. 1.7. Material flows for an enrichment plant

give any improvement in burn-up over the natural uranium systems if the fuel is used on a once-through cycle only.

A similar calculation for the Advanced Gas-Cooled Reactor (AGR) with an average feed enrichment of 2.4% and a burn-up of 18 000 MWd/t shows that about 0.6% of the initial uranium is consumed on a once-through cycle.

Since in all existing thermal reactor types, the numbers of plutonium atoms created is less than the number of fissile atoms destroyed, even recycling the plutonium would not give a great increase in fuel utilization. For a more detailed exposition of the reactor physics see references 1, 2, and 3.

1.4 THERMAL REACTOR SYSTEMS

Nearly all of the electricity generated from nuclear fission is at present from thermal reactors. A world-wide survey of reactor specifications can be found in reference 4. Initially reactors were made for producing plutonium for weapons purposes. Since enrichment processes were still being developed, the only available fuel was natural uranium. Heavy water was also not available in sufficient quantities and graphite had to be used as the moderator. The first reactor, CP1 (zero energy) went critical at Chicago University in December 1942 and this was followed by the plutonium producing reactors at Hanford in Washington State, which were water cooled. Later, in the UK, two natural uranium graphite moderated air-cooled reactors were constructed at Windscale. These used metallic fuel clad in aluminium.

1.4.1 Magnox Reactors

It was soon realized in the UK that the waste heat could be used to raise steam and generate power by using pressurized CO_2 as coolant and by changing the cladding to a magnesium alloy which has a higher melting point than aluminium. This led to the Magnox reactor design, the name coming, somewhat oddly, from the type of canning used for the fuel. Initially the four-reactor station at Calder Hall, with a total electrical output of 200 MW was built, the first reactor beginning power generation in 1956. This was followed by a similar station at Chapelcross. Both these stations are still operational.

Following a generally similar design, nine two-reactor power stations were constructed for the CEGB. The power outputs range from 250 MW(e) to 840 MW(e) per station, the total output being 3800 MW(e). This is less than the design figure of 5200 MW(e) since the operating temperature of the later stations has been lowered significantly to reduce corrosion problems arising from the hot CO_2. Although these reactors were constructed by five different organizations the core designs are all similar. Fuel rods of metallic uranium (density 18 700 kg/m^3) are used. Their diameter is about 29 mm and they are canned in Magnox about 2 mm thick with deep transverse or helical fins to aid

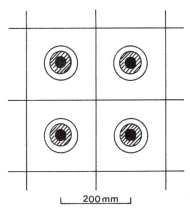

200mm

Fig. 1.8. Section of Magnox core

heat transfer. The rods are approximately 1 m long and are arranged in a regular square lattice of about 200 mm pitch in channels about 100 mm diameter through the graphite moderator. The carbon dioxide coolant flows upwards through these channels.

Figure 1.8 shows a typical section through a few fuel channels in a Magnox core. Since the core is about 8 m high there is about 100 kg of uranium in each channel and an average thermal power output of 250 kW per channel.

A more detailed example is provided by the reactors used at the Sizewell station. Each reactor has a thermal output of 810 MW and an electrical output of 250/210 MW (gross/net) with a thermal efficiency of 26%. The fuel elements are 28 mm diameter and 1.067 m long and are placed eight to a channel in a cylindrical core of diameter 13.7 m and height 8.1 m. This gives an average power density of 0.68 kW/l of core. The total uranium loaded into each reactor is 321 t, and occupies 3225 channels. About 0.8 m of graphite surrounds the core as a reflector and the weight of graphite per reactor is 2237 t. The coolant is CO_2 at 19.6 bar pressure. In order to contain the coolant pressure, the reactor is inside a mild steel spherical pressure vessel 19.4 m diameter and 105 mm thick. The CO_2 enters the bottom of the fuel channels at 210 °C and emerges at 360 °C. Four separate coolant circuits, each with a mass flow of about 1.3 t/s, are used to convey the hot CO_2 to steam generators and recirculate it to the core.

The steam generators are situated outside a concrete biological shield, as shown in Fig. 1.9. A dual pressure steam cycle is used and there is one turbine per reactor.

With the exception of the Calder Hall and Chapelcross reactors the Magnox reactors have been designed for on-load refuelling, which means changing two or three channels per day. The elements are removed and loaded singly from refuelling machines which couple on to standpipes on the reactor top serving

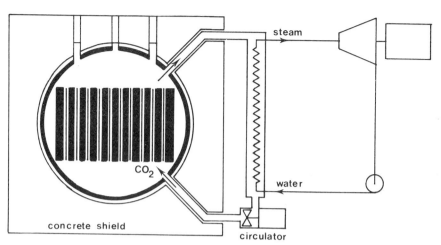

Fig. 1.9. Magnox reactor

several fuel channels, one standpipe per channel being impossible with a pitch of only 200 mm.

Apart from the corrosion problems, the ultimate operating temperature limit for this type of reactor is the use of metallic uranium. It is necessary to restrict the temperature at the centre of the most highly rated fuel element to less than 670 °C, the temperature of the $\alpha-\beta$ metallurgical phase change. This limits the surface temperature of the rod to about 440 °C and, with further temperature drops across the canning and from canning to gas coolant, the effective maximum gas outlet temperature is therefore about 410 °C. If a higher power generation efficiency is required a higher gas outlet teperature has to be used and significant changes in materials and core design must be made.

1.4.2 Advanced Gas-Cooled Reactor

These limitations to the Magnox system led to the design of the Advanced Gas-Cooled Reactor (AGR) in the UK. The most important change is the use of oxide fuel instead of metallic uranium. Since sintered UO_2 has an effective density of about 8700 kg/m³, the uranium density is about half that of metal. It also has a poorer thermal conductivity than the metal so smaller diameter fuel elements must be used, 14.5 mm being the chosen size for the standard fuel rod used in the AGR stations. Instead of a single fuel rod in each channel a cluster of 36 is used. A cross-section of the fuel cluster is shown in Figure 1.10. Increased fuel temperature also necessitates a change in the canning material and stainless steel 0.38 mm thick with shallow transverse ribs is used. Each fuel element is 1.039 m long and the fuel consists of a series of pellets 14.9 mm long inside each can, rather than a single long rod. The outer part

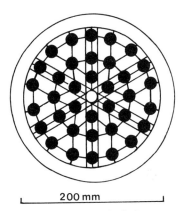

Fig. 1.10. AGR fuel element

of the fuel element is a graphite sleeve 190 mm internal diameter which is changed each time the fuel is changed. This is necessary because of corrosion and erosion at the higher gas coolant temperatures. Eight elements are linked together and are suspended from a gas duct and shielding plug making a fuel stringer of length about 23 m which is replaced as a single unit. In order to include sufficient graphite for moderation between fuel channels a pitch of

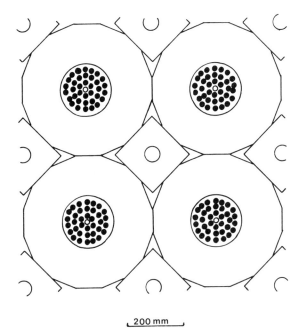

Fig. 1.11. Section of AGR core

460 mm is used. Figure 1.11 gives a section through part of the core and shows how the moderator is made up of 16-sided and square section graphite blocks.

Compared with the Magnox reactor, the large oxide fuel cluster in the AGR increases the resonance capture of neutrons, and the use of stainless steel canning increases the parasitic absorption of thermal neutrons. The penalty for these changes is the need for enriched uranium. Two enrichment zones are used, an inner zone of 2.1% and an outer zone of 2.6%. The total number of fuel channels is just over 300 and is therefore much smaller than in a Magnox core. The active core is 9.3 m diameter and 8.2 m high and can produce a total thermal power of 3000 MW. This is equivalent to an average power density of 2.7 kW/l, that is about four times the power density in a Magnox reactor. The total fuel per reactor core is about 120 t and the graphite requirement 1200 to 1600 t.

An interesting feature of the AGR (introduced in the last two Magnox stations) is the use of a prestressed concrete pressure vessel which also serves as the biological shield. This contains the core, steam generators and gas circulators and so has only steam, water, and refuelling penetrations, making a loss of coolant unlikely as there are no external gas ducts which could fracture. Figure 1.12 gives the general arrangement of an AGR. The coolant is CO_2 pressurized to about 42 bar with an outlet temperature of about 650 °C and inlet temperatures ranging from 286 °C to 321 °C in the different designs. Part of the cool gas is circulated through the moderator to keep the bulk graphite temperature down to about 350 °C and then mixes with the main inlet stream to the fuel channels. A single cycle steam system is used, providing steam at

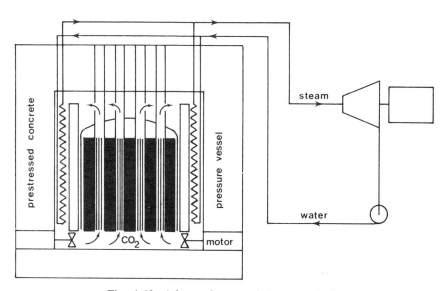

Fig. 1.12. Advanced gas-cooled reactor (AGR)

538 °C and 163 bar to a single 660 MW(e) turbogenerator. Five two-reactor stations are now complete or almost complete and work has begun on two more. One notable feature is in the siting of three of these seven stations close to large centres of population. This is regarded as acceptable because of the high degree of protection from loss of coolant accidents afforded by the design. Although the reactors have been designed for on-load refuelling, there are problems of vibration of fuel stringers when withdrawing or inserting into a channel with a large coolant mass flow (average 12 kg/s per channel). On average, one channel needs to be changed every five days so intermittent refuelling off-load would not seriously affect the running of an AGR.

Two layouts for the steam generators or boilers inside the pressure vessel have been tried, pod boilers which are inserted in vertical holes in the concrete pressure vessel, and annular boilers which surround the core, with an intermediate boiler radiation shield to limit radiation damage to the boilers. This second arrangement seems to be preferred in the latest stations[5].

A total of 8.4 GW(e) installed capacity will be available from AGR stations in the UK when the present construction programme is complete.

1.4.3 The Pressurized Water Reactor

The pressurized water reactor, or PWR, was originally developed in the USA as a small compact reactor for the propulsion of submarines. Nuclear powered submarines have the great advantage of being able to run at high speed for long distances underwater as they do not need any external oxygen supply. Water moderated reactors, which use enriched fuel, can be made compact and the use of water as moderator as well as coolant means that high power densities can be achieved compared with the bulky and heavy graphite moderated reactors. The reactor is pressurized sufficiently to suppress any boiling in the core and so avoid possible damage to the highly rated fuel by local overheating. It was soon realized that the small submarine reactors could be scaled up and used as the basis of land-based power stations. Assisted by the large home market in the USA the PWR has become the most widely used type of power reactor.

Single PWRs with an output of up to 1300 MW(e) exist but, by way of illustration, the standard French 900 MW(e) reactor of Westinghouse design will be described. Thirty-two of these reactors are completed, under construction, or planned and two 1300 MW(e) reactors are also under construction, making a total of 31.4 GW(e) capacity in France. By 1985 it is intended that 77% of the electricity generated in France will be from nuclear stations (including a small contribution from natural uranium graphite moderated reactors and fast breeder reactors).

The main features of the standard French reactor are as follows [6]. The total thermal power is 2785 MW and the electrical output is 925 MW net giving a thermal efficiency of 33%. Compared with gas-cooled reactors, the core is small having a height of 3.66 m and a diameter of 3.04 m. Four radial enrichment zones are used, 1.8%, 2.4%, 3.1%, and 3.25% and the core loading is 70.4 t. With such a small core the power density is high, at 100 kW/l average and 245 kW/l peak. The small size of core arises from the use of water as the moderator. Slowing down and thermal diffusion distances for neutrons are low and, to reduce parasitic neutron capture in hydrogen, the moderator to fuel volume ratio has to be low (about 1.5 to 1). To allow adequate heat removal, the fuel rods are small in diameter, comprising pellets of UO_2 8.2 mm in diameter and 13.5 mm long inserted in a zirconium alloy tube of wall thickness 0.57 mm. Each fuel rod is the full 3.66 m length of the core and 264 rods are arranged on a 12.5 mm square pitch in a square fuel element. Figure 1.13 shows a section through such an element. Some spaces are left for measuring devices or for control rods, which are made of silver, indium, and cadmium. A total of 157 elements make up a complete core.

The light water moderator also acts as the coolant and enters the core at 286 °C to leave at 326 °C. To prevent boiling the core is pressurized to 155 bar, the pressure being contained by a strong pressure vessel made of Mn–Mo–Ni steel. The pressure vessel is 13 m high, 4 m diameter, and 200 mm thick and the upper part may be removed for refuelling after reactor shut-down and depressurization. A shut-down time of 60 days is necessary every 12–15 months for refuelling, one-third of the fuel being replaced.

Three coolant loops are used, each having one steam generator 20 m high, one circulating pump, and a mass flow of 4.4 t/s. Water is fed to each steam

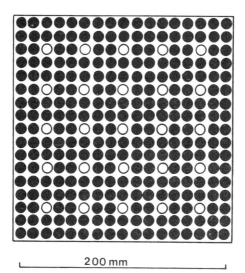

200 mm

Fig. 1.13. PWR fuel element

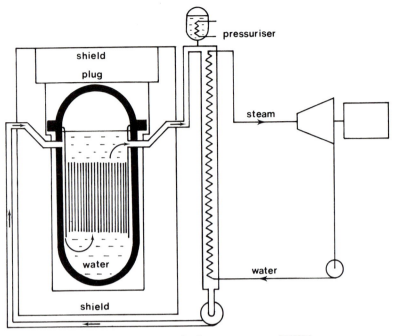

Fig. 1.14. Pressurized water reactor (PWR)

generator at 220 °C and 61 bar and steam produced at 270 °C 56 bar. A turbine consisting of one high pressure and three low pressure sections drives a single 975 MW alternator. For those stations sited near the sea, sea water is used to feed the condensers which are then fabricated of pure titanium tubes. Figure 1.14 shows the general arrangement of a pressurized water reactor system. For such a highly pressurized reactor containing over 100 t of hot water, which would flash off into steam if the pressure vessel were to fail, some form of secondary containment is needed. This takes the form of a prestressed concrete, steel lined shell, cylindrical in shape with a hemispherical top, 40 m in diameter and a total height of 57 m which can withstand an overpressure of 5 bar.

1.4.4 Boiling Water Reactors

A variant of the pressurized water reactor is the boiling water reactor (BWR). Instead of using an external heat exchanger the moderator is allowed to boil in the upper part of the core and so raise steam directly. This makes for a simpler system but the volume fraction of water in the core must be greater than the PWR to prevent steam bubbles blanketing the fuel rods and so the power density is lower. A result of these changes is a larger core for a given power, but since the operating pressure is lower the pressure vessel can be made thinner.

The Pilgrim 1 reactor at Plymouth Mass., USA is an example of a BWR power plant. This installation has been operational since 1972,[4] and produces 1998 MW of thermal power and 664 MW net of electrical power with a thermal efficiency of 33%. Its core is 3.64 m high and 4.14 m diameter and contains 112 t of fuel at an enrichment of 2.1%. The average power density is 40 kW/l. As with the PWR the fuel rods are made from pellets of UO_2, but are 12.7 mm in diameter because the power density is lower. The cladding is a zirconium alloy 0.81 mm thick. The maximum can temperature is 294 °C and the fuel centre temperature is 2300 °C. A single fuel element contains 49 rods on a square pitch of 18.7 mm and 508 elements make up a complete core. Control rods move in spaces between the fuel elements and are made of boron carbide. This material can be used instead of the expensive silver/indium/cadmium control rods used in the PWR since the neutrons are moderated more effectively by the larger water volume in the BWR and absorption of incompletely slowed down neutrons is therefore not so important.

Water enters the core at 276 °C and boils in the core at 288 °C, 72 bar, feeding steam to the turbine at 283 °C and 68 bar. A steel pressure vessel 19.7 m high and 5.7 m diameter with a wall thickness of 145 mm contains the

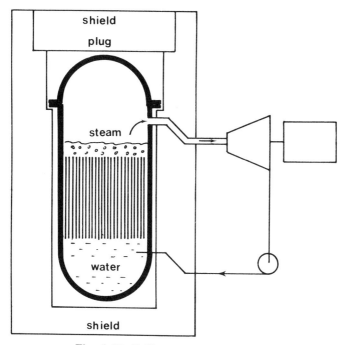

Fig. 1.15. Boiling water reactor (BWR)

core and, as in the PWR, the upper section is removable for off-load refuelling. One-third of the core is replaced every 12 months with a shut-down period of 30 days. The whole reactor (but not including the 690 MW turbogenerator) is enclosed in a steel lined prestressed concrete shell designed to withstand an internal pressure of 5 bar. Figure 1.15 shows the general arrangement of a BWR.

A BWR is much simpler than a PWR since there are no separate steam generators. However, the steam is radioactive with a 7 s half-life because of fast neutron activation of the oxygen in the water. Also, if a fuel can leaks, fission products will be carried to the turbine which will become contaminated, making maintenance more difficult.

1.4.5 CANDU Reactors

Canada has an extensive nuclear power programme with about 6 GW(e) already installed and 14 GW(e) planned for 1990. The reactor type used is different from any of the previous types in that it uses heavy water as the moderator and coolant and natural uranium dioxide as the fuel. This type of reactor, known as CANDU, is a pressurized heavy water reactor, but the coolant is separated from the moderator by pressure tubes. A large pressure vessel is not therefore needed and increasing the power output only needs an increase in the number of pressure tubes. Steam is raised from ordinary light water using conventional steam generators in the heavy water coolant circuit. Outputs per reactor range from 206 MW(e) to 740 MW(e) in existing designs and a larger reactor with 881 MW(e) output is proposed.

The Bruce reactors on the shores of Lake Huron may be taken as examples of the CANDU design.[7] Four reactors are working and another four are planned. Each reactor has an electrical output of 740 MW net and a thermal output of 2520 MW. The thermal efficiency is 29.4%.

Figure 1.16 shows the general arrangement of the CANDU system. The fuel elements are 493 mm long and contain 37 pins per element. Each pin consists of a zirconium alloy tube of 0.42 mm wall thickness containing natural UO_2 pellets 11.3 mm diameter and 15.3 mm long. Strings of twelve fuel elements are contained by horizontal pressure tubes made from a zirconium/niobium alloy having an internal diameter of 103.8 mm and a wall thickness of 4.2 mm. There are 480 such tubes arranged on a square pitch of 286 mm. The total core loading of uranium is 108 t and the core size is 7.06 diameter and 5.94 m length.

The moderator is contained in a stainless steel tank 5.95 m long and 8.52 m in diameter with horizontal zirconium alloy tubes. These tubes have the pressure tubes running through them so that the coolant is thermally isolated from the bulk of the moderator (338 t), which is only at 38 °C. Outside the core

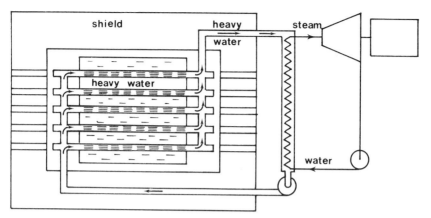

Fig. 1.16. CANDU reactor

vessel is a shielding tank of light water and this extends upwards to give better shielding around the vertical control rods.

The heavy water coolant is pressurized to 93.6 bar and enters the core at 249 °C, leaving at a temperature of 300 °C. The average power density is 10.9 kW/l and is therefore intermediate between the gas-cooled reactors and the light water moderated reactors. Steam is raised in two steam generators each fed by two pumps with a capacity of 3 t/s and is supplied to a single 791 MW electrical turbogenerator at 43 bar and 254 °C.

As with the gas-cooled reactors on-load refuelling is used, but with horizontal fuel channels fuel elements can be fed in at one end and removed at the other. About 105 t of uranium is replaced per year with a burn-up of 8167 MWd/t.

Containment is not by the cylindrical domed pressure building of the light water reactor in this instance, although that method has been used for some of the CANDU reactors. A rectangular vault made of concrete and steel is used instead and can withstand up to 7 bar pressure.

CANDU reactors require large quantities of heavy water, which has to be isotopically separated from light water, being present to only 0.015%. About 0.8 t of heavy water is needed for each MW (electric) taking into account the moderator, coolant, and reserves. About 11 000 t total of heavy water is needed for the Canadian Nuclear programme, and for this reason, large scale enrichment plants have been constructed in Canada[8] for heavy water production. At present, there is excess capacity and some of the plant has been put into reserve. It would therefore be possible for a significant number of heavy water reactors to be constructed in other parts of the world. One of the main advantages of the CANDU reactor is that it uses natural uranium at burn-ups in excess of 1% of the uranium feed, i.e. rather better than any other system on a once-through cycle.

1.4.6 Other Thermal Reactor Types

Many other reactor types have been studied but either have had only limited use or have been found to be unsatisfactory. One rather unusual system used in the USSR is the boiling channel reactor, a graphite moderated reactor with light water as the coolant. Reactors of 1000 MW electrical (3200 MW thermal) output are in existence,[9], producing steam at 280 °C and 65 bar. A 1000 MW(e) reactor has quite a large core, 7 m high and 11.8 m diameter. The fuel is uranium dioxide with a pellet diameter of 13.5 mm and an enrichment of 1.1% to 1.5%. It is clad in zirconium, 18 rods being used per fuel cluster, giving a loading of 192 t of uranium in 1693 channels. The pressure tubes containing the fuel are also made of zirconium. Single reactors of up to 2400 MW(e) are planned with a rectangular core 7.0 m × 7.5 m × 27 m. In these more advanced designs some channels will use stainless steel clad fuel to act as superheaters, giving steam at 450 °C.

A similar reactor design but using a heavy water moderator is the steam generating heavy water reactor (SGHWR). This was tested in prototype form in the UK where it has given satisfactory service for a number of years. The system could be regarded as a combination of CANDU and BWR, using heavy water as the moderator in an arrangement like CANDU, but passing light water through the pressure tubes where it is allowed to boil. All the advantages of pressure tube reactors exist, particularly the elimination of the need for a large pressure vessel and the ability to scale up in power output simply by increasing the number of coolant channels. The Winfrith SGHWR[10] has an electrical output of 94.5 MW net and a thermal power of 309 MW, giving a thermal efficiency of 35%. The average power density is 10.9 kW/l. A total of 104 channels are arranged on a square lattice pitch of 260 mm. In the heavy water moderator tank vertical zirconium tubes of 184 mm outside diameter and 3.5 mm wall thickness contain the pressure tubes. These are also made of zirconium and have an inside diameter of 130 mm and a wall thickness of 5.1 mm. The active core size is 3.12 m diameter and 3.66 m high and the moderator tank is 3.7 m diameter and 3.96 m high. Because of the neutron absorption properties of the light water in the coolant channels, it is necessary to use uranium enriched to 2.3%. Each fuel element comprises 36 fuel rods 3.66 m long made from 14.5 mm UO_2 pellets (similar in diameter to AGR fuel) clad in 0.66 mm zirconium alloy. The central position in each element is a zirconium tube pierced by holes that would be used for an emergency spray of cooling water in the case of loss of coolant. The maximum can temperature is 307 °C and the fuel centre temperature has a limit of 1905 °C. Light water coolant at a pressure of 67 bar enters the core at 275 °C and leaves at 282 °C producing steam to the 100 MW turbine at 280 °C and 63 bar. The method of control is interesting, using moderator height for normal control during running and varying the concentration of boric acid dissolved in the moderator to compensate for fuel burn-up. Emergency shut-down can be carried out by dumping the moderator and injecting boric acid into the light water coolant channels. Core design has been carefully carried out so that an increase in steam voidage slightly increases multiplication of neutrons. This enables some degree of automatic loading in the following way. An increase in steam demand at the turbine valve lowers the pressure in the coolant tubes. More water is consequently turned into steam and the increased steam voidage causes the reactor power to rise until pressure is restored. Decreases in steam demand are followed in an equivalent manner.

In the USA and Europe there has been interest in a high temperature gas-cooled reactor. The OECD reactor DRAGON, situated at Winfrith began operation in 1966 but is now shut down. This had an unusual fuel to enable temperatures much higher than those in the AGR to be used. A small sphere of UO_2 surrounded by layers of pyrolitic carbon, silica, and pyrolitic carbon with a total diameter of about 800 μm is the basic unit of fuel, the carbon and silica layers forming an impervious pressure vessel around the fissile kernel and therefore acting as a can. In the Dragon reactor the graphite core was only 1.6 m high and 1.07 m diameter and was made up of graphite blocks containing annular rings of coated particles bound in a graphite matrix. The coated particles had a maximum surface temperature of 950 °C and a maximum centre temperature of 1500 °C. Helium gas was used as the coolant with an outlet temperature of 750 °C and pressure of 20 bar, enabling 20 MW of heat to be removed from the core. No use was made of the heat, the reactor serving solely as a test facility for the design of large power reactors of this type.

The main problem with this type of reactor is that the coolant has too high a temperature for raising steam and too low a temperature for driving a gas turbine. It could, however, provide process heat and electricity. It has, for example, been suggested that the process heat is used to reduce iron ore to metallic iron and the electrical power is employed in the arc furnace that produces the steel.[11] A total of 60% of the heat would then be used. No coal or coke would be necessary but hydrogen, probably derived from methane, would be needed for the reduction process.

1.5 FAST BREEDER REACTORS

As has already been stated, it is not necessary to slow neutrons down in order to obtain a self-sustaining chain reaction provided that the fissile material content in the fuel is made sufficiently high. For ^{235}U as fissile materials about 20% enrichment would be needed even for large reactors. Therefore considerably more energy would need to be expended in isotope separation than for enriched thermal reactors and the advantages of employing fast fission in such a system would be uncertain. If, however, plutonium is used as the fissile material, the circumstances change markedly since it then proves feasible to design reactors that produce more fissile material than they consume. This may be seen by considering the average number of neutrons released per fission as a function of neutron energy. Table 1.4 gives a comparison between ^{239}Pu and ^{235}U and includes the ratio of non-fissile capture to total capture in these isotopes. It can be seen that ^{239}Pu is not significantly different from ^{235}U in thermal reactors since more of the neutron interactions lead to non-fissile capture and this compensates for the higher number of neutrons per fission. For fast neutrons (see Fig. 1.2), however, ^{239}Pu yields almost 0.5 neutron extra per fission than ^{235}U and has a lower fraction of interacting neutrons leading to non-fissile capture. Since also the fission cross-section for ^{239}Pu is generally about 50% higher than for ^{235}U, there will be more spare neutrons available from fast fission of ^{239}Pu after the demands of the chain reaction have been met. If these neutrons can be captured in ^{238}U then plutonium can be bred, just as in thermal reactors. By careful design more than enough plutonium can be produced to keep the reactor working on a feed of just ^{238}U. Separation of plutonium from uranium can be achieved by normal chemical methods. Some plutonium is already available from the spent fuel rejected from thermal reactors and this, like the plutonium produced by fast reactors, is not pure ^{239}Pu but also contains decreasing percentages of ^{240}Pu (half-life 6600 years), ^{241}Pu (14.3 years), and ^{242}Pu (3.87×10^5 years). Of these isotopes ^{240}Pu and ^{242}Pu are neutron absorbers below about 1.5 MeV and, like ^{238}U, can undergo fission only with neutrons above this energy. The other isotope, ^{241}Pu, can be made to fission by neutrons of all energies.

Ideally, therefore, it is possible to sustain a reactor indefinitely on a feed of ^{238}U even if it produces only just enough plutonium to feed back to itself. As a result almost complete burn-up of natural uranium could be achieved in time by continuous recycling of the plutonium back into the reactor. This cannot be achieved in practice partly because of build-up of the higher plutonium isotopes which reduces the multiplication and breeding and partly because of plutonium losses during reprocessing. As a result, it is estimated[12] that the

Table 1.4. Neutrons per fission and non-fissile capture in ^{239}Pu and ^{235}U

Neutron energy (MeV)	^{239}Pu neutrons per fission	Non-fissile capture (%)	^{235}U Neutrons per fission	Non-fissile capture (%)
10	3.90	0.1	3.90	0.5
6	3.52	0.1	3.33	1
3	3.23	0.5	2.80	2
1.35	3.06	2	2.57	4
0.5	2.97	6	2.49	14
0.2	2.92	11	2.45	18
thermal	2.90	38	2.42	15

overall utilization is limited to a maximum of about 70%. This is still about 100 times better than for thermal reactors. There is, therefore, a very strong incentive to use the fast breeder reactor in order to use the limited supply of uranium to the fullest extent.

The name 'fast breeder reactor' can be misunderstood since the system does not breed at a fast rate, only with fast neutrons. Production of plutonium is remarkably slow, most reactor designs requiring at least twenty-five years in order to produce enough excess plutonium to start up another reactor system. Whereas with thermal reactors fuel reprocessing becomes necessary only when the cans of the spent fuel elements become corroded in storage, with a fast breeder reactor it is essential to reprocess the fuel and breeder elements as quickly as possible in order to minimize the total amount of plutonium held up outside the reactor. With a typical fuel life of the order of one year and a one year hold-up in the reprocessing and fuel element manufacturing plant the total plutonium inventory would be twice the core loading. Figure 1.17 shows the fuel flows associated with a complete fast reactor system. Because of the short time allowable for reprocessing, the plant must handle much more highly active fuel than plants designed for thermal reactor fuel. A minimum hold-up of about six months is needed in order to allow the volatile and highly radioactive ^{131}I fission product to decay since this could cause severe handling hazards. Likewise, the activity of the discharged fission products and used fuel cans will be much higher than for thermal fuel which has been stored for a long time.

In order to limit the fuel inventory, fast reactors must also be highly rated. Fast reactors are in operation with average power densities up to 500 kW/l, that is five times that for a pressurized water reactor. The core of a fast reactor is therefore small and a substantial fraction of the neutrons can escape from its surface. In order to utilize these neutrons the core is completely surrounded by ^{238}U in the breeding blanket, where additional plutonium is created.

Plutonium is also bred from the uranium present in the core, but at a rate which is less than the consumption rate of plutonium. Core enrichment

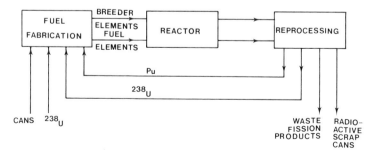

Fig. 1.17. Fuel flows in a fast reactor system

therefore gradually falls and the only way of compensating for this is to transfer plutonium from the blanket to the core via the fuel reprocessing and refabrication route.

Production of plutonium can be characterized by the breeding ratio (BR) defined, as for the conversion ratio in thermal reactors, by the number of fissile atoms created per fissile atom destroyed. Excess plutonium production can be characterized by (BR − 1) but the breeding gain (BG) is commonly used. This is the number of excess fissile atoms created per fission. The breeding gain is proportional but not equal to (BR − 1) and enables spare plutonium production to be calculated from reactor power, which is proportional to the fission rate. Another parameter which indicates the performance of a breeder reactor is the doubling time (T_D). This is the time taken to create enough spare fuel to start up another similar reactor and is dependent on breeding gain, total fissile fuel inventory (I), enrichment (e), and specific fuel rating R (e.g. kW/kg) in the following way:

$$T_D \alpha \ \frac{I \times e}{BG \times R}$$

Apart from low inventory and high breeding gain as already established, a low enrichment and high specific fuel rating are also desirable for a short doubling time. High breeding gain and high fuel rating may necessitate specific design features for the reactor system that do not correspond to the most economic generation of power so inevitably the design must be a compromise between these factors. A further complication is the small percentage of plutonium lost in reprocessing.[13] With a breeding gain of 1.1 only 10% of the reprocessed plutonium is spare, so a loss of 2% in reprocessing is actually 20% of the spare plutonium.

With the high density required in a breeder reactor, the method of cooling is an important design feature. So that the average neutron energy is kept as high as possible the coolant should not have good moderating properties but must have good heat transfer properties. Water obviously cannot be used since it is a good moderator and although gas cooling is feasible and has been proposed in some design studies, the only coolant currently in use is liquid sodium. With a melting point of 97.8 °C and a boiling point of 883 °C it is suitable for use unpressurized although to prevent oxidation it must be covered by an inert gas. Formation of sodium oxide can cause plugging of the narrow coolant channels with consequent overheating of the fuel so that on-line sodium purification is necessary, usually by diverting part of the flow to a freezing trap. Another potential problem area is leaking tubes or welds in steam generators which permits the reaction between sodium and water and the release of hydrogen. Careful quality control in the manufacture of steam generators is therefore essential and, to prevent any leaks affecting the reactor core, intermediate sodium/sodium heat exchangers are used so that only the secondary coolant circuits could become contaminated.

Oxide fuel clad in stainless steel is used in the reactor core and blanket so there are substantial amounts of light materials present, which tend to moderate the neutrons although not as effectively as conventional moderators (Table 1.1 indicates the effect of the oxygen and sodium). The neutron spectrum will, therefore, have a mean energy lower than that of fission neutrons. It is therefore important that the amount of sodium in the core is minimized since breeding is dependent on neutron energy. Fuel pins (i.e. rods) in fast reactors are small in diameter, typically about 6 mm, in order to facilitate heat transfer. The gaps between the pins must also be small to obtain a low sodium fraction. Blocking of a coolant channel must be avoided since it could lead to local burn-out of the highly rated fuel. In the blanket, some power is generated from fission of ^{238}U and from fission of the low percentage of bred plutonium. However, the power density is much lower than in the core and larger pins can be used in the radial section of the blanket, but not in the axial blanket where the breeder region is an extension of the fuel pins.

Large quantities of liquid sodium must be pumped around the reactor circuit. In early fast reactors electromagnetic pumping was tried but this was found to be inefficient and the speed of such a pump is inferior to that of mechanical pumps. Conventional submersible type pumps with bearings designed to work in liquid sodium are now used. Two types of sodium circuit layout have been tried. The pool type of reactor has the core, primary pumps, and intermediate heat exchangers all immersed in one large tank of sodium. This pool type of reactor has low leakage probability and provides a large heat sink in the sodium in case of total circulation failure. The alternative design is the loop type of reactor which has all the components separated and linked by sodium ducts. This second type gives easier access to the individual components and reduces the total amount of sodium needed but sodium leaks from the system are more likely.

At present there are only four fast breeder power reactors operational in the world, with electrical outputs from 150 MW to 600 MW. These are the PFR (250 MW) in the UK, Phénix (250 MW) in France, and BN–350 (150 MW + desalination) and BN–600 (600 MW) in the USSR. West Germany, the USA, and Japan also each have a reactor of this general size under construction or planned, and a 1200 MW electrical output fast reactor, Super Phénix, is under construction in France. In spite of the small number of reactors, fast breeders have been in existence for over thirty years now and there is considerable experience in the use of liquid metal coolants in high power density cores.

Reactors in the 300 MW electric range are regarded as prototype reactors to provide experience and component testing for full scale reactors of 1000 MW

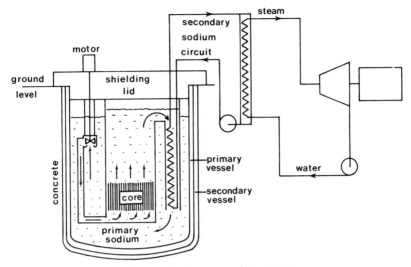

Fig. 1.18. Pool type fast reactor

electric or greater. An example is the Prototype Fast Reactor (PFR) at Dounreay in Scotland,[14] see Fig. 1.18. This reactor is typical of the pool type and has an electrical output of 250 MW and a thermal power of 600 MW. The core is 0.9 m high and 1.4 m in diameter, and has an average power density of 415 kW/l. Fuel pins are assembled from PuO_2/UO_2 pellets 5 mm in diameter inside stainless steel cans of wall thickness 0.38 mm and outside diameter 5.85 mm. These are arranged on a triangular pitch of 7.4 mm, separated by helical stainless steel wire wrappers. One hexagonal fuel element contains 325 pins and has an outer stainless steel jacket 140 mm across the flats. Also in the fuel pin are upper and lower axial breeder zones of depleted uranium in oxide form. The total loaded length of the fuel pin is 2.04 mm, of which 0.9 m is the active core length. The inner 30 of these elements are enriched to 22% plutonium and the outer 48 to 29%. Six live fuel elements are grouped around a dummy fuel element, termed a 'leaning post' which contains either a control rod or an irradiation experiment. The complete group, which occupies seven positions in the core, is removed and replaced as a single unit and the complete core and radial reflector is made up from these units.

In the core region, the volume occupied by the fuel is 35%, the sodium coolant 42%, and stainless steel (cans, etc.) 23%. Radial breeder elements surround the core and contain depleted UO_2 pins at a larger diameter than in the core. In these the sodium volume fraction is reduced to 19% since the heat generation is much lower than in the core. The Dounreay PFR is designed to demonstrate that breeding is possible but is not designed to produce excess plutonium. For this reason the radial blanket is three elements thick (0.42 m) around one-third of the core and only one element thick (0.14 m) around the remainder of the core (a total of 54 elements). Surrounding the breeder is a steel neutron reflector.

The whole reactor is immersed in a stainless steel tank surrounded by a steel leak jacket, the main vessel being 12 m diameter and 15 m high. As this is a pool type reactor, the three primary pumps and six intermediate heat exchangers are also in the tank, which contains 920 t of sodium. A shielding lid covers the tank and contains a rotateable plug for aligning the refuelling machine over individual leaning posts. The space above the sodium is filled with an inert gas to prevent oxidation of the coolant. Each of the three pumps has a flow rate of 1 t/s of sodium. This enters the bottom of the core at 400 °C and emerges at 585 °C to flow downwards through the six intermediate heat exchangers and so back to the pumps. There are three secondary sodium circuits, one to two intermediate heat exchangers, and three steam generators. Steam at 530 °C and 162 bar is provided to the single 260 MW turbogenerator. Fuel changes involve replacing one-sixth of the core every seven weeks at 100% load factor and spent fuel is stored in the core tank for some time until the decay heat has dropped sufficiently for transfer to an external storage system.

The core tank is sunk into a concrete vault so that the top of the reactor shield plug is at ground level. No special containment is used over the reactor but the reactor building is run at a slight under-pressure as a safety precaution in the event of a leak of radioactivity.

Designs have been produced for a commercial demonstration fast reactor of about 1200 MW electrical output but no decision has yet been taken on this proposal. This reactor would use similar fuel to PFR but at about 17% enrichment. Core height would be about the same but the core diameter would be over 3 m. Since this size will probably be typical of fast reactor power stations some idea of the technology involved can be gained from the French Super-Phénix reactor at present under construction at Creys-Malville.[15,16] This has a thermal power of 3000 MW and a net electrical power of 1200 MW and is rather more conservatively designed than Phénix or PFR. The fuel pins are made from 7 mm diameter hollow pellets of PuO_2/UO_2 clad in stainless steel 0.7 mm thick. There are only 271 pins in each fuel element. All the 193 inner fuel elements are enriched to 15% Pu and the outer 171 elements to 18%. making up an active core 1.0 m high by 3.66 m diameter. The core contains 32 t of Pu + U and has an average power density of 285 kW/l. Both upper and lower axial breeder regions of the fuel elements and pins are 0.3 m in length. Three rows of radial breeder elements surround the core, giving a radial blanket about 0.5 m thick and these use breeder pins 15.8 mm outside diameter. The total breeder uranium content is about 40 t. Around the breeder there are steel reflector elements and steel shielding elements to reduce radiation dose to the other components in the reactor pool. The main vessel is 21 m in diameter and 15.5 m high with an inner vessel 20 m in diameter and 10 m high and an outer safety vessel of 22.4 m diameter and 17.4 m high.

The total sodium in the primary circuit is 3200 t, the sodium being pumped through the core by four mechanical pumps immersed in the tank, each with a flow of 4.1 t/s. The core inlet temperature is 395 °C and mean outlet temperature 545 °C. Eight intermediate heat exchangers are also in the pool and there are four secondary coolant loops with a flow of 3.3 t/s of sodium each. The total amount of sodium in the secondary circuits is 1500 t. Steam is raised at 487 °C, 177 bar and feeds two 660 MW(e) turbogenerators. Fuel burn-up is estimated to be 70 000 MWd/t or about 7% of the fuel in the core half of which will be replaced every 13 months at a load factor of 75%. The whole reactor is enclosed in a prestressed concrete pressure shell similar to the containment of a light water reactor.

The total fuel inventory for this system would be about 10 t of plutonium. Even with an estimated breeding gain of over 0.2, the fuel doubling time at 75% load factor would be of the order of twenty-five years. To start a fast reactor programme, therefore, the plutonium from thermal reactors is essential. Taking, for example, the French system with about 31 GW of PWR producing about 200 kg of plutonium per year per GW at 75% load factor a total of 6 t of plutonium per year is made available for fast reactors. Over a reactor life of thirty years there will be 180 t of plutonium available, sufficient to start up 18 fast breeder reactors similar to Super Phénix, or about 22 GW(e) of installed capacity.

Since some fast breeder reactors will be installed during the thirty-year thermal reactor lifetime the installed capacity of breeders can be greater than indicated above. If a new fast reactor is commissioned as soon as the fuel is available from thermal and fast reactor sources then, assuming as above 6 t of plutonium per year from thermal sources and a fuel doubling time for the fast reactors of twenty-five years, after thirty years there could be 28 fast reactors or 34 GW(e). This would allow only for a moderate growth rate in the demand for nuclear electricity but would replace the thermal reactors with fast reactors that have a much higher utilization of fuel. The fuel requirements for the twenty-eight fast reactors would be a feed of natural or depleted uranium of about 32 t/a and there would be about 3500 t of uranium permanently committed in the reactor and reprocessing systems.

The implications for the UK are even less optimistic. Since the Magnox stations are nearing the end of their life the major source of plutonium will be the 8.4 GW(e) of AGR stations. These only produce 60% of the amount of plutonium compared with equivalent PWR stations. There is a stock of plutonium created by the Magnox reactors but unless additional plutonium can be imported or the rapid introduction of thermal reactors such as PWR or CANDU is initiated, then there will be insufficient plutonium available for a fast reactor based generation system to have a significant impact within thirty to forty years. The only alternative is to design fast reactors with much more highly rated fuel than that achieved by Super Phénix in order to reduce the fuel inventory.

1.6 REACTOR SAFETY

There is a common but mistaken idea that nuclear reactors could run out of control and explode like a nuclear weapon. In designing a nuclear weapon the main problem is holding the component parts together for a sufficiently long time to enable the majority of the fuel to undergo fission. A nuclear reactor core would be disrupted into a non-critical geometry long before this condition was reached. Great care is taken in the design of control rod mechanisms and in instrumentation to ensure that rapid rises in power do not occur. The rate of withdrawal of control rods, taken with the multiplication controlled by each rod or group of rods is deliberately limited so that the rate of increase in multiplication is slow. Also instrumentation monitoring the neutron flux (which is proportional to power level) measures the power level and its rate of rise. Shut-down is initiated when either of these quantities exceeds a preset level. To avoid unnecessary shut-downs and yet allow for possible instrument failure, two out of three separate measuring channels have to indicate the fault before automatic action is taken. Shutdown is achieved by inserting absorbing rods (similar to control rods) rapidly into the core, generally under gravity.

Apart from these mechanical means all reactors, thermal and fast, have an inherent safety feature arising from the neutron absorption resonances of ^{238}U. Since the ^{238}U is intimately mixed with the fissile material its temperature increases with an increase in fission rate. The neutron absorption resonances have a high cross-section over a narrow energy range. An increase in fuel temperature raises the thermal energy of the uranium atoms in the fuel and so increases the vibrational velocities about their fixed positions in the crystalline lattice. Hence neutrons over an increased velocity range can be absorbed and resonance capture increases, lowering neutron multiplication. This effect on multiplication, know as the Doppler coefficient, gives automatic negative feed-back as the temperature of the fuel increases.

An increase in moderator temperature affects the Maxwellian neutron energy distribution to give relatively more neutrons at the high energy end of the spectrum. Most materials in a reactor have a neutron absorption cross-section that varies as the inverse of the neutron velocity in the thermal region. The most marked exception is ^{239}Pu fission (Fig. 1.19). This has a broad resonance at 0.3 eV, bringing the cross-section in this region well above a $(1/v)$ type of variation. As the number of neutrons of higher energy in the thermal region increases with temperature, the fission rate in ^{239}Pu therefore increases relative to other neutron reactions and so the neutron multiplication is increased. This increase is, however, slow and is easily controlled.

With light water moderated reactors an increase in moderator temperature has two effects that oppose each other. Moderator density decreases, increasing neutron leakage and so reducing multiplication. However, neutron absorption also decreases, resulting in an increase in multiplication. In pressurized water reactors the net effect is a reduction in multiplication but in

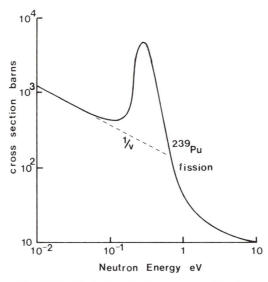

Fig. 1.19. Plutonium fission cross-section for thermal neutrons

boiling water reactors, including the SGHWR, the net effect is a small increase.

Formation of steam voids has a similar effect to a decrease in moderator density. An increase in steam demand lowers pressure and so increases the steam voidage. This, in turn, increases multiplication and the power rises until pressure restores equilibrium conditions. There is, therefore, a moderate degree of automatic load following.

The most important factor which might lead to a release of radioactive materials is the heat released by the decay of the radioactive fission products in the fuel since this continues even when the release of energy from fission has ceased. In Table 1.5, the decay heat from a nominal reactor of 1000 MW thermal power is given for times between one minute and one hundred days after shut-down. It can be deduced that the decay heat of a 1000 MW(e),

Table 1.5. Decay heat from 1000 MW thermal reactor after shut-down

Time after shut-down	Radioactive decay heat (MW)
1 minute	57
1 hour	28
1 day	7
100 days	1.7

3000 MW(th) PWR is 170 MW(th), immediately after shut-down, and that the average power density arising from decay heat is almost $6 \, kW/l$ at this time.

For any reactor the most hazardous conditions that might result in the release of radioactive materials follow a loss of coolant. Magnox reactors enclosed in a steel pressure vessel could develop coolant leaks, particularly where the large diameter gas ducts are welded on to the pressure vessel. Since the power density due to fission product decay is low ($35 \, W/l$ at maximum), a supply of carbon dioxide at atmospheric pressure is sufficient for emergency cooling. Advanced gas-cooled reactors are unlikely to suffer a loss of coolant accident since they use a prestressed concrete pressure vessel containing the steam generators and gas circulators. The only penetrations through the vessel are therefore for steam and water.

Pressurized water reactors operate at high pressures (up to 160 bar) and the reactor vessel typically contains about 100 t of water at $300 \, °C$. If a break in the pressure vessel were to develop about thirty tonnes of water could flash off as steam. Assuming that the reactor shuts down there would still be an average heat release of $6 \, kW/l$ initially. This could be sufficient to result in the partial melting of canning and fuel if the core became uncovered by water. Emergency supplies of cold water are therefore available for spraying the core for emergency heat removal. In addition, the whole of the reactor and primary circuit are enclosed in a large volume, prestressed concrete, steel-lined containment shell that can withstand up to five atmospheres over-pressure. This will retain the released steam and any volatile and gaseous fission products that might escape from the fuel if a partial meltdown were to occur.

The accident with the PWR at Three Mile Island that occurred in 1979 began with a faulty valve in the pressurizer of the reactor and was compounded by maloperation and unfamiliarity with emergency procedures.[17] Activity was only released because some radioactive water was pumped out of the containment building. With more thorough training of operating staff such a sequence of events should not recur.

Similar safety considerations apply to the Boiling Water Reactor but this has a lower average power density due to fission product decay of $2.3 \, Kw/l$ immediately after shut-down.

The CANDU pressure-tube type of reactor is not prone to large leaks since cracks in the relatively thin walled pressure tubes do not propagate into large breaks. In addition the use of a liquid moderator that is separated from the coolant allows use of an effective shut-down mechanism by draining the moderator rapidly into a dump tank.

From the foregoing remarks it is apparent that loss of coolant is the most serious hazard to be considered, except for exceptional circumstances. A related hazard to loss of coolant is loss of coolant flow, resulting for example from loss of electrical power. It is therefore essential to provide standby power supplies such as diesel or gas-turbine driven generators. Since the reactor would automatically shut down on loss of power because the safety rods are held out of the core on electromagnetic clutches, the auxiliary power supplies need only be sufficient to allow decay heat removal.

Fast breeder reactors have the highest power density of all reactor systems and consequently the highest power density on shut-down, in the range 20 to $30\,kW/l$. Large quantities of liquid sodium are contained in the primary circuit, especially in pool type reactors. The PFR, for example, has 920 t of primary sodium at a mean temperature of about $400\,°C$. The liquid sodium will circulate by natural convection through the core when shut down and during the first hour after shut-down from full power, the sodium temperature would rise by about $80\,°C$ if no heat removal were possible. A thermal syphon system filled with a sodium-potassium alloy that is liquid at just above room temperature is provided in the PFR to discharge waste heat to air in the event of prolonged total loss of electrical supplies.

The mechanical and civil engineering of reactors must also take into account abnormal accidents such as being struck by a crashing aircraft. Also the possibility of floods or earthquakes must be considered. Reactors with solid moderators in earthquake zones pose the greatest problem since the graphite blocks may be shaken out of position and distort fuel and control rod channels. To overcome this, the Magnox reactors in Italy and Japan have been designed with extra core constraints to prevent distortion and an additional emergency shut-down system of boron-loaded steel balls is available for filling distorted channels.

A further safeguard is the existence of independent inspection and licensing authorities such as the Nuclear Installations Inspectorate in the UK. These examine all aspects of a reactor including construction, nuclear design, control systems, and operational procedures and have the power to insist on changes before operation is permitted.

An important aspect of the safety philosophy applied to reactors that is not well appreciated by the public is the concept of risk. Daglish[18] reports on the almost complete lack of correlation between perceived (or imagined) risk and actual risk (loss of life expectancy) of various common activities and occupations. This attitude is a major difficulty in the general acceptance of the benefits of nuclear power. In normal operation, the additional radiation exposure above natural background and medical irradiation arising from occupational exposure and disposal of radioactive waste is shown by Fremlin[19] to be only 0.6% of the total exposure. Obviously, if a serious reactor accident were to occur then the general public would suffer additional exposure. This would occur mainly downwind of the reactor and would be attributable to the radioactive fission product ^{131}I (half-life 8 days) which is readily taken up by the thyroid gland. The form of the relationship between the total release and the probability of occurrence is shown in Fig. 1.20 and is known as the Farmer criterion. This type of relationship is also shown by natural disasters and

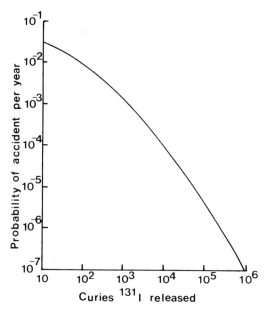

Fig. 1.20. Farmer risk curve

accidents, as discussed by Farmer.[20] Serious accidents occur only infrequently and the safety design of a reactor system has to ensure that the risk is sufficiently low as to be acceptable.

A safety analysis associated with normal operation of nuclear power system is given by Hill[21] who estimates the deaths due to accidents associated with extraction, transport, and power generation for coal, oil, gas, and nuclear power sources. The analysis concludes that nuclear power is about seven times safer than coal. Other studies quoted in this paper also indicate that, in normal operation, the risks due to nuclear power are much less than those of conventional power sources.

1.7 RADIOACTIVE WASTE

An increasingly important consideration in planning a nuclear power programme is the disposal of the waste in the form of fission products in the spent fuel. For 1 GWa(e) the amount of waste will be between 0.9 t and 1.5 t depending on the reactor design and its thermal efficiency. This is equal to the mass of fuel fissioned. Although the amount of waste is small compared with the water, carbon dioxide, and ash resulting from combustion of fossil fuels to provide a similar amount of energy, the activity is high. Many different radioactive species are present, and the half-lives cover a wide range. Most of the fission products have half-lives of less than one year so the activity drops rapidly after the fuel elements are removed from the reactor. A few, however, have half-lives between 30 years and 10^5 years so there is a long term waste disposal problem. At present, the fuel from thermal reactors is stored for as long as possible, the limit being corrosion of the fuel can. Some Magnox fuel has been reprocessed but at present there is little reprocessing of PWR/BWR fuel. When a significant quantity of electricity is generated from fast breeder reactors, storage of spent fuel for an indefinite time will not be possible since the fuel will have to be quickly reprocessed in order to feed the plutonium back to the reactors.

Even for thermal reactors it will eventually prove necessary to reprocess the fuel because of can failure in the spent fuel store. The problem of safe storage or disposal of the waste fission products then arises. At first sight disposal into the sea appears attractive because of the large volume available for dilution. Because of the successive food chains from plankton through to fish there is an important concentration effect and significant quantities of activity could occur in food for human consumption. Some of the radioactive elements can be rapidly eliminated from the body but others are concentrated in certain organs or parts of the body. For example ^{90}Sr with a 28-year half-life tends to be concentrated into the skeleton with only a low loss by excretion compared with the natural radioactive decay. Long-term storage is mainly determined by the ^{90}Sr and the ^{137}Cs content (^{137}Cs having a half-life of 30 years). After about 500 years these will have decayed to negligible proportions and the remaining long half-life actinides (^{241}Am, ^{243}Am, and ^{239}Pu) which are α-particle emitters will then have an activity comparable with uranium ore. At the Windscale processing plant, for example, the high activity waste is stored in tanks that are cooled to remove the decay heat, low activity liquid waste is discharged into the sea and low activity solid waste is sealed into containers and dumped in the ocean.

Work is in progress on methods for long term safe storage of the high activity waste. The most promising of these is the incorporation of the activity in glass, which is then cast into cylindrical blocks in stainless steel outer jackets.[22] There would be about 4 m^3 of highly active waste in vitreous form per GWa(e). These glass blocks, known as HARVEST containers, would have to be stored for a number of years with suitable cooling. Eventually they would be buried in suitable geological formations such as igneous rocks, salt beds, or deep clay layers where they would be completely inaccessible. In the unlikely case of the glass becoming exposed to the surroundings there would be no leakage of activity into surface waters.

1.8 ECONOMICS

The cost of electricity generation is not the same as the cost to the consumer. At present in the UK distribution costs are approximately equal to the cost of

Table 1.6. UK generation costs (p/kWh)

Year	Magnox	Coal	Oil
1971/72	0.43	0.43	0.39
1972/73	0.48	0.49	0.40
1973/74	0.52	0.53	0.55
1974/75	0.48	0.74	0.88
1975/76	0.67	0.97	1.09
1976/77	0.69	1.07	1.27
1977/78	0.76	1.23	1.42
1978/79	1.12	1.33	1.35
1979/80	1.30	1.56	1.93
1980/81	1.65	1.85	2.62

generation. Generation costs are mainly composed of the capital cost of construction, fuel costs, and operating costs. A comparison of the generation costs of Magnox, coal and oil fired stations since 1971 is given in Table 1.6.[23-26] This shows clearly the effects of inflation and particularly the dramatic increase in oil prices. Table 1.7 shows the split between fuel costs and capital costs for the three main fuels in 1980/81.[26] Although the capital costs in all these cases are given at historic prices, it illustrates that generation costs in fossil fuelled stations are mainly dependent on fuel costs since capital costs are low.

In nuclear stations, the reverse is true; costs are high and fuel costs low. Nuclear stations should therefore be run at high load factors; that is they should be used for base load generation. Fossil fuelled stations equivalent to perhaps 25% of the total installed capacity should therefore be available to handle peak load demands if the cost of generation overall is to be kept as low as possible.

Generation costs for AGR stations can be illustrated by a comparison between Hinkley Point B AGR and Drax A coal-fired stations,[26] both of similar age, for the year 1980/81. The AGR station has a total generation cost of 1.45 p/kWh (fuel 0.70 p/kWh) and the coal-fired station a total generation cost of 1.87 p/kWh (fuel 1.45 p/kWh) including a component for eventual

Table 1.7. Fuel, operating and capital costs (p/kWh) in 1980/81

	Magnox	Coal	Oil
Fuel	0.74	1.54	2.15
Operating	0.43	0.21	0.24
Capital	0.40*	0.10*	0.23*
Total	1.65	1.85	2.62

* Including eventual decommissioning.

Table 1.8. French capital and generation costs at 1979 prices

Station type	Size (MW(e))	Capital (£/kW)	Generation (p/kWh)
Coal	2 × 600	260	2.3
Oil	2 × 600	260	2.5
PWR	1300	390	1.4
Super-Phénix	1200	890	2.7

decommissioning. Projected costs for stations under construction are also given in reference 26 and the total costs are (AGR) 2.65 p/kWh, (coal) 3.84 p/kWh, and (oil) 8.72 p/kWh. The large increases over working stations are due to higher capital and decommissioning costs and forecast increases in fossil fuel prices.

Comparisons between coal, oil, PWR, and Super-Phénix have been made in France,[27] and these are summarized in Table 1.8. These 1979 prices are not directly comparable with the UK prices in Tables 1.6 and 1.7 but show that oil is the most expensive and thermal reactors the cheapest source of electricity. Capital costs for the fast reactor Super-Phénix are high but it is to be expected that future fast reactors could be reduced to capital costs of 655 £/kW and generation costs to 1.8 p/kWh also at 1979 prices. It therefore seems that although fast reactors give up to 100 times the fuel utilization, thermal reactors give the lowest generation costs.

A further point to be considered is station reliability. This can be measured by the percentage annual load factor. Table 1.9 gives the cumulative load factor and the 1980 load factor for various reactor types[28] where the load factors for individual stations have been weighted according to the rated power output. The Canadian CANDU reactors appear so far to be the most reliable.

Magnox are the next most reliable, although the load factor for these has dropped markedly in recent years because of the discovery of cracks in gas duct bellows supports. These cracks appear to have been present for a long time without causing problems but there have been various lengthy shut-downs to correct the defects. Apart from this the annual load factors are showing a gradual improvement.

Table 1.9. Load factors for the main reactor types

Reactor type		Cumulative load factor (%)	1980 load factor (%)
CANDU	(12 at 6306 MW(e))	72.3	80.8
MAGNOX	(26 at 8513 MW(e))	58.7	50.9
PWR	(78 at 61389 MW(e))	54.0	59.0
BWR	(52 at 37052 MW(e))	53.7	59.9

1.9 FUEL SUPPLIES

How much of the installed power generation capacity in the future will be nuclear depends partly on the availability of fossil fuels. As mentioned in the section on pressurized water reactors, in France it is planned that about 77% of the installed capacity will be nuclear by 1985, the incentive being that it has little coal and wants to reduce its dependence on imported oil. If the proportion of electricity from nuclear power is to increase, the growth rate of nuclear installation has to be considerably greater than the overall growth rate. Various predictions have been made as to the growth of nuclear capacity and the assumed capacity in AD 2000 is shown by Hunt[29] to range from 2500 GW(e) to 500 GW(e). This lower figure is generally similar to predictions by Ion[30] who assumes that nuclear power will contribute only about 10% of the total power generation capacity by the end of the century.

Whether such capacities are feasible also depends on the uranium consumption, enrichment capabilities, rate of uranium extraction, and total uranium resources. Thermal reactors operating at about 75% load factor require 100 to 150 t/a of natural uranium per GW(e) and a figure of 120 t/a per GW(e) will be taken for illustrative purposes. Table 1.10 shows the annual and cumulative uranium requirements for thermal reactors at the year 2000 for three installed capacities, 500 GW(e) and the lower and upper OECD predictions of 1977[29] which are 800 GW(e) and 1600 GW(e). As with all mineral resources the amount of uranium available is uncertain. At present uranium costing less than $130 per kg to produce is considered to be economic for use in thermal reactors. Such reserves occur non-uniformly around the world, as shown in reference[31] and Table 1.11. The reasonably assured category is that already prospected and additional reserves refer to those expected from the similarity of the geological areas to known uranium reserves. Also given is a more speculative estimate of uranium resources at lower concentrations in ores that would cost in excess of $130 per kg to extract. Potentially the largest source of uranium is the sea, which contains about 10^{13} t. However, it occurs at such a low concentration that it may require more energy to extract it than would be obtained by fission even using fast reactors.

In order to estimate how long the 2190 kt of reasonably assured uranium

Table 1.10. Uranium requirements for thermal reactors (year AD 2000)

Installed capacity GW(e)	Annual supply (kt/aU)	Cumulative supply (kt/aU)
500	60	600
800	96	960
1600	192	1900

Table 1.11. Uranium resources (world outside communist area) (Unit 1000 t U)

	Reasonably assured < $130/kg	Additional < $130/kg	Speculative > $130/kg
Africa	570	200	1 300– 1 400
N. America	830	1711	2 100– 3 600
S. and Central America	60	14	700– 1 900
Asia and Far East	37	24	200– 1 000
Australia/Oceania	296	49	2 000– 3 000
W. Europe	388	91	300– 1 300
Total	2190	2100	6 600–14 800

reserves or the 4290 kt reasonably assured plus additional reserves might last it is necessary to assume a growth rate. Ilustrative estimates assuming a linear growth rate of 5% per annum of the capacity at AD 2000 are given in Table 1.12 for the three capacities used in Table 1.10. The calculations indicate that uranium will not last long if used only in thermal reactors. One of the more fortunate countries is Canada which has 587 kt of uranium measured, indicated, and inferred. With an installed capacity of 14 GW(e) by 1990, 10% of the reserves would be depleted over the reactor life of thirty years.[32]

Another fuel for thermal reactors could be ^{233}U bred from thorium. The reserves of thorium outside the communist area[33] are given in Table 1.13. Total reserves of 1133 kt are less than for uranium. The main difficulty is the production of enough ^{233}U to start a significant reactor programme. This would involve loading thorium into reactors fuelled with uranium, thereby reducing the amount of plutonium available for fast reactors.

In order to make uranium reserves last sufficiently long to provide a substantial contribution to world supplies, fast breeder reactors are required. With fast breeder reactors, not only will the lower cost uranium last longer but it may also prove economic to use some of the lower concentration ores for which the uranium cost exceeds $130 per kg. This is because the lower fuel requirements of fast reactors makes them less sensitive to fuel cost. The main problem is the need for a substantial thermal reactor programme to produce

Table 1.12. Date of exhaustion of reasonably assured and additional uranium reserves

Installed capacity GW(e) (AD 2000)	Linear growth GW(e)/a	Year of exhaustion	
		Reasonably assured	Reasonably assured + additional
500	25	2020	2035
800	40	2010	2020
1600	80	2002	2010

Table 1.13. Thorium resources (world outside communist area) (Unit 1000 t Th)

	Established reserves	Estimated additional resources
Brazil	1.2	31.8
Canada	80	80
Egypt	14.7	280
India	—	300
South Africa	20	—
USA	52	265
Others	8	—
Total	176	957

enough plutonium to provide the initial fuel charge for the fast reactor cores. As discussed in the section on fast reactors, the balance may prove difficult to achieve unless fast reactors with a low plutonium inventory and a high breeding gain (in the initial stage) can be produced.

1.10 REFERENCES

1. Bennet, D. J. (1972) *The Elements of Nuclear Power,* Longman, London.
2. Glasstone, S., and Edlund, M. C. (1952). *Nuclear Reactor Theory,* Macmillan, London.
3. Murray, R. L. (1957). *Nuclear Reactor Physics,* Prentice-Hall, Englewood Cliffs, N. J.
4. *Nuclear Engineering International* (1980), **25**, 302.
5. *Nuclear Engineering International* (1981), **26**, 310, 27–42.
6. Wilkie, T. (1979). *Nuclear Engineering International* **24**, 291, 54–7.
7. *Nuclear Engineering International* (1976), **21**, 244, 58–63.
8. Lumb, P. R. (1976). *Nuclear Engineering International,* **21**, 244, 64–9.
9. Petrosyants, A. M. (1981). *Problems of Nuclear Science and Technology* the Soviet Union as a World Nuclear Power, Ch.6, Pergamon, London.
10. Rippon, S. (1974). *New Scientist,* **63**, 910, 398–401.
11. Valery, N. (1973). *New Scientist,* **59**, 863, 610–15.
12. Seaborg, G. T. (1967). *Journal of British Nuclear Energy Society,* **6**, 10–23.
13. Marshall, W. (1980). *Atom,* **282**, 88–103.
14. Evans, A. D., Laithwaite J. M. and Nunn D. A. (1974). *Proceedings of International Conference on Fast Reactor Power Stations,* 1–12, BNES, London.
15. Simpson, H. C. (1977). *The Breeder Reactor* (ed. Forrest, J. S.), 37–48, Scottish Academic Press, Edinburgh.
16. *Nuclear Engineering International* (1978), **23**, 272, 43–60.
17. *Atom,* (1980). **280**, 38–42.
18. Daglish, J. (1981). *Atom,* **291**, 9–13.
19. Fremlin, J. (1980). *Atom,* **283**, 128–31.
20. Farmer, R. (1980). *Atom,* **282**, 108–109.
21. Hill, J. (1981). *Atom,* **293**, 64–9.
22. Roberts, L. (1979). *Atom,* **267**, 8–20.

23. *Atom* (1978), **259** 145.
24. *Atom* (1979), **270,** 112.
25. *Atom* (1980), **288,** 271–2.
26. *Atom* (1981), **299,** 229–31.
27. Rippon, S. (1980). *Atom,* **281,** 60–2.
28. *Nuclear Engineering International* (1981), **26,** 314, 15–16.
29. Hunt, S. E. (1980). *Fission, Fusion and the Energy Crisis* 2nd Edn. Ch. 9. Pergamon, Oxford.
30. Ion, D. C. (1980). *Availability of World Energy Resources* 2nd Edn. p. 310, Graham and Trotman, London.
31. *World Uranium Resources—an International Evaluation* (1979), OECD, NEA/IAEA, OECD Paris.
32. *Energy Report* (1981), **8,** March (2), 6.
33. Bowie, S. H. U. (1974). *Phil. Trans. Roy. Soc.* **A1276,** 503.

Energy—Present and Future Options, Volume 2
Edited by D. Merrick
© 1984 John Wiley & Sons Ltd

T. D. BEYNON
University of Birmingham, UK

2

Thermonuclear Fusion as an Energy Source

2.1 INTRODUCTION

A fusion reaction is that which occurs when two nuclei combine to produce more tightly bound heavier nuclei with a subsequent release of energy. Fusion reactions can in principle occur with all light elements up to about mass number 56 (iron). Nuclei with masses heavier than this may be considered as potentially fissionable nuclei, in which two lighter nuclei of approximately equal masses are produced, again with a subsequent energy release.

An example of a fusion reaction is that which occurs between the two heavier isotopes of hydrogen, deuterium (D) and tritium (T). These, under suitable conditions, can combine to form a helium nucleus (α − particle) and a neutron; the reaction releases 17.6 MeV (1 MeV $\equiv 1.60 \times 10^{-13}$ J) of energy which is carried away as kinetic energy by these reaction products.

In practical terms, the potential of such energy release is enormous. A few milligrammes of deuterium could release as much fusion energy as the combustion of about 25 tons of coal, whilst the natural deuterium in one litre of normal water has the energy equivalent of about 250 litres of automobile fuel. Despite these incentives, nuclear fusion as a practical energy source has proved to be an elusive goal, particularly when making comparisons with the rapid technological development of fission reactors following the first demonstration of the neutron chain reaction by Fermi in 1942. Indeed, the striking success of fission reactor development has reduced the economic incentive to pursue expensive alternative nuclear sources and has been largely responsible for the extended timescale of fusion development. Recently, various socio-political pressures have brought a wider awareness of the advantages of fusion power.

In some cases, this has led to increased funding of national programmes and, in other cases, to international collaborative research programmes to study various fusion devices.

This chapter presents a broad overview of the aims and achievements of the development of the science and technology of fusion power for the last dozen or so years. Consideration is given to both magnetic and inertial confinement approaches, to the resource demands of such reactors and to the many limitations imposed by the materials problems. Whilst fusion energy should in no way be considered as a cheap energy source, it does offer promise as a plentiful source without the serious problems of the long-lived radioactive by-products which are formed during the operation of fission reactors.

2.2 GENERAL CONSIDERATIONS

For near-term applications, the isotopes of hydrogen and helium will be the preferred fuels. In particular, the hydrogen isotopes deuterium ($^2_1$H or D) and tritium ($^3_1$H or T) and the helium isotope ^{3}He are the most likely candidates. The basic reaction cycles are

$$D + T \rightarrow {}^4_2\text{He} \ (3.52\,\text{MeV}) + {}^1_0 n \ (14.06\,\text{MeV})$$

$$D + D \quad \begin{array}{l} \xrightarrow{50\%} \quad T \ (1.01\,\text{MeV}) + {}^1_1\text{H} \ (3.03\,\text{MeV}) \\[2ex] \xrightarrow{50\%} \quad {}^3_2\text{He} \ (0.92\,\text{MeV}) + {}^1_0 n \ (2.45\,\text{MeV}) \end{array}$$

and

$$D + {}^3_2\text{He} \rightarrow {}^4_2\text{He} \ (3.67\,\text{MeV}) + {}^1_1\text{H} \ (14.67\,\text{MeV})$$

Note that the ^{3}He $(D,p)\alpha$ reaction produces only charged particle reaction products which could therefore lead to quite low levels of induced radioactivity compared, for example, with the 14.1 MeV neutrons produced in the D–T reaction. The energy available in these reactions appears in the form of kinetic energy in the reaction products, and its availability will depend on the form these reaction products take. For example, in the D–T reaction, the 3.52 MeV α-particle could be retained in the plasma to provide for self-heating which could give rise to sustained operation. The neutrons, however, would have no reactions in the low-density plasma of a magnetic system and would lose only about 10% of their energy in the high-density plasma of an inertially confined system. These neutrons appear outside the plasma essentially unmoderated and their relatively long mean free path (i.e. the average distance between collisions) is 100–200 mm. Entering bulk material around the plasma, the so-called 'blanket' region, they exchange their energy with the blanket nuclei through collisions, to provide thermal energy for steam raising. We shall see later that blankets are required to perform functions other than those of heat exchangers. Neutrons from the D–D reaction would perform a similar role, although at a lower energy.

To use fusion as a practical source of electricity, two fundamental demands arising from the physics of the system must be met simultaneously. For a deuterium–tritium plasma these are:

(1) The plasma must be confined for a time τ seconds so that the product $n\tau$, where n is the D–T ion density, is about 10^{21} s/m^3. This is essentially the *Lawson criterion* and guarantees ignition and net energy output against energy losses by radiation from the plasma and steam plant losses in converting thermal energy to electrical energy.

(2) The temperature of the plasma should be in the region of $kT \sim 20\,$keV ($\equiv 2.3 \times 10^8\,$K) to ensure adequate fusion reaction rates.

These two conditions, which we shall consider in more detail later, define our concept of 'engineering breakeven'. An earlier milestone to be achieved is 'scientific breakeven', when the net energy gain in the system is just unity. At this point $n\tau$ is about 2×10^{20} s/m^3 and a simultaneous temperature of about 10 keV is adequate.

In the magnetic confinement programme, these scientific breakeven conditions have been met with the toroidal geometry devices called tokamaks, though not yet simultaneously. The first has occurred in ALCATOR at MIT and the second in the Princeton Large Torus (PLT). Simultaneous achievement is expected in the Tokamak Fusion Test Reactor (TFTR) at Princeton and, at a later stage, in the Joint European Torus (JET) at Culham. Similar experiments are planned in Japan (JT–60) and in the Soviet Union (T–15). For inertially confined plasma, scientific breakeven may occur with the large Nova glass laser planned for the Lawrence Livermore National Laboratory, or the light ion driven assembly (PBFAII) at the Sandia National Laboratory, New Mexico, although both these devices possess a number of formidable technological obstacles which have yet to be overcome.

2.3 ENERGY BALANCES IN FUSION REACTORS

Although the fusion reactions we have listed are exoergic, the probability of a reaction occurring depends on whether the colliding nuclei have sufficient energy to penetrate the repulsive Coulomb barrier, which exists between the positively charged ions, in order to come within range of the attractive nuclear forces. For D–T, for example, the cross-section for the fusion reaction becomes significant only for collision energies in excess of about 125 keV, with higher energies required for the D–D reactions and the ^{3}He $(d, p)^4$He reaction. This is illustrated in Fig. 2.1, where the cross-sections for these reactions

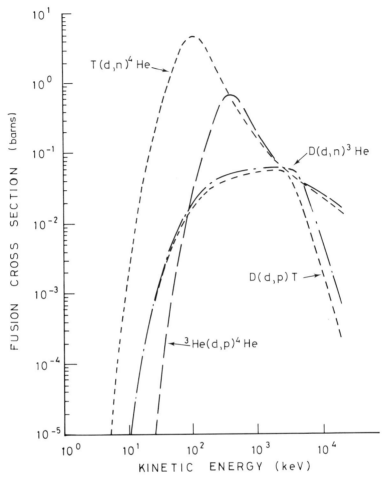

Fig. 2.1. Cross-sections for fusion reactions as a function of the kinetic energy of the colliding nuclei

are plotted as a function of the total kinetic energy of the colliding nuclei. The optimum arrangement for obtaining a fusion reaction is obtained first by using a neutral plasma of ions and electrons, rather than a gas of just ions, so that Coulomb scattering interactions occur as statistical fluctuations in an otherwise neutral plasma. Secondly, the plasma should be allowed to come into collisional equilibrium so that a Maxwellian distribution of energies is attained. The quantity of interest now becomes not the cross-section, σ, but the Maxwellian-averaged product of the relative velocity of the ions and σ, namely $<v\sigma>$. Once we know $<v\sigma>$, then the number of fusion reactions per second per unit volume is simply $n_D n_T <v\sigma>$, where n_D and n_T are respectively the deuterium and tritium ion densities. In Fig. 2.2, the averaged reaction rates $< v\sigma >$ are plotted as a function of the temperature T (in keV) of the ions in the equilibrium plasma.

Two points are of immediate importance. First, although the fusion cross-sections, which represent the interaction probabilities between *two* colliding particles, required relatively high energies to be significant, the function $< v\sigma >$ peaks at much lower temperatures when an equilibrium ensemble of particles is used. The reason for this is simply that the ions in the high energy

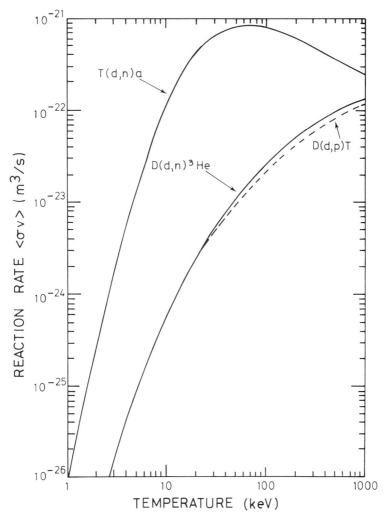

Fig. 2.2. The Maxwellian-averaged reaction rate $< v\sigma >$ as a function of ion temperature for the main fusion reaction cycles

tail of the Maxwellian distribution are responsible for the fusion reactions, since they alone are sufficiently energetic to overcome the Coulomb barrier. The role of the lower energy ions is simply to maintain the equilibrium state of the plasma.

Secondly, we see that at lower temperatures the $< v\sigma >$ values for the D–T cycle are two orders of magnitude higher than the D–D reaction or any of the other reactions. This is the reason why so much attention has been given to the D–T reaction for near-term fusion devices.

Fusion reactions which proceed via an equilibrium plasma condition such as that described above are referred to as thermonuclear fusion. The fusion products will now not be produced monoenergetically, but with a spread of energies arising because of the Doppler broadening produced by the Maxwellian distribution of ion energies. Thus, in the D–T reaction, the 14.1 MeV reaction is a distribution peaked at the 14.1 MeV energy but with a spread $\Delta E/E$ proportional to $(kT)^{1/2}$. This amounts to about ± 10% at 10 keV.

We can now obtain some simple criteria which allow the energy balance and ignition conditions in thermonuclear plasmas to be examined. The simplest approach is to obtain a minimum plasma condition, known as the Lawson criterion, which we shall derive for an equimolar D–T plasma. The major energy loss from the plasma occurs as bremsstrahlung radiation. Bremsstrahlung—the German for 'braking radiation'—arises from the Coulomb collisional deceleration of charged particles and, in particular, from the two-body collisions between particles. The energy lost in the process appears as a photon. In the non-relativistic case, the radiation fields produced by two like particles (i.e. electron–electron or ion–ion) just cancel each other, so that only electron–ion collisions produce bremsstrahlung radiation. For an ion species of charge $\mathcal{Z}$, density n_i, interacting with an electron gas of density n_e ($\mathcal{Z}n_i = n_e$) we have a radiated power density P_x given by

$$P_x = 4.81 \times 10^{-37} \, \mathcal{Z}^2 n_i n_e T_e^{1/2} \quad (\text{W/m}^3)$$

where T_e is the electron temperature in keV.

For a pure hydrogenic plasma of electron density n_e, this reduces to

$$P_x = 4.81 \times 10^{-37} n_e^2 \, T_e^{1/2} \quad (\text{W/m}^3)$$

The thermal energy content of the plasma is simply $\frac{3}{2} n_e \, kT_e$ in electrons and $\frac{3}{2} n_i \, kT_i$ in ions, where n_i and kT_i refer respectively to the ion density and temperature. Now assume that $n_e = n_i = n$, $T_i = T_e = T$ and that the plasma is contained for a time τ seconds. Fusion reactions between D and T ions proceed at the rate of $(\frac{1}{2}n)^2 < v\sigma >$ per unit volume per second, with an energy release of E_f per reaction. Let us now assume that the radiation, the thermal energy and the fusion energy are all available for conversion to electrical energy using power plant of efficiency η. This electrical energy is then simply used to reheat the plasma and balance radiation losses. Hence, during our confinement time τ,

$$\left(\frac{n^2}{4} < \nu\sigma > E_f\tau + P_x\tau + 3kTn\right)\eta = P_x\tau + 3kTn$$

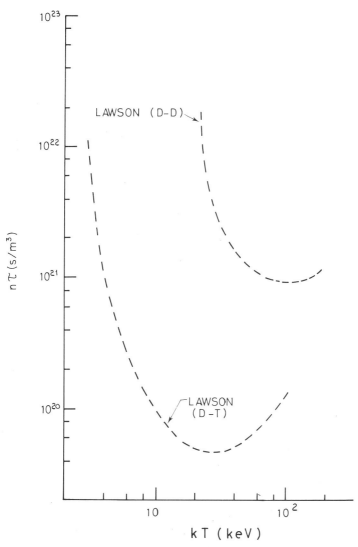

Fig. 2.3. The Lawson ignition conditions for the D–T and D–D cycles

Putting $P_x = cn^2 T^{1/2}$, c being a constant, we obtain the simple Lawson criterion

$$n\tau = \frac{3kT\,(1-\eta)}{\eta\left[\left(\dfrac{<v\sigma>}{4}\right)E_f + c\sqrt{T}\right] - c\sqrt{T}}$$

The product $n\tau$ is simply dependent on the plasma temperature, T, and the plant conversion efficiency, η, for a given plasma species.

Figure 2.3 plots $n\tau$ for D–T and D–D plasmas as a fraction of temperature for $\eta = 0.35$. We see that the minimum value of $n\tau$ occurs at a temperature of about 20 keV for D–T and about a 100 keV for the D–D cycle, at values of $n\tau \sim 5 \times 10^{19}$ s/m^3 and 10^{21} s/m^3, respectively. More detailed calculations of the energy balance are required in practice, in order to include the effects of charged particle plasma heating and externally supplied heating. Generally, the Lawson criterion serves to illustrate well the minimum-plasma requirement, namely that for a D–T reactors a minimum $n\tau$-value of order 10^{20} s/m^3 is required, with ion temperatures ranging from 10–40 keV and electron temperatures ranging from 10–20 keV.

2.4 THE BASICS OF FUSION REACTORS

Whether we are referring to magnetic fusion energy (MFE) reactors or inertial confinement fusion (ICF) reactors, the broad details of the design of the reactor chamber are the same. The major differences are that, for MFE, we require magnetic field configurations to confine the plasma which may have dimensions of the order of a few metres, whereas for ICF the ignited thermonuclear plasma may have dimensions of the order of 10^{-5} to 10^{-6} m. Figure 2.4 illustrates the required engineering components of a thermonuclear reactor and we consider these now, in more detail.

2.4.1 The First Wall and Blanket

The first wall and the blanket are the first materials which the fusion reaction products encounter. As well as these products, the first wall and blanket see electromagnetic radiation (x-rays) from the hot plasma, neutral atoms derived from charge-exchange processes in the plasma and, in the case of ICF, the debris from the target. The first wall must therefore be designed to withstand the resultant high surface and volume heating rates, the pressure exerted by the coolant required to remove this heat and the radiation and particle damage produced. Heat fluxes in the first wall can vary from 100 kW/m^2 to 2000 kW/m^2 for surface ratings, and from 50 MW/m^3 to 100 MW/m^3 for volumetric ratings.

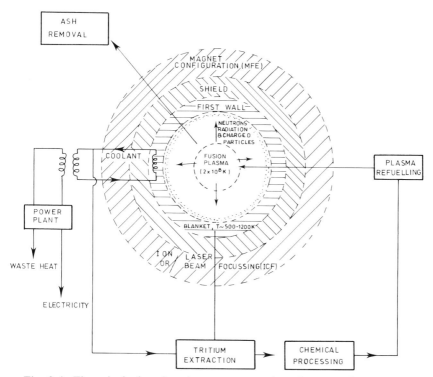

Fig. 2.4. The principal engineering components in a thermonuclear reactor

The materials problems of reactor first walls have been studied in detail and we can only briefly mention here the main points. The sputtering away of atoms from the first wall, due to neutron bombardment, charged particle and fast neutral atom erosion all cause serious problems which severely limit the lifetime of the first wall and possibly even the operation of the reactor. For example, the erosion can cause the release of high-Z atoms into the plasma which, in the case of MFE, would greatly enhance the bremsstrahlung radiation losses and shut down the thermonuclear reaction. Similar problems for ICF would arise when these atoms could cause possible defocusing of the driver beams focused on to the small D–T target in the centre of the chamber, particularly for the case of accelerator ion-beam drivers.

The role of the blanket in a D–T reactor is twofold. First, it is the bulk region in which the 14.1 MeV neutrons deposit their energy to provide the primary thermal energy source for the steam raising plant. Secondly, it provides the means of breeding the isotope tritium to maintain the fuel cycle. Let us consider the second point first.

Tritium ($^3_1$H) does not occur naturally. It has a half-life of 12.35 years, so that any of this material formed at the creation of the Universe has long since disappeared. Traces of tritium which appear now are due to recent atmospheric weapons testing or releases from heavy-water moderated fission reactors. Consequently, there is a need to breed tritium and this can be relatively easily done using the abundant material lithium. Lithium occurs naturally as two isotopes, ^{6}Li (7.4%) and ^{7}Li (92.6%). The following reactions between these isotopes and neutrons produce tritium:

$$^6\text{Li} + n \longrightarrow {}^4\text{He} + T + 4.8 \text{ MeV}$$
$$^7\text{Li} + n \longrightarrow {}^4\text{He} + T + n - 2.5 \text{ MeV}$$

The first reaction will occur at all neutron energies since it produces 4.8 MeV of kinetic energy. The second reaction absorbs energy and has a threshold neutron energy ($E_n = 2.8$ MeV) below which the reaction will not proceed. This has an important effect on the design and performance of blankets. It should also be noted that the ^{7}Li reaction produces, in addition to tritium, an extra neutron which could react with ^{6}Li to increase the tritium breeding.

For each D–T reaction in the plasma a triton is lost for each neutron produced. Thus each neutron entering the blanket must produce something slightly in excess of one triton, to allow for losses in tritium gas during production, reprocessing, and chemical handling. We can define a *breeding ratio* as the number of tritium nuclei produced per neutron incident in the first wall. Typically, breeding ratios would need to be 1.15 to 1.20.

If we were to consider a blanket containing just the isotope ^{6}Li, then the breeding ratio must necessarily be less than unity. Although in the ^{6}Li reaction, one neutron produces just one triton, there are a number of competing neutron-capturing reactions in Li and structural materials in the blanket which do not yield tritium as a product. An extra source of secondary neutrons is therefore required to balance this neutron loss. This can be achieved most easily by using the neutron produced in the ^{7}Li reaction, that is, use natural lithium as a breeding material. There may be a cost penalty involved, however, in having a large breeding zone necessary to accommodate the relatively large bulk of natural lithium required to give a good breeding ratio. Consequently lithium enriched in ^{6}Li can be used, with another material added to provide the necessary additional neutron. There are a number of suitable candidates for this role, each using $(n, 2n)$ reactions. The most likely are beryllium, ^{9}Be$(n, 2n)$, and lead, Pb$(n, 2n)$. Lead is a particularly useful possibility since, alloyed as a euctectic with lithium, it can also have advantageous heat transfer properties, a point we discuss next.

The mean free path between collisions of the thermonuclear neutrons is of the order of 0.1 to 0.15 m in the blanket. Since the heating process is the result of such collisions, the spatial distribution of the thermal heating is relatively flat across the blanket region. This is in complete contrast to a fission reactor, where the energy production is localized within the fuel pin. There are two basic ways in which the 14.1 MeV neutron gives its energy to the blanket. The first is simply to have elastic collisions with the nuclei when, at each collision, a fraction of the neutron's kinetic energy is given to the target nuclei which thus acquires thermal (kinetic) energy. The neutron continues to have collisions with other nuclei, and is thus moderated. The second mechanism for the neutron moderation is that of inelastic scattering, where part of the neutron's energy is converted to γ-rays, of energy 1 to 10 MeV. These photons then diffuse through the blanket, having collisions with the atomic electrons and consequently producing heating. This effect also contributes to the flattening of the heating distribution in the blanket. A total energy balance for a 14.1 MeV neutron would show that it is responsible for producing typically 16 to 17 MeV of energy in the blanket, the extra energy arising from exoergic reactions. Of this energy, about 50% arises from direct neutron heating and 50% from γ-ray heating, as discussed above. This again should be compared with the case of the fission reactor, where only a small percentage of the total heat production arises because of γ-radiation.

Having deposited this energy in the blanket, it must then be removed by a coolant. The choice of coolants is limited by chemical and materials compatibility, and pumping power requirements. Coolants for a fusion reactor can be either a liquid or a gas. The primary function is the removal of bulk-heat from the blanket. Secondary roles include maintaining the first wall at as low a temperature as possible to maximize performance in the radiation environment, and the minimization of temperature variations during the thermonuclear burn cycle. Conceptually, the simplest liquid coolant is possibly liquid lithium which, as well as providing the necessary moderation–breeding zone, acts as a heat transport medium. Liquid lithium is compatible with stainless steel and most nickel-based alloys up to temperatures of about 500 °C, above which temperature significant corrosion arises from the resultant nickel-leaching from the surfaces of the alloys. The radioactive corrosion material then contaminates the coolant. Another major disadvantage with this coolant, for magnetically confined plasmas, is that high pumping powers and pressures are required to overcome the magnetohydrodynamic losses associated with pumping such a highly conductive material in a magnetic field. For ICF systems, on the other hand, liquid lithium has been proposed as a first wall material, flowing down the inner walls of the reaction chamber. However, the primary role of the liquid lithium in this case is to provide a shock-absorbing buffer for the explosively released debris following the fusion of the D–T pellet. This is the 'waterfall' concept and will be discussed later. Other possible coolants include water, organic liquids, and various fused salts of which Li–Be–F or FLIBE has received the most attention.

A particularly interesting concept involves the use of a lithium–lead eutectic in the blanket. Apart from its tritium breeding properties, with tritium formed via the lithium with the lead providing neutron multiplication via the $(n, 2n)$ reaction, this material can act as an energy storage unit which can maintain the blanket structure at near constant temperatures to minimize thermal cycling and fatigue. A typical eutectic might be $Li_{62} Pb_{38}$ which melts at $464\,°C$ and is partly in a liquid phase and partly in a solid phase. The concept is applicable to a MFE system operating in a pulsed mode, where the thermo-nuclear burn occurs over a period from a few seconds to a few tens of seconds. During the burn phase of the cycle, the fraction of the eutectic in the liquid phase increases, with an adsorption of heat as latent heat of fusion. During the subsequent down-phase of the burn, the solid fraction of the eutectic in-creases, yielding latent heat to maintain a constant heat flow to the cycle. The material, although not a coolant in the conventional sense, provides a smoothing mechanism for pulsed-mode magnetic fusion reactors.

Water-cooled blankets have been studied as an obvious extension of the technology based on fission reactors. Dry-steam cooling possesses much the same properties as helium gas, except that tritium gas contamination is a prob-lem and its use would appear to give little advantage. Pressurized-tube water designs, similar to the Canadian CANDU reactor concept using Zircaloy pressure tubes, would appear to be feasible particularly for blankets which are capable of breeding fission material (^{239}Pu or ^{233}U) with the thermonuclear neutrons.

For nearly all designs using a gas as a coolant, the inert gas helium has been proposed. This gas is compatible with most structural materials provided the concentration of oxygen is kept low. Thus stainless steels, nickel-based and aluminium-based alloys, and some molybdenum alloys are compatible with helium at an oxygen level of a few p.p.m. Helium cooling with niobium or vanadium alloys would require technologically impossible oxygen levels of a few parts per billion. Finally, it should be noted that carbon dioxide is not compatible with most proposed fusion reactor structural materials.

2.4.2 The Shield Region

Returning to Fig. 2.4, it can be seen that a shield region follows the blanket. The function of such a shield is to attenuate the neutron fluxes and γ-fluxes to a sufficiently low level that, for components outside the shield, induced radioactivity, heating rates, and radiation damage are kept to a minimum. In magnetically confined reactors, it is most important to shield the supercon-ducting magnets. Failure to do so would result in heating in the cryogenic components, increased electrical resistance in the copper or aluminium components and damage to the superinsulation components.

For ICF reactors, the shield's region would fulfil a similar requirement. Thus, for ion-beam driven fusion, a high degree of protection would be required for the magnets in the final quadrupole focusing system particularly if they were of a superconducting design. For a laser driven system, the highly

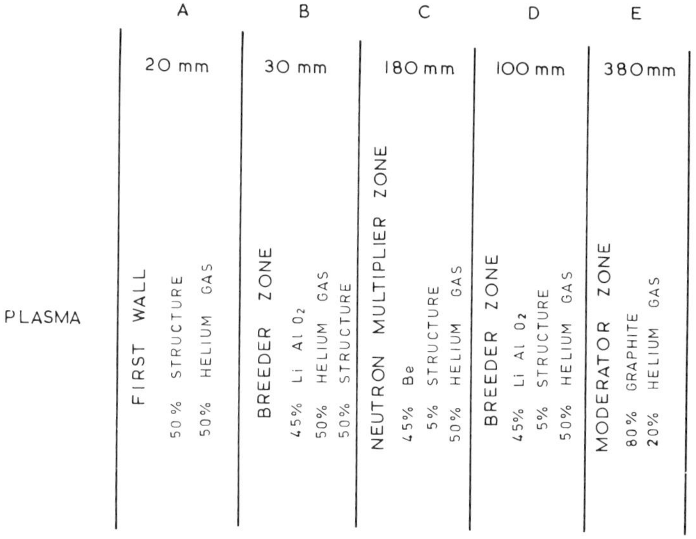

A   B   C   D   E

20 mm   30 mm   180 mm   100 mm   380 mm

PLASMA | FIRST WALL | STRUCTURE 50% | HELIUM GAS 50% | BREEDER ZONE | Li Al O₂ 45% | HELIUM GAS 50% | STRUCTURE 5% | NEUTRON MULTIPLIER ZONE | Be 45% | STRUCTURE 5% | HELIUM GAS 50% | BREEDER ZONE | Li Al O₂ 45% | STRUCTURE 5% | HELIUM GAS 50% | MODERATOR ZONE | GRAPHITE 80% | HELIUM GAS 20%

Fig. 2.5. Schematic of a solid breeding material gas-cooled blanket

susceptible laser components would require probably a higher degree of radiation protection than the systems mentioned above.

The design of a typical shield presents no critical problems. Boron carbide is a suitable material for the absorption of the now largely moderated neutron flux, with materials like iron and lead providing sufficient attenuation for both the high energy component of the neutron flux and the γ-ray flux. Only about 1% of the total heat production is deposited in the shield and this can be dissipated easily.

To illustrate a typical design for a complete blanket assembly, Fig. 2.5 shows a schematic of a helium gas-cooled system. The first wall (zone A) is gas-cooled and is followed by a breeding zone (B) of solid lithium aluminate (Li AlO₂) contained in clad rods cooled by gas at 50 bar. A larger zone (C) follows and contains beryllium which acts as a neutron enhancer via the (n, 2n) reaction. This zone enhances the breeding in (B). A further breeding zone (D) is followed by a large graphite region (E) which thermalizes the neutrons to enchance their capture in the ⁶Li component of the breeder zone (D). Note that this zone is also gas cooled. This blanket would have a breeding ratio of 1 : 11 for a stainless steel structure, with 18.5 MeV nuclear heating per thermonuclear neutron produced.

2.5 STRUCTURAL MATERIALS FOR FUSION REACTORS

For the past decade, a series of engineering design studies and experimental measurements has resulted in a relatively short list of materials which are thought suitable for fusion reactor operation. The basis for selection falls approximately into two categories, namely near-term suitability and long-term suitability. In the first category, the criteria are those essentially for demonstration plants, with relatively low irradiation exposures. In both cases, the first wall materials will have the largest demands made on them. We can summarize the materials requirements as follows, in order of priority.

2.5.1 Radiation Damage

Radiation damage is by far the most important consideration. Such damage occurs principally through the interaction of the fast thermonuclear neutrons with the blanket materials. The neutrons transfer their energy collisionally with the nuclei in these materials with sufficient energy to displace them from their equilibrium positions in the solid lattice structure, leaving voids in the lattice. If the energy of the atoms disturbed is sufficiently high, they will move through the lattice, colliding with other atoms to produce a displacement cascade.

Some of the nuclear reactions which initiate such cascades are $(n, 2n)$ and reactions, leading to nuclear transmutations and gas-producing reactions such as (n, p) and (n, α). The transmutation reactions produce impurities whilst the (n, p) and (n, α) produce hydrogen and helium gas. These gases may not diffuse out of the materials, but instead form bubbles. The overall result of void formation and gas bubble formation is swelling and dimensional changes in the bulk materials.

Materials least susceptible to the resultant swelling and dimensional instability of radiation damage are titanium, vanadium, molybdenum, and stainless steel. Associated embrittlement is higher for molybdenum than the other materials, but graphite and niobium could now be considered.

2.5.2 Materials Compatibility

Coolant and tritium compatibility are only marginally less important than radiation damage. Titanium, vanadium, niobium, molybdenum, and stainless steel are all compatible with lithium whereas, for the reasons previously stated, helium gas cooling is compatible only with stainless steel, graphite, aluminium, molybdenum, and titanium. Water, as a coolant, is known to be compatible with stainless steel, aluminium, and titanium.

The use and production of tritium in D–T fusion reactors poses a particular set of radiological and materials problems. Tritium is a weak (mean energy 5.7 keV) β-emitter with a half-life of 12.35 years. Uptake in the human body produces a significant radiological and biological hazard. In the case of the former, a steady state concentration of 1 millicurie (1 mCi) of tritium in the body produces a dose of 0.1 rem per week, corresponding to a tritium concentration in body fluids of about 0.2 μCi/ml. As a gas, the maximum permitted level of tritium in air for occupational exposure is 2×10^{-3} μCi/ml compared with 5×10^{-6} μCi/ml in water. These figures indicate that tritium forms a much larger hazard when ingested than when inhaled in the form of gas. The external dose to the body from tritium is generally regarded as negligible and shielding is not required. The biological hazard from tritium arises primarily from the decay, in tritiated water, of HTO to form the radical (OH):

$$HTO \longrightarrow (OH) + {}^0\beta + {}^3_2He.$$

This radical can then take part in abnormal biochemical reactions.

The materials problems associated with tritium must be assessed in terms of the permeability, solubility, and diffusivity of the gas through a metal. Diffusivity is the mobility of tritium atoms through the crystal lattice of a metal by jumping from one site to another. Solubility is a general term for the property of tritium to form a solution in a metal, believed to occur by the initial formation of a metallic hydride which is itself soluble in the metal. In summary, the properties restrict the number of favoured structural materials to stainless steel, aluminium, and molybdenum.

2.5.3 Materials Strength

The mechanical and thermal properties of irradiated materials must be evaluated in terms of their creep strength, yield strength, fracture toughness and various thermal stress parameters. Molybdenum, vanadium, titanium, and stainless steel meet the required creep strength and yield strength demands. However, only stainless steel, titanium, and aluminium meet the demands for fracture toughness. Stainless steel will not possess the full range of thermal stress parameters, although molybdenum, aluminium, niobium, and vanadium will.

2.5.4 Induced Radioactivity

An important issue which must be considered for the operation of any reactor, whether it be a fission or a fusion system, is the level of induced radioactivity after a given operational time. In particular, it is necessary to know the time required for the radioactivity to decay to a given acceptable level following a shut down of the reactor, and the level of afterheat associated with the decay. Obviously, a fusion reactor does not possess many of the problems associated with fission reactors, particularly the build-up of the very long lived higher actinide isotopes and fission products. In fusion systems, the radioactive chains of isotopes are driven largely through $(n, 2n)$ and (n, γ) reactions in structural materials, particularly in the isotopes of nickel, iron, molybdenum, and manganese.

Figure 2.6 shows the decay of the induced radioactivity in various structural materials, in Ci/kW(th), following the shutdown of a fusion system which has been operating for two years with a first wall loading of $1.25\,MW/m^2$ (the first wall loading is defined as the flow rate of 14.1 MeV neutron kinetic energy through the first wall). Also shown for comparison is the total actinide and fission product activity from a liquid metal-cooled fast breeder reactor (LMFBR). A number of points are noticeable. First, at shutdown, a number of structural materials produce comparable activities with the LMFBR, particularly 316 stainless steel and the molybdenum alloy TZM (99% Mo with 1% stainless steel). These two materials also have the highest decay rates after 100 years. At periods greater than this, TZM as a structural material would constitute a worse source than the fast breeder, with stainless steel still

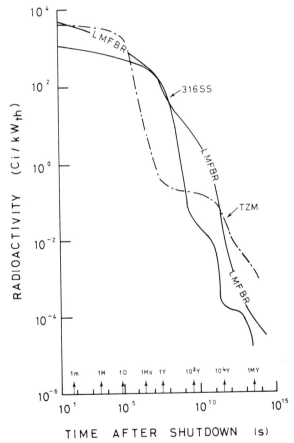

Fig. 2.6. Decay of induced radioactivity in various structural materials

representing a major problem. Choices of other alloys would present few problems, but as indicated earlier other criteria must be considered in choosing them.

2.5.5 Materials Resources and Costs

The final set of criteria for materials selection embraces the broader issues of cost, resource availability, industrial manufacturing ability, and data base. In each of these areas, stainless steel and graphite are dominant, together with aluminium and titanium. However, because of the widespread use of stainless steel in the fission reactor industry, relatively little data exist for radiation damage in high neutron flux exposures for aluminium and titanium, particularly with accompanying helium contamination. Although excessive corrosion limits the use of stainless steel to temperatures of about 500 °C, this material together with graphite look like strong contenders for the first generation of power generating fusion reactors.

The demands of fusion reactors for lithium will obviously be high and the long-term resource availability of this material has been the subject of several studies. In the US alone, assuming a 1000 GW(e) user rate from fusion reactors, there is sufficient supply for about 5000 years, with comparable availability for Western Europe. Hopefully, long before such a period has elapsed, the need for tritium breeding from lithium will have disappeared with the advent of D–D reactors or even systems working on more advanced fuel cycles.

2.6 MAGNETIC FUSION REACTORS

The description of fusion power covered so far in this article has been applicable, with a few minor exceptions, to either MFE or ICF concepts. In this and the following section, we shall discuss specific reactor systems, starting with magnetic systems.

Two main reactor types have emerged as the forerunners for possible commercial power producers, namely the tokamak and the magnetic mirror machine. Other concepts exist, including the reversed-field pinch and the ELMO Bumpy torus.

2.6.1 The Tokamak Reactor

A tokamak is a concept with its origins in the Soviet Union in the early 1950s. The geometry is a toroid which is essentially an axisymmetric magnetic container working on the principle of an electrical transformer. The basic configuration is illustrated in Fig. 2.7. A toroidal magnetic field, B_ϕ, is generated by the toroidal field magnet. The curvature and gradient of this field produces a magnetohydrodynamic instability in the toroidally shaped thermonuclear

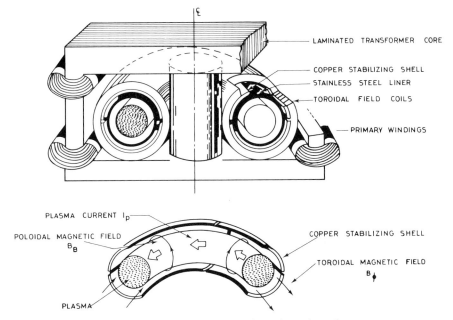

LAMINATED TRANSFORMER CORE

COPPER STABILIZING SHELL
STAINLESS STEEL LINER
TOROIDAL FIELD COILS

PRIMARY WINDINGS

PLASMA CURRENT I_p
POLOIDAL MAGNETIC FIELD
B_θ
COPPER STABILIZING SHELL
TOROIDAL MAGNETIC FIELD
B_ϕ
PLASMA

Fig. 2.7. Basic configuration of a tokamak

plasma and, in order to achieve confinement, it is necessary to superimpose a poloidal (B_θ) field to cause the effects of the curvature and gradient of the B_θ field to cancel out, on average. The poloidal field can be produced by a set of poloidal coils which carry a toroidally directed current or by a toroidal current flowing in the plasma. Devices using the first method are known as stellarators. Tokamaks employ the latter method, a time-varying current in the primary windings generating an electric field that induces a current flow in the plasma.

Because of the transformer principle, a tokamak will have a pulsed burn cycle, of duration from a few tens of seconds up to about 1000 seconds, depending on the control of impurities and helium ash removal. Continuously operated tokamaks have been proposed, using radio frequency heating or neutral particle heating. Such forms of auxiliary heating are also required to provide the required thermonuclear temperatures, as discussed in Section 2.3.

Figure 2.8 illustrates a conceptual tokamak fusion reactor with a neutral beam injector (I) and a poloidal coil structure (G). The poloidal field is essentially a device for producing a null region in the weak poloidal field which serves as a magnetic diverter for the control of impurities in the plasma. Also illustrated is the neutral beam injector which, typically, would provide auxiliary heating using the injection of neutral hydrogen atoms of energies of several hundred keV. The neutral particles traverse the vacuum and magnetic fields unhindered and eventually become trapped in the confinement region by

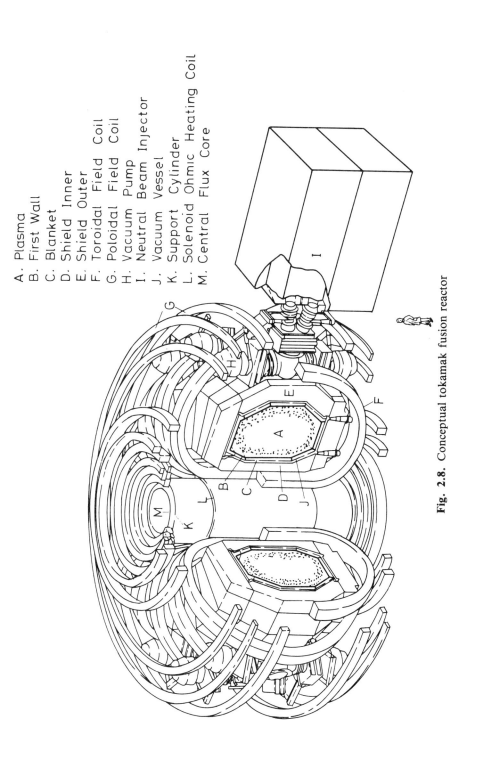

A. Plasma
B. First Wall
C. Blanket
D. Shield Inner
E. Shield Outer
F. Toroidal Field Coil
G. Poloidal Field Coil
H. Vacuum Pump
I. Neutral Beam Injector
J. Vacuum Vessel
K. Support Cylinder
L. Solenoid Ohmic Heating Coil
M. Central Flux Core

Fig. 2.8. Conceptual tokamak fusion reactor

charge exchange, sharing their energy with the plasma by Coulomb collisions. Notice the transformer structure embodied in L and M in Fig. 2.8.

Figure 2.9 shows the Culham Mk IIB whole-reactor design which has been optimized for blanket maintenance. This conceptual design has a neutron wall loading of $4.5\,MW/m^2$ with a thermal power of 3.4 GW, generating 1.2 GW of electricity. It is a helium gas-cooled system with natural lithium blanket breeding.

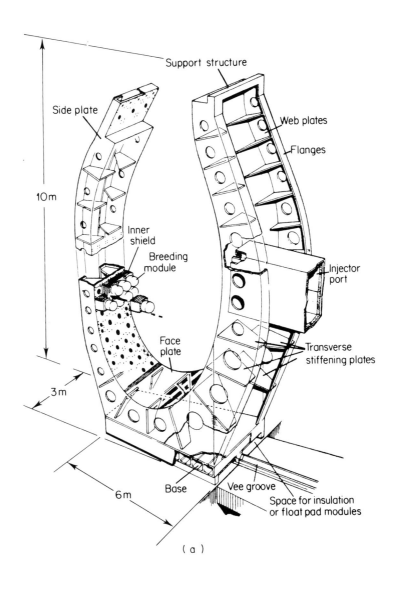

(a)

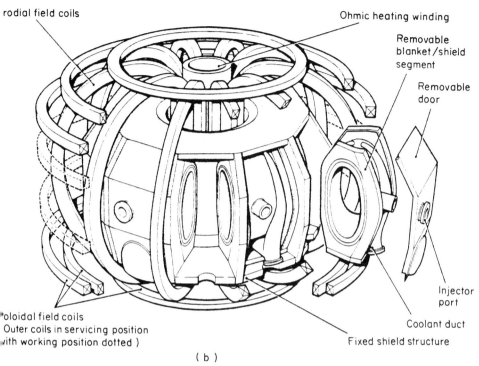

radial field coils

Ohmic heating winding

Removable
blanket/shield
segment

Removable
door

Injector
port

Coolant duct

Poloidal field coils
Outer coils in servicing position
with working position dotted)

Fixed shield structure

(b)

Fig. 2.9 The conceptual Culham MkII B reactor: (a) the details of a segment (b) the overall arrangement illustrating a blanket segment removal and replacement

2.6.2 The Tandem Mirror Reactor

If, in contrast to a toroidal geometry, we consider an open system, we are led initially to the configuration shown in Fig. 2.10, namely a simple magnetic mirror. Particles which have a sufficient ratio of transverse to parallel velocity (relative to the field lines at each point) will become trapped in the centre of the system since they are reflected as they approach the higher B-fields in the region of the field coils. Such a system is, unfortunately, subject to magnetohydrodynamic instabilities which limit plasma confinement.

However, these MHD instabilities, so-called 'flute' modes, can be controlled by producing field-lines which are concave into the plasma, so that a minimum B-field is produced. The most promising, and most complicated, devices for securing minimum-B regions are Yin-Yang coils, illustrated in Fig. 2.11, where the resultant plasma shape is shown. These enhanced mirror devices can now be used to plug the ends of a long solenoidal plasma of the type shown in Fig. 2.10.

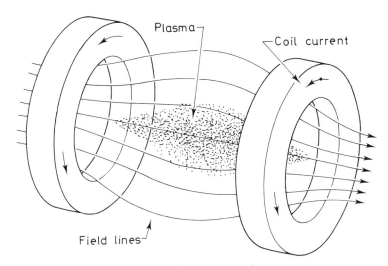

Fig. 2.10. A simple magnetic mirror system

A conceptual design of a tandem mirror reactor (TMR) has been made at the Lawrence Livermore National Laboratory and is illustrated in Fig. 2.12. Original studies for a 1 GW(e) reactor showed a requirement for a B-field of about 17T in the end cells and negative-ion, neutral beam injection with particle energies of 1.2 MeV. The 500 MW(e) design in Fig. 2.12. utilizes direct

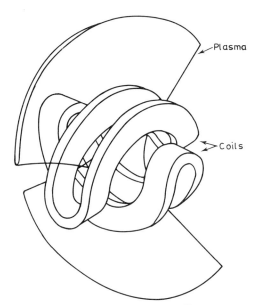

Fig. 2.11. Yin-Yang coils

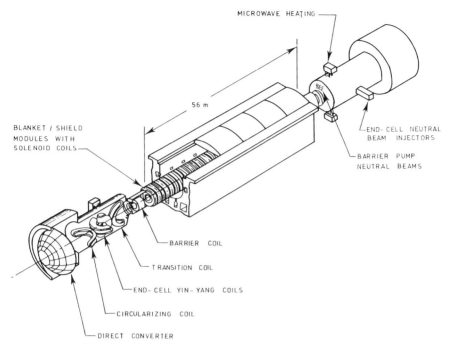

Fig. 2.12. Conceptual design of a tandem mirror reactor (TMR)

electron heating in the central cell, thermal barriers to insulate end-cell electrons from central-cell electrons and the addition of auxiliary mirror cells to improve the confinement. The thermal barrier concept promises a higher power gain factor with much less demanding neutral beam and magnet technology than indicated in earlier designs. The concept is simply that the provision of a mirror cell with a depressed plasma potential between the central cell and the end cell serves as a thermal barrier to the transport of electrons between these two cells. These and other concepts are presently being tested in TMX, the Tandem Mirror Experiment, at Livermore. Such a power reactor design, although technologically very difficult, promises considerable advantages over the tokamak concept. This is apparent by comparing Figs 2.9 and 2.12; the linear nature of TMR should lend itself to easier blanket maintenance than the toroidal geometry of the tokamak.

2.7 INERTIAL CONFINEMENT REACTORS

2.7.1 Some General Considerations

An attractive alternative to plasma confinement by magnetic fields is inertial confinement. In this approach, a small pellet of frozen D–T is rapidly heated and compressed to a density and temperature where fusion occurs, as defined by the requirements for ignition discussed earlier. The Lawson criterion is equally valid for inertial confinement fusion (ICF) as for magnetic confinement.

The qualitative physics of ICF is essentially as follows. Energy is delivered to the target, a small pellet a few millimetres in diameter, by a suitable driver at the rate of 5 to 10 MJ in a few tens of nanoseconds. This corresponds to a target illumination of about 10^{19} W/m^2. The energy rapidly heats the outer regions of the pellet and is transferred into the resulting plasma. The energy is then converted into kinetic energy and transported to what is called the ablation front. Here, heated dense matter is rapidly transformed to a lower density, depositing momentum in the remaining material and forming an imploding shock wave which compresses and heats the D–T fuel. If all goes well, the minimum $n\tau$ value is attained and the thermonuclear energy output is sufficient to produce a net energy gain. The principle is illustrated in Fig. 2.13.

We can make a number of simple calculations to illustrate the magnitudes of the pellet dimensions required for ICF. First, it should be noted that, unlike MFE, the pellet design should guarantee a high burn-up of the D–T fuel, i.e. 30–40%, in order to obtain a high net energy gain. The original ignition arguments now need modifying to allow for this fuel depletion. Accordingly,

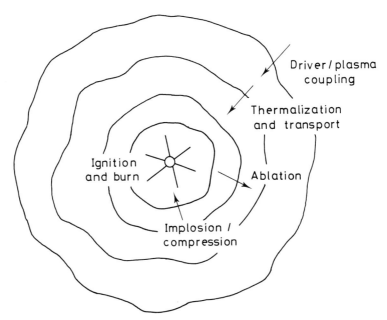

Fig. 2.13. Principle of inertial confinement fusion (ICF)

we can write (for an equimolar D–T mixture of ion density n)

$$\frac{dn}{dt} = - <v\sigma> (n(t)/2)^2$$

where we have assumed that ignition has occurred. Integrating this equation over the confinement time, τ, and defining the fractional burn-up at a time, t, $\phi(t)$, as follows:

$$\phi(t) = 1 - \frac{n(t)}{n_0}$$

where n_0 is the initial density, we obtain the result appropriate to a temperature of 20 keV (cf. Fig. 2.2) that

$$n_0\tau \sim 0.9 \left(\frac{\phi}{1-\phi}\right) \times 10^{22}\,\text{s/m}^3$$

For a 30% depletion, $n_0\tau$ is approximately 0.4×10^{22} s/m^3. The confinement time τ can be approximated by (r/v) where v is the thermal speed of the ions in the compressed pellet of radius r.

If the mass of the D–T fuel is M kg, then we can write directly that

$$n_0\tau \sim \frac{3M}{2\pi v r^2 (M_D + M_T)}$$

where M_D and M_T are the deuterium and tritium ion masses respectively, so that

$$\left(\frac{M}{r^2}\right) \sim 140\ \text{kg/m}^2$$

where a value of $v \sim 2 \times 10^6$ m/s has been assumed. About 10^{-6} kg of D–T fuel is appropriate for an ICF pellet, which therefore implies a compressed density of $n_0 \sim 9.5 \times 10^{31}$ m^{-3}. The uncompressed density of solid D–T ($\varrho_0 \sim 0.2 \times 10^3$ kg/m^3) is 4.8×10^{28} m^{-3}, so that a density compression of about 2×10^3 is required, corresponding to a factor of about 10 in the radius. It is important to note that, at these plasma densities, careful consideration must be given to the flow of energy, both radiation (x-rays) and fusion-product particles (neutrons and α-particles), in the design of a pellet.

Sometimes, a more useful rule for ICF than the Lawson criterion is the 'ϱr rule'. If ϱ is the density of the compressed pellet of radius r, we can rewrite the modified 20 keV Lawson criterion given above so that

$$\phi \sim \frac{\varrho r}{75 + \varrho r}$$

Thus at about 30% burn-up, we require that $\varrho r \sim 32$ kg/m^2. Another practical advantage of this approach is that some plasma diagnostic techniques can measure ϱr directly.

2.7.2 Requirements for Drivers and Pellets

At the present time, likely candidates as drivers for ICF are high power lasers, light ion accelerators, and heavy ion accelerators. Before discussing these further, it is necessary to define the parameter space of an ICF power producing reactor and see how the various constraints limit the choice of drivers.

Suppose we postulate a generalized driver which has a wall-plug efficiency, η_D, defined simply as the output energy divided by the input energy. This device is used to drive a pellet which has a gain G, defined as the ratio of the total output energy to the input energy. The pellet is placed at the centre of the reactor chamber and the exoergic reactions in the blanket and structural materials produce a further gain of η_R. This thermal energy is now converted into electrical energy with a plant efficiency η_P. A fraction E of this final energy is recirculated to the driver and a fraction $(1-\epsilon)$ is available for external use. Representing this loop in Fig. 2.14 we see that

$$\epsilon \eta_D \, G \eta_R \, \eta_P = 1$$

Assuming a plant efficiency $\eta_P \sim 0.3$, a 30% feed back of energy and $\eta_R \sim 1.2$, we have the requirement that $\eta_D \, G \sim 10$.

In Fig. 2.15 some typical performance gain characteristics for simple single-shell target designs driven either by short wavelength laser light (e.g. KrF) or an ion beam are displayed. The range of input energy chosen, 1 to 20 MJ, is probably the most optimistically achievable by either system in the foreseeable future. A projected efficiency for a KrF system is $\eta_D = 5\%$, whereas an efficiency $\eta_D = 25\%$ is easily foreseen for a light or heavy ion accelerator. Thus the intersections with the appropriate gain curves of the two lines $G = 10/\eta_D$ ($\eta_D = 0.05$ and $\eta_D = 0.25$) in Fig. 2.15 indicate the required target performance in each case. We see that in the case of the short wavelength laser, a target gain of about 200 with an input energy of about 30 MJ is necessary, whereas for the ion-driven system a target with a gain of 40 at 2 to 3 MJ input is sufficient. Apart from the difficulty in designing such a high-gain pellet, the explosive yield at 6000 MJ output is equivalent to about 1.5 t of high explosive per shot which puts a considerable constraint on the design of a suitable reaction

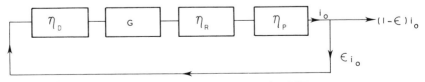

Fig. 2.14. Energy balance loop for ICF

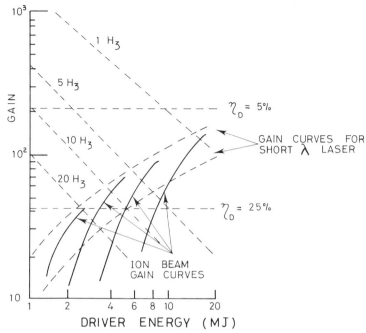

Fig. 2.15. Typical performance gain characteristics for simple single-shell target designs

chamber. The equivalent yield for the low gain pellet is about 30 kg of explosive.

Figure 2.16 shows a typical design for a single shell target. Energy is deposited in the outer layer, part of which ablates and part of which acts both as a shield against radiation present and as a compressor for the inner layer, a zone of frozen D–T.

The demands on the repetition rate of a given driver can be obtained by specifying a given power station output. To be economical, an ICF reactor system would require an output of at least 2 GW(th). If pellets are burned at a frequency of f Hz then by plotting in Fig. 2.15 the curves $G = 2000/fE$, where E is the input energy to the pellet in MJ, we can see whether our choice of $\eta_D G = 10$ can be achieved. At $\eta_D = 5\%$, a frequency of 0.3 Hz is required. This is an extremely high repetition rate for a laser of this power. At $\eta_D = 25\%$, a frequency of 5 to 15 Hz is required, a figure which could be easily met by an accelerator system.

In summary, therefore, we see that for commercial power production the higher efficiency and repetition rates of ion-beam accelerators compared with lasers make them the more likely candidates for ICF. Nevertheless, laser driven systems will continue to play the major role in contributing to our knowledge of target physics for some time, as we now see.

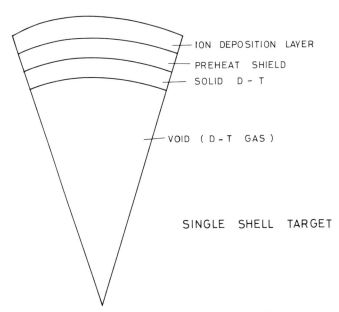

ION DEPOSITION LAYER

PREHEAT SHIELD

SOLID D - T

VOID (D - T GAS)

SINGLE SHELL TARGET

Fig. 2.16. Typical design for a single-shell target

2.7.3 Laser Driven Targets

For any driver there are two basic issues which must be carefully examined. These are: (1) coupling efficiency and (2) hot electron production.

A useful definition of coupling efficiency is the ratio of the useful energy imparted to the D–T fuel to the input energy of the driver beam. The coupling efficiency embraces the hydrodynamic efficiency and the absorption efficiency. In the former, we are simply referring to the effectiveness of the target in compressing and shock-heating the D–T. The latter measures how well the incoming beam energy is converted to mass ablation in the outer region of the target, and depends very much on the type of driver and the nature of the beam, i.e. photons or ions.

Hot electron production is a crucial factor since it can lead to excessive preheating of the D–T fuel, with a resultant poor performance of the target.

For lasers, energy absorption proceeds by three mechanisms:

(1) *Inverse bremsstrahlung.* This, as the name implies, is the inverse of the radiation-producing process discussed in Section 2.3. A photon is absorbed following a collision with an ion-electron pair, the free electron absorbing the photon's energy. This collisional process dominates at low laser intensities and produces only low-temperature electrons.

(2) *Resonance absorption.* With an increase in intensity the laser light will begin to interact collectively with the whole plasma, rather than with its constituents. Resonance absorption occurs near the point in the ablated plasma where the density is equal to a critical value (n_{crit}) such that the light frequency equals the fundamental plasma (i.e. plasma wave) frequency

(3) *Parametric instabilities.* This term refers to a further collective process which involves non-linear coupling of the light to plasma waves.

The relative strengths of the three processes depend on both the wavelength and intensity of the light beam. However, the modes (2) and (3) have in common the undesirable result of generating hot electrons (i.e. electrons with a temperature of more than 20 keV) and subsequent penetrating x-rays with strong pre-heat properties.

The effects on absorption efficiency and hot electron production of irradiance and laser wavelength are illustrated in Figs 2.17 and 2.18. Figure 2.17 shows that, at a given wavelength, the absorption efficiency falls off with increasing irradiance (W/m^2) on the target. Similarly, at a given irradiance, the absorption fraction increases significantly as the wavelength decreases and the collisional processes dominate. In Fig. 2.18 we see how the x-ray spectra produced by hot electrons have fluxes and mean energies which decrease for shorter wavelengths.

Reflection of the incident laser beam can occur, particularly at high intensities. The scattering occurs from ion acoustic waves (known as 'stimulated Brillouin' scattering) and from electron plasma waves (known as 'stimulated

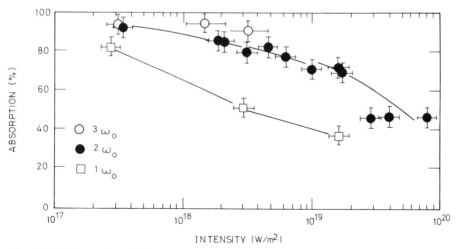

Fig. 2.17. Absorption efficiency for laser light as a function of irradiance, showing the advantage of using wavelengths less than 1 μm at high irradiance

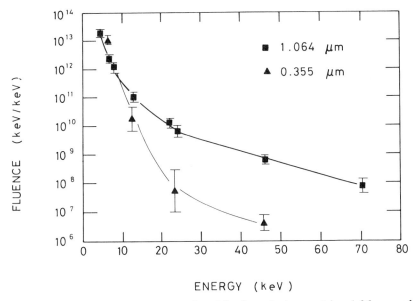

Fig. 2.18. The x-ray spectra produced by hot electrons at $\lambda = 1.06\,\mu$m and $\lambda = 0.35\,\mu$m. Both the fluxes and mean energies decrease for shorter wavelengths

Raman' scattering). Such scattering can generate hot electrons and give rise to convective instabilities in the plasma. The fraction of light reflected can be from 30% to 90%, with a significant reduction in the effective coupling efficiency of the driver resulting. Shorter wavelength lasers favour enhanced collisional losses but give reduced collective losses and weaker stimulated Brillouin and Raman scattering.

Figure 2.19 shows a schematic of the idealized density and temperature distribution in a laser driven target. A major area of intensive research at the present time is understanding the mechanism of thermal conduction from the critical layer into the ablation front.

For ICF laser-plasma interaction studies, there are two major high-power laser types available, the neodymium-glass laser ($\lambda = 1.06\,\mu$m) and the carbon dioxide gas laser ($\lambda = 10.6\,\mu$m). Much research is also being devoted to the development of krypton fluoride (KrF) gas lasers, at a wavelength $\lambda = 0.25\,\mu$m. Most of the Nd-glass lasers have also been modified to generate light at higher harmonics, so that as well as the infra-red generation ($\lambda = 1.06\,\mu$m), green light ($\lambda/2 = 0.53\,\mu$m), blue light ($\lambda/3 = 0.35\,\mu$m), and ultraviolet light ($\lambda/4 = 0.26\,\mu$m) can be used for experiments.

A number of countries now have major Nd-glass laser installations. In the USA, the Lawrence Livermore National Laboratory is constructing a ten-beam system called Nova, with each beam having a capability of 10 to 15 kJ

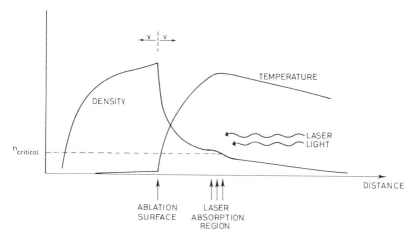

Fig. 2.19. A schematic of the density and temperature distribution near the critical and ablation layers for a laser driver

and 5 to 15 TW. Initially, the system will have only two beams and is referred to as Novette. Other programmes in the USA include those at KMS Fusion, the Naval Research Laboratory and the University of Rochester. In Europe, there are programmes at the Ecole Polytechnique (France), the Lebedev Institute (USSR), and the Rutherford-Appleton Laboratory (UK). There is also work at Osaka University in Japan. The associated research programmes are largely concerned with the detailed physics of the laser–plasma interaction. The predecessor to Nova, Shiva (a 20-beam system with a total output of 10 kJ and 20 TW), for example successfully compressed fuel ablatively to 100 times liquid density. However, the Nd-glass system is not seen as a possible ICF reactor driver. A pulse rate of about 1 per hour is about the maximum attainable due to heating effects in the system. Also, the overall efficiency, η_D, is about 0.1%, and therefore two orders of magnitude less than that required for a commercial system, as discussed in Section 2.7.2.

The long wavelength ($\lambda = 10.6\,\mu\text{m}$) CO_2 gas laser, on the other hand, would appear to have some of the required properties, i.e. $\eta_D \sim 10\%$ with a high achievable repetition rate. However, we have seen that at long wavelengths and high target irradiance, the absorption efficiency and the hot electron production are major problems; there is therefore considerable doubt as to whether successful ablative implosion and ignition of reactor-type targets can be achieved with this approach. These doubts have let to a significant reduction in funding for the partially built Antares CO_2 system at Los Alamos, a 24-beam system designed to deliver 40 kJ at 40 TW. Nevertheless, the intrinsic high efficiencies and high repetition rates of gas lasers could well be realized if the successful development of the KrF system is achieved. Due to the comparatively short wavelength of $\lambda = 0.25\,\mu\text{m}$, the absorption and hot electron problems are significantly reduced. Severe practical difficulties still remain, however, in developing optical coatings to handle the power densities in this ultra-violet region.

2.7.4 Ion-beam Driven ICF

Present studies on the use of ion beams for ablatively imploding ICF targets are derived from two important properties of ion sources:

(1) the high repetition rate, f, and high efficiency, η_D, of such devices.
(2) the present extensive technological base of accelerators, largely due to the rapid and successful growth of high energy physics.

Add to these factors the evidently well understood deposition mechanism of ions in matter, via coulomb interactions, and there is apparently an excellent scenario for an ion beam driver for ICF.

Let us first examine some of the parameters necessary for an ion beam to implode a target. Whatever the driver system, the hydrodynamics of a target demand that the energy deposition per unit mass of the target, ϵ, should be about 30 to 40 MJ/g and the irradiance, I, about 10^{19} W/m^2. These figures assume an optimized pulse shape which should be of about 30 to 40 ns duration.

Consider a simple single shell target consisting of a hollow spherical shell, radius r, thickness Δr, and mean density ϱ. The energy deposited by an ion beam will be almost entirely due to Coulomb interactions between the incident ions and the electrons in the (now) plasma shell of the target. We can assume with sufficient accuracy for the present argument that the ions of energy E slow down as

$$\frac{\mathrm{d}E_i}{\mathrm{d}x} \simeq \frac{-4\pi N Z_i^2 M_i e^4 \ln \Omega}{2mE_i}$$

where m is the electron mass of charge e, M_i is the ion mass, Z_i is the effective charge on the ion whilst slowing down, N is the electron number density in the target and $\ln \Omega$ is the Coulomb logarithm which is assumed constant during the slowing down. By integrating this expression we see that, for an ion range $\overline{X}$, the product $\overline{X}\mathrm{p}$ behaves essentially as

$$\overline{X}\mathrm{p} \propto \frac{E_i^2}{Z_i^2 M_i}$$

If it is assumed that $\Delta r \sim 2\overline{X}$, the energy per unit mass, ϵ, can be calculated directly as

$$\epsilon \sim \frac{I\,\Delta t}{2(\overline{X}p)}$$

where I is the irradiance (W/m^2) and Δt is the pulse length (s). For a pulse length of 30 ns, we require ($\overline{X}p$) $\sim$ 2 to 3 kg/m^2.

If we now specify an ion, we can define its kinetic energy. Thus for a ^{238}U ion, the required energy is $E_i \sim 10$ GeV. For a proton we can use the $\overline{X}p$-scaling law given above (assuming $\mathcal{Z}_i \sim 85^+$ for the ^{238}U ion) to see that an energy of about 7.5 MeV is required.

A cylindrical current i of radius r_b (the target radius) carrying ions of energy E_i in charge state $\mathcal{Z}'_i$, would be of magnitude,

$$i = \left(\frac{I}{E_i}\right)\mathcal{Z}'_i(\pi r_b^2 e)$$

Thus, for U^{+1} ions at 10 GeV the current would be 20 kA, and for 7 MeV protons 30 MA, assuming $I = 10^{19}$ W/m^2 and $r_b = 2.5$ mm. This corresponds to 4×10^{15} U-ions per pulse and 6×10^{18} protons per pulse.

Table 2.1 summarizes typical values for the parameters necessary for an ICF power plant driver.

Considerable work is now known to be underway in the United States, Europe, and Japan to answer two fundamental questions for ion-beam driven ICF. First, can currents of these magnitudes be produced and transported in a stable mode to the edge of the reactor chamber? Secondly, can these currents then be focused over a distance of 1 to 5 m down to the target of diameter about 5 mm? No definitive answers exist yet for either of these questions but detailed computational modelling and some small scale experiments provide an optimistic framework for expanding the ion beam concept.

The focusing of intense light ion beams is limited by the radial space charge spreading of the beam and by the diverging effect of the beam's self-magnetic field. Schemes to minimize these effects and thereby permit sufficient focusing

Table 2.1

Beam energy	3 MJ per pulse
Ion kinetic energy (ϵ_i)	10 Gev U^{+1}
	7 MeV p
Ion range (λ_0)	$\sim$ 2 kg/m^2
No. of ions per pulse	4×10^{15} U^{+1}
	6×10^{18} p
Pulse length at target (Δt)	$\sim$ 30 ns
Power at target (I)	$\sim 10^{19}$ W/m^2
Beam current at target	20 kA U^{+1}
	30 MA p
Beam spot radius at target (r_b)	2.5 mm
Specific energy deposition in target (ϵ)	30 GJ/kg
Energy released (assuming $G \sim 80$)	480 MJ ($\equiv$ 0.7 mg of fuel consumed per pulse)

rely on space-charge and current neutralization of the beam in the chamber. Heavy ion beams, on the other hand, are not limited by these two effects because of their inherent 'stiffness'. However, their focusability is determined by lateral momentum components in the beam which lead to a dilution of the beam's brightness.

At the Sandia National Laboratory, an intensive programme of light ion ICF research is underway using so-called magnetically insulated power diodes. These are essentially intense ion sources in which the electron flow has been suppressed. Such devices form the basis of the PBFA I (Particle Beam Fusion Assembly) which is now operational at 2–4 MV, 1 MJ, 30 TW (30 beams at 1 TW each). The successor, PBFA II, is scheduled for operation in the middle to late 1980s and should be working at 2–16 MV, 3.5 MJ, and 100 TW. Both PBFA I and PBFA II are 'one shot' devices in contrast to reactor-type repetitive-shot machines. As with the laser programme, the objective is to provide information on the basic physics of ion-plasma interactions, intense particle beams and target design, with the ultimate aim of achieving scientific break-even conditions.

For heavy ion fusion, detailed studies indicate that two possible drivers, both employing linear accelerators (linacs) have promise. These are the radiofrequency (rf) linac and the induction linac.

With the rf linac design, an ion current of about 100 mA is accelerated to the final ion energy of 5 to 10 GeV. The initial part of the system is designed as a parallel set of low frequency linacs arranged in a diverging tree geometry (see Fig. 2.20). As the full ion energy is attained, the total beam is transferred to charge-accumulating storage rings, 10 to 20 in number, via a stacking ring to allow multi-turn stacking in both vertical and horizontal planes. In these storage rings, which are similar to those for e–p colliding beam systems, the storage time must be of the order of a few ms to avoid serious beam loss from collisionally-driven ion–ion charge exchange. The current is finally further increased by strong bunching and eventually delivered to the target in multiple beams ('beamlets').

The induction linac is a single-pass device where the current amplification occurs continuously during the accelerating process. The current is injected at several amps and the entire beam is then accelerated in a single long sausage-like bunch. Initially, the voltage pulses to the induction cores are ramped upwards slightly with time. This has the effect of differentially accelerating the tail of the bunch with respect to the head. The ion velocity increases, the bunch length decreases and the current rises to several kA by the end of this stage. In the transport system to the reactor chamber, a strongly ramped voltage is applied to produce a strong longitudinal compression. The current rises sharply and the beam is split so that the current in each transport line does not exceed about 1 kA. Figure 2.21 shows a conceptual design for an induction linac driver.

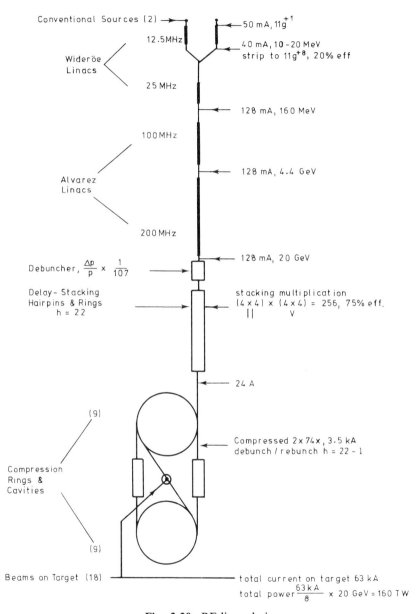

Fig. 2.20. RF linac design

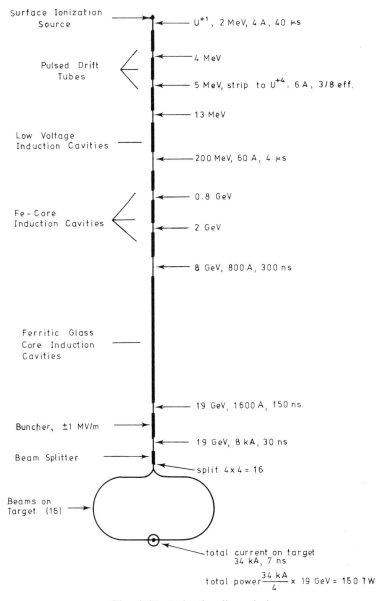

Fig. 2.21. Induction linac design

Although considerable expertise presently exists for rf linacs in nuclear physics research, the demands of heavy ion fusion require extensions of these designs to much higher currents, lower velocities, and smaller charge–mass ratios. Induction linacs have been widely used for 1 kA electron beams and the extension of this technology to the lower velocity ions required for ICF is also underway.

2.7.5 ICF Reactor Design Concepts

As discussed earlier, much of the conceptual reactor design for ICF is similar to MFE. There are, however, two major differences. First, ICF systems deliver energy to the reactor chambers in short (< 0.1 ns) pulses, each pulse containing several hundred MJ of energy, at repetition rates of 5 to 15 Hz. Secondly, and this difference is a major advantage, the reactor chamber design is essentially independent of considerations of the thermonuclear ignition and burn of the plasma. This permits a wide choice of reactor design, scale, shape, and refuelling facilities. For a tokamak system, by contrast, these parameters must be consistent with the constraints associated with being in the toroidal and poloidal field coils and operating in a high magnetic field whose source must be shielded.

The pulse in an ICF chamber has a complex energy and time distribution. As ignition occurs, about 70% of the thermonuclear energy is in the form of neutrons with 30% delivered mainly by X-rays and hot ionized target debris. The neutrons and hard X-rays (~ 100 keV) are emitted promptly. Shortly after, the target debris which has been heated by absorbing energy from the neutrons, α-particles and hard X-rays radiates soft X-rays into the chamber. It then expands and cools with a resultant downward shift in the wavelength of the emitted radiation to the optical and thermal regions. Two major engineering problems arise. The first problem results from the thermal transients which arise in the bulk materials; however, these can probably be overcome by conventional engineering techniques. The second problem is more serious and arises because the X-rays are absorbed in a shallow surface layer, unlike the neutrons which deposit their energy almost uniformly in the bulk containment medium. In stainless steel, for example, this can give rise to significant erosion by sputtering and evaporation at rates which could be as high as tens of millimetres per year.

A number of ways have been proposed for protecting the surfaces from such erosion. One, the dry wall concept, considers the use of pyrolytic graphite or niobium or even the use of a replaceable sacrificial liner. For laser-driven ICF designs, the reactor could be filled with a gas that is relatively opaque to both x-rays and debris ions. The gas pressures required would make such a concept unsuitable for ion beam systems. Alternative ways include the wet wall concept in which a waterfall of liquid lithium is constructed inside the chamber wall.

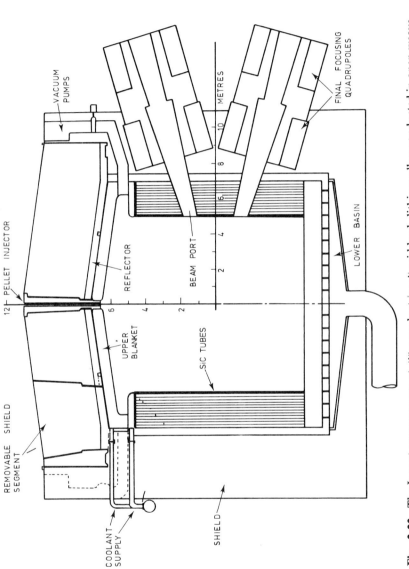

Fig. 2.22. The Inport reactor concept. The coolant is a liquid lead–lithium alloy conducted in open-weave silicon carbide tubes. Ten ports through which the heavy ion beams enter are arranged in five pairs around the chamber; one pair is shown

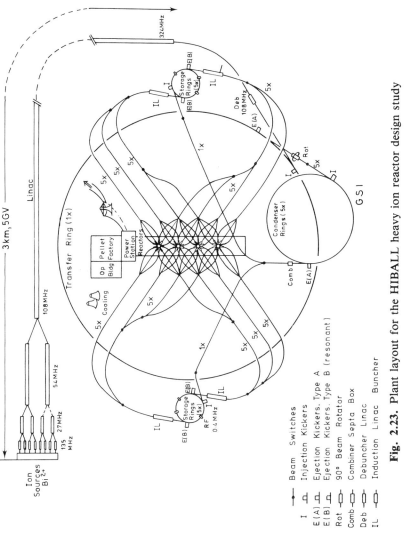

Fig. 2.23. Plant layout for the HIBALL heavy ion reactor design study

The liquid removes the energy as well as contributing to the tritium breeding. A particularly interesting idea which could well be suitable for heavy ion reactors is the Inport concept. Here, low-speed streams of lead–lithium alloy move vertically through flexible tubes of loosely woven silicon carbide. Sufficient seepage through the tubes occurs to produce a protective layer on the outside. This scheme, which was developed for the joint West German–University of Wisconsin heavy ion reactor design study, HIBALL, is illustrated in Fig. 2.22.

To illustrate the scale of an rf linac/storage ring reactor, Fig. 2.23 shows the ground plan of the HIBALL design. The plant would have an output of 3.7 GW(e) and would use bismuth ions at 10 GeV, with 4.8 MJ input energy per pulse and a repetition rate of 20 Hz. Such studies are not intended as engineering blueprints for construction but rather as a bench-mark to focus on the magnitude and priorities of the problems which arise and to provide a discussion and growth point for further studies. Figure 2.23 clearly illustrates the physical size of such a power station. However, to keep matters in perspective, the accelerator complex required for HIBALL is not significantly different from the high energy accelerator LEP presently under construction at CERN, Switzerland.

2.8 CONCLUDING REMARKS

The last decade or so has seen a remarkable advance in both the science and engineering of fusion as an energy source. There is little doubt that the MFE tokamak devices at Princeton, Culham, in the Soviet Union, and in Japan will attain the scientific breakeven condition in the next few years. However, it may well be that the main function of such experiments using tokamaks will be to demonstrate this scientific feasibility rather than to provide a growth point for a tokamak-based commercial power programme. The technological difficulties associated with toroidal geometry concepts certainly tend to favour the ultimate development of linear devices such as tandem mirror reactors.

Inertial confinement fusion has, of course, long since demonstrated its scientific feasibility but in the uncontrolled form of thermonuclear weapons. For power plant applications, considerable work remains to be done to develop and assess the relative merits of the various options for drivers, i.e. short wavelength gas lasers, light ion accelerators or heavy ion accelerators. The next five to ten years should see the successful demonstration of scientific breakeven for ICF, using either lasers or light ion accelerators, as the basic physics of target design becomes better understood. The attainment of this milestone, as would be the case with MFE, could be expected to lead to significant increases in funding for fusion research and development.

For both MFE and ICF, the materials and design parameter data bases have become reasonably well founded as a result of the existing fission reactor programmes and specialized fusion-orientated experiments. Nevertheless, the technological problems should not be underestimated. Quantitatively, they are an order of magnitude worse for fusion power reactors than for fission power reactors of the same generating capacity. The potential rewards, however, are more than commensurate with the effort.

2.9 BIBLIOGRAPHY

Relevant Plasma Physics
Rose, D., and Clarke, M. (1961). *Plasmas and Controlled Fusion,* Wiley, New York.
Stacey, W.M. Jr. (1981). *Fusion Plasma Analysis,* Wiley, New York.
Duderstadt, J. J., and Moses, G. A. (1982) *Inertial Confinement Fusion,* Wiley, New York.
Motz, H. (1979). *The Physics of Laser Fusion,* Academic Press, London.
Brueckner, K. and Jorna, S. (1974). *Rev. Mod. Phys.,* **46**, 325.

Reactor Design
Ribe, F. L., *Rev. Mod. Phys.,* **47**, 7 (1975); *Nucl. Technol.,* **34** (2) 179 (1977).
Steiner, D. (1975) *Nucl. Sci. Eng.,* **58** (2), 107.
Conn, R. W. (1977). *Adv. Nucl. Sci. Technol.,* **10**, 405.
HIBALL—A Conceptual Heavy Ion Beam Driven Fusion Reactor Study, Karlsruhe rep. no. KFK-3202 (1981).
Badger, B. *et al. SOLASE—A Conceptual Laser Fusion Reactor Design,* Univ. of Wisconsin report no. UWFDM-220 1977.
See also Special Edition of *Nuclear Engineering and Design* on Fusion Design (1981).

Materials Problems
Thompson M. W. (1969). *Defects and Radiation Damage in Metals,* Cambridge University Press, London.
Kulcinski G. L. (1979). *Contemp. Phys.,* **20**, 417.

Laser Fusion Studies
Probably the best source of information is the current annual report of the Laser Division, Lawrence Livermore National Laboratory.

Light Ion Studies
Buzzi, J. M. (ed.) (1981). *Proc. 4th Int. Top. Conf. High-Power Electron and Ion beam Res. Technology, Palaiseau* (Paris, École Polytech.).

Heavy Ion Studies
Hermannsfeldt, W. B. (ed.) (1979). *Proc. Heavy Ion Fusion Workshop, Berkeley 1979,* Lawrence Berkeley Lab. Report no. LBL-10301.
Böhne, D. (ed.), *Proc. Symp. on Accel. Aspects of Heavy Ion Fusion, Darmstadt 1982;* Darmstadt report no. GS1-82-8.
Beynon T. D. (ed.), *Heavy Ion Beam Fusion,* Plenum Press (to be published).

Energy—Present and Future Options, Volume 2
Edited by D. Merrick
©1984 John Wiley & Sons Ltd

F. C. TREBLE
Consulting Engineer

3

Photovoltaic Solar Energy Conversion

3.1 INTRODUCTION

To many of us, conditioned as we are to thinking of electrical power generation in terms of large central stations and extensive distributing networks,
photovoltaic solar cells, those expensive devices used to power satellites, would
appear at first sight to have no prospect of making any significant impact on
energy supplies. Compared with coal, oil, or nuclear power stations, the
capital cost of photovoltaic generation is still about fifty times too high and
the storage necessary in many applications to offset the effects of the daily and
seasonal variations of solar radiation is prohibitively expensive.

Yet, before writing off photovoltaics as a future energy option, it would be
wise to look at the capabilities of this unique technology in a world-wide
context and see how it has developed in recent years. In particular, it is
relevant to consider not only what it might do for industrialized countries but
also how it might help the millions of people in scattered communities who are
at present without electricity and pumped water or who are facing escalating
fuel costs for diesel or gasoline generators.

Solar cells convert daylight into direct current electricity at an efficiency of
10 to 15%. Made of silicon, one of the most abundant elements on earth, they
are silent and clean in operation and leave no harmful waste products. Being
mechanically simple with no moving parts to wear out, they are capable of
operating for many years without degradation and with little or no
maintenance, provided they are properly protected from the weather and
accidental damage. Because they work effectively in cloudy weather, are more
efficient at low temperatures and respond rapidly to sudden changes of
illumination, they are an attractive possibility not only for sunny lands but

also for places like north-west Europe, where over half of the annual solar input comes in the form of diffuse radiation.

Photovoltaic generators of any size and voltage can be constructed quickly and easily from mass-produced modules and their efficiency, unlike that of thermodynamic generators, is practically independent of size. This modularity means that potential users of large generators can gain valid experience with a smaller version. Systems can grow as more funds become available or as demand increases. One or more modules can fail and the system continue to operate until replacements are available. But perhaps the most important characteristic of solar cells is that, because sunlight is freely available everywhere, they can generate power as and where it is needed and thus save the cost and losses of transmission lines. The long lead times, commonly ten years, associated with the planning and construction of central power stations can be avoided and with them the necessity for accurate long-term forecasting of future energy requirements.

The so-called 'photovoltaic effect' was first observed by Becquerel[1] in 1839, when he directed sunlight on to one of the electrodes in an electrolyte. Thirty-eight years later, Adams and Day[2] observed the effect in selenium. The pioneering work of Lange,[3] Grondahl,[4] and Schottky,[5] with other solid-state workers in selenium and cuprous oxide photocells, led to the development of the photographic exposure meter and many other applications. But it was not until 1954 that solar cells with an acceptably high conversion efficiency for electrical power generation were produced. In that year, Chapin, Fuller, and Pearson[6] reported that they had made silicon solar cells with an efficiency of 6% and Reynolds *et al.*[7] made a similar breakthrough with a cadmium sulphide device.

Silicon solar cells soon found their first application as a power source for spacecraft when Vanguard I, the first solar-powered satellite, was launched in 1958. Since then, practically every one of the hundreds of scientific, military, meteorological, communications, and other applications satellites launched into orbit have been powered by silicon cells. Indeed, were it not for the timely invention of these devices, the exploration and exploitation of space would not have developed as they have. Progressive improvements in technology have raised conversion efficiencies to 15% and increased spacecraft power from the few milliwatts of Vanguard I to tens of kilowatts for the latest communication satellites. Satellite life has been extended from a few months to several years and photovoltaic power generation in space has become a well-established technology with proven reliability. In the fifteen years up to 1975, the annual production of silicon cells for space programmes was about 100 kW.

Until the 1973 fuel crisis, interest in the terrestrial applications of photovoltaics was muted and sporadic but, since then, the possibilities have become more widely recognized and photovoltaic conversion is now an important element in most renewable energy research, development, and demonstration programmes. Although the capital cost is still high, it has been brought down by an order of magnitude compared with space systems and the downward trend is expected to continue, helped by the R & D effort and a rapidly expanding market, which has long since overtaken sales for space applications. In the USA, annual production of silicon cells is expected to rise from 4 MW in 1980 to 7 MW in 1981 and 14 MW in 1982. Japan's targets are 50 MW/year by 1985 and 3 GW/year by 1990. According to the Commission of the European Communities,[8] the total productive capacity in the European Community is expected to reach about 6 MW in 1982 and, by the year 2000, the total installed capacity of photovoltaic generators in Europe will have reached 10 GW.

Admittedly, these figures represent only a few per cent of the total generating capacity of the countries concerned but they indicate that photovoltaics will become a major new industry in the course of the next decade.

In this chapter, we look in some detail at the construction and operation of the silicon solar cell, the design of modules and systems, cost targets, and the many approaches to cost reduction. We then review the applications for which there is already a commercial market and discuss the near- and long-term prospects of photovoltaic technology. However, in order to understand the problems facing the designers of photovoltaic systems, it is necessary to begin with some background information on solar energy.

3.2 SOLAR ENERGY

3.2.1 Physical Characteristics

The total radiant power from the sun falling on unit area of a surface is termed the 'irradiance'. Outside the Earth's atmosphere, normal to the solar beam, it averages 1.35 kW/m^2, with a ± 3.4% variation due to the change in the distance from the sun with time of year. The irradiance within a particular spectral waveband is called the 'spectral irradiance' and, when this is plotted as a function of wavelength, the resulting curve is the 'spectral irradiance distribution' (sometimes referred to as the 'spectral energy distribution'). The uppermost curve in Fig. 3.1 shows the spectral irradiance distribution of extra-terrestrial sunlight. The radiation extends over the ultraviolet, visible, and infrared parts of the spectrum, peaking in the visible at about 500 nm. Note that the distribution closely follows the Planckian curve for a black-body radiator at 5900 K.

In its passage through the atmosphere, a considerable amount of the radiation is lost by scattering and absorption, some wavelengths being more affected than others, with the result that the distribution is substantially

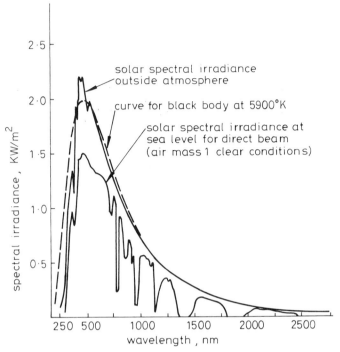

Fig. 3.1. Spectral irradiance distributions (Reproduced by permission of the Institution of Electrical Engineers)

modified. The loss of energy depends on the path length through the atmosphere and the composition of the atmosphere at the time. The term 'air mass' is commonly used to denote the path traversed by the direct solar beam, expressed as a multiple of the path traversed to a point at sea-level when the sun is directly overhead. Thus, the path length at sea-level with the sun at zenith is 'Air Mass 1' or 'AM1' and above the atmosphere it is 'AMO'. The air mass depends on latitude, altitude, time of year, and time of day.

The main absorbents in the atmosphere are ozone, which removes much of the harmful ultraviolet radiation, water vapour, which absorbs strongly in the near infrared, and carbon dioxide, which absorbs strongly in the middle infrared. The principal natural scattering agents are the air molecules and water vapour but small water droplets, dust and man-made aerosols, especially smoke, contribute to the attenuation both by scattering and absorption. The atmospheric content of these absorbents and scattering agents varies widely with location and the weather.

The lowest curve in Fig. 3.1 is a typical spectral irradiance distribution of direct terrestrial sunlight at AM1 on a clear day with a clean atmosphere. There is almost total absorption by ozone below 300 nm and by carbon dioxide

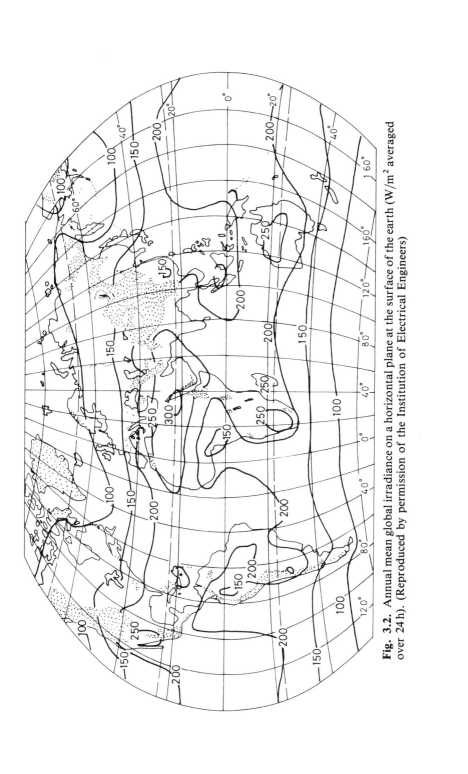

Fig. 3.2. Annual mean global irradiance on a horizontal plane at the surface of the earth (W/m² averaged over 24h). (Reproduced by permission of the Institution of Electrical Engineers)

above 2500 nm. Day-to-day variations in the distribution and the total irradiance, the area under the curve, can be considerable.

About half of the scattered energy from a clear sky is returned to the Earth in the form of diffuse radiation. The sum of the direct and diffuse components of the irradiance on a horizontal surface is termed the 'global irradiance'. The diffuse component can vary from under 20% of the global on a clear day to 100% in heavily overcast conditions. As most of the scattered light is at the blue/violet end of the spectrum, global sunlight is bluer than direct sunlight. On a clear summer day with the sun near its zenith the global irradiance at sea-level can be as high as 1000 W/m^2. In cloudy conditions, it is an order of magnitude less, rarely exceeding 100 W/m^2 even in high summer.

3.2.2 Availability

Figure 3.2, based on data from Budyko,[9] shows the availability of solar energy in various parts of the world in terms of the annual mean global irradiance in W/m^2, averaged over 24 hours. These units can be converted to the more useful units of 'insolation', that is to say the mean solar energy received by unit surface area during a specified time period, by reference to Table 3.1. The official meteorological unit of insolation is the MJ/m^2 per day, month, or year but the kWh/m^2 per year is useful to the photovoltaic system designer.

The sunniest parts of the world are seen to lie in the continental desert areas around latitudes 25 °N and 25 °S. The annual insolation falls off towards the Equator because of cloud and towards the Poles because of low solar elevations. But the difference between the Red Sea area with 300 W/m^2 and the United Kingdom, which averages about 110 W/m^2, is considerably less than might be expected.

3.2.3 Regional Distribution

Over a particular region, characterized on Budyko's map by a single figure, there can be substantial differences in insolation. For example, Fig. 3.3, reproduced from a Meteorological Office map, shows the variation of annual

Table 3.1 Insolation equivalents

Annual mean global irradiance (W/m^2)	Equivalent mean daily global insolation (MJ/m^2 per day)	Equivalent mean annual global insolation (kWh/m^2 per year)
100	8.64	876
150	12.96	1314
200	17.28	1752
250	21.60	2190
300	25.92	2628

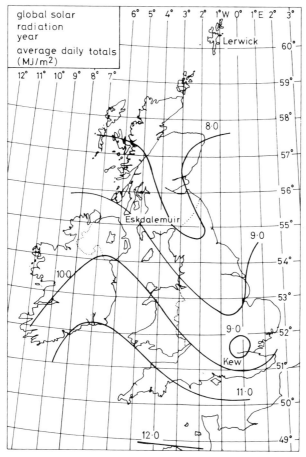

Fig. 3.3. Annual mean global insolation over UK and Ireland. (Reproduced by permission of the Institution of Electrical Engineers)

mean global insolation in MJ/m² per day over the United Kingdom and Ireland. The most favoured areas are SW England, SW Wales, and S. Ireland. Northern Ireland has high values, considering its latitude. In Scotland, the east and west coasts fare better than inland areas. Similar variations exist in other countries.

3.2.4 Seasonal Variation

Generally speaking, equatorial regions experience little seasonal variation in insolation, in contrast to higher latitudes, where the summer/winter ratios are large. To illustrate this, Figs. 3.4 and 3.5, derived from Meteorological Office

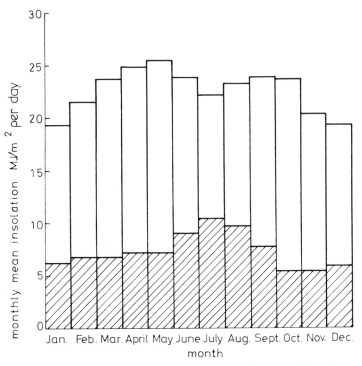

Fig. 3.4. Variation of monthly mean insolation at Aden (mean 1958–1967). (Reproduced by permission of the Institution of Electrical Engineers)

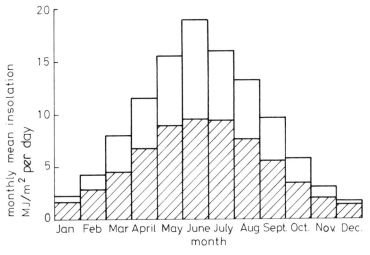

Fig. 3.5. Variation of monthly mean insolation at Kew (mean 1955–1970). (Reproduced by permission of the Institution of Electrical Engineers)

data,[10] show the variations of the monthly mean insolation at Aden (12.5 ° N) and Kew (51.5 ° N) respectively. The diffuse element is shown hatched and the direct element clear, the two together making up the global insolation. At Kew, which is fairly typical of the United Kingdom, the seasonal variation is marked, about 77% of the annual input being obtained in the six summer months. Some 60% of the annual insolation is diffuse and in winter the proportion can be as high as 70% of the monthly total. In contrast, the seasonal variation at Aden is less than ± 14% and only about a third of the global insolation is diffuse.

3.2.5 Influence of Surface Orientation

So far, the discussion has been confined to the radiation falling on a horizontal surface, the parameter measured by most meteorological stations. However, if a flat-plate photovoltaic array is tilted towards the sun, the direct irradiance is increased in proportion to the cosine of the angle of incidence, the diffuse element is reduced and some ground reflection is added. The bonus from ground reflection can be substantial where there are highly reflective surfaces such as concrete, sand, or snow. The array will collect the maximum amount of solar energy if it is kept facing the sun on clear days (the tracking need not be highly accurate) and moved to the horizontal when the sky is overcast. But for the simpler and more reliable static array, the optimum tilt angle will depend on latitude, the proportion of diffuse radiation, seasonal variations of insolation, and load requirements.

In an ideal situation, with clear sunny weather throughout the year, a fixed array will receive maximum input at an inclination equal to the latitude. But a lower inclination will be better for sites with a high proportion of diffuse radiation and a steeper angle will increase output during sunny winter days and thus help to reduce the seasonal variation. More will be said on this subject in Section 3.6.

3.3 THE SILICON SOLAR CELL

3.3.1 The Photovoltaic Effect

To understand the photovoltaic effect, it is necessary to go back to some basic atomic concepts. In the simplest model of the atom, negatively charged electrons orbit a central nucleus and have discrete kinetic energy levels which increase with the orbital radius. When atoms bond together to form a solid, the electron energy levels merge into bands. In electrical conductors, these bands are continuous but in insulators and semiconductors there is an 'energy gap' of width E_g, in which no electron orbits can exist, between the inner or 'valence' band and the outer or 'conduction' band. Valence electrons help to bind the atoms in a solid together, while conduction electrons, being less closely bound to the nucleus, are free to move in response to an applied voltage or electric field. The fewer conduction electrons there are, the higher the electrical resistivity of the material.

In semiconductors, the materials from which solar cells are made, the energy gap is fairly small and so electrons can be thermally excited from the valence to the conduction band. This explains why the resistivity of semiconductors, unlike that of conductors, decreases as the temperature is raised.

At room temperatures, pure ('intrinsic') semiconductors have high resistivities but the resistivity can be greatly reduced by 'doping', i.e. introducing a small amount of impurity. There are two types of dopant. Those which have one more valence electron than the semiconductor are called 'donors' and cause extra electrons to appear in the conduction band. Semiconductors so doped are termed '*n*-type'. The other type, with one less valence electron per atom than the semiconductor, are called 'acceptors'. They leave so-called 'holes' in the valence band and render the semiconductor '*p*-type'. Holes act like positive charges and, although less mobile than electrons, move when a voltage is applied and contribute to the electric current. The effect of doping on resistivity is pronounced, even if only one impurity atom is added to every million atoms of the semiconductor.

Figure 3.6 shows the essential features of a solar cell. It is made from a wafer of crystalline silicon, the best semiconductor for the purpose at present, which has been doped with boron, an acceptor impurity, to a resistivity of 1 to 10 ohm cm. The cell, which is normally one-third of a millimetre thick, can

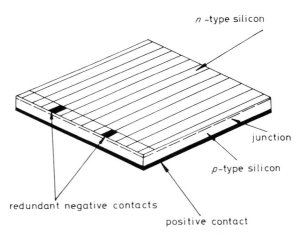

Fig. 3.6. Silicon solar cell. (Reproduced by permission of the Institution of Electrical Engineers)

be square, round, or semicircular in shape and up to 100 mm across. Phosphorus, a donor impurity, is diffused into the wafer at about 850 °C to form a $p-n$ junction a fraction of a micrometre below the front (active) surface. This is the so-called 'n-on-p' cell. The alternative p-on-n type can be made by diffusing boron into a phosphorus-doped wafer. The front contact is in the form of a narrow-fingered grid, while the back contact usually covers the entire back surface. A single-layer antireflective coating is usually applied to the front surface.

At the junction, conduction electrons from the n-region diffuse into the p-region and combine with holes, cancelling their charges. The opposite action also takes place, holes from the p-region crossing into the n-region and combining with electrons. This results in the disappearance of electrons and holes from the vicinity of the junction, the so-called 'depletion area' or 'barrier layer', and leaves behind layers of charged impurity atoms, which are positive on the n-side and negative on the p-side. Thus a *reverse* electric field is set up around the junction.

When light falls on the active surface, photons with energy exceeding the energy gap E_g of the semiconductor (1.1 eV in the case of silicon) interact with valence electrons and lift them to the conduction band. This movement leaves behind holes, so the photons are said to generate 'electron-hole pairs'. In crystalline silicon, these 'carriers' are generated throughout the thickness of the cell in concentrations depending on the intensity and spectral composition of the light. Photon energy is inversely proportional to wavelength. The highly energetic photons in the blue/violet part of the spectrum are absorbed near the surface, while the less energetic, longer wavelength photons in the red and infrared parts are absorbed deeper in the crystal and further from the junction. The electrons and holes diffuse through the crystal, the electrons moving directly and the less mobile holes by valence electron substitution from atom to atom. Some recombine after a 'lifetime' of the order of one millisecond, neutralizing their charges and giving up energy in the form of heat. Others reach the junction before their lifetime has expired and are there separated by the reverse field, the electrons being accelerated towards the negative contact and the holes towards the positive. If the cell is connected to a load, there will be a drift of electrons from the negative contact through the load to the positive contact, constituting an electric current. In silicon, the generated current in any particular type of irradiation is directly proportional to the irradiance and the active area of the cell. For all practical purposes, it is equal to the short-circuit current.

3.3.2 Spectral Response

The generated current is composed of increments produced by photons of various wavelengths within the effective range. The highly energetic short

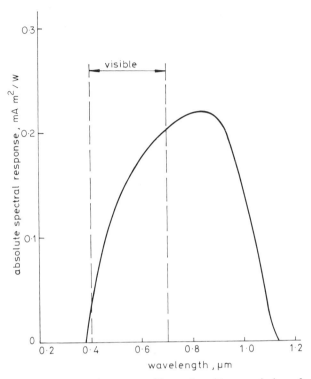

Fig. 3.7. Spectral response. (Reproduced by permission of the Institution of Electrical Engineers)

wavelengths in the blue/violet part of the spectrum are absorbed near the surface, while the longer ones penetrate deeper into the silicon, most being absorbed within a thickness of 100 μm. If the incremental current generated by unit irradiance is plotted as a function of wavelength, the resulting curve is called the 'absolute spectral response'. Figure 3.7 shows a typical example for a modern cell. Note that it covers the whole of the visible spectrum and part of the infrared, peaking at about 0.9 μm.

The short-circuit current of a solar cell in radiation of known intensity and spectral composition can be computed by multiplying the ordinates of the absolute spectral response by the corresponding ordinates of the absolute spectral irradiance distribution (e.g. Fig. 3.1) and integrating the resulting products.

3.3.3 Voltage-current Characteristic

Figure 3.8 shows the voltage-current characteristic of a typical modern silicon solar cell at 25 °C in sunlight at an irradiance of 1000 W/m². Maximum power,

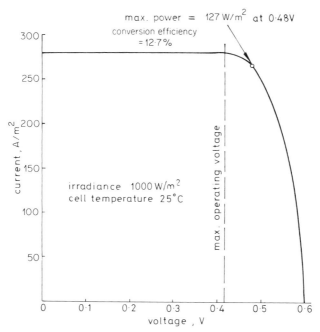

Fig. 3.8. Voltage-current characteristic

represented by the area of the largest rectangle that can be fitted under the curve, is produced at a voltage between 0.4 and 0.5 V, depending on the quality and base resistivity of the cell. In the case illustrated, it is $127\,\text{W/m}^2$ at 0.48 V.

The conversion efficiency is the maximum output power, expressed as a percentage of the input power:

$$\text{Conversion efficiency} = \frac{\text{maximum power}}{\text{irradiance} \times \text{area}} \times 100\%$$

The accepted convention among engineers is to use the total cell area, inclusive of the front contact grid, in this calculation, but some research workers calculate their efficiencies on the basis of the active area and exclude the grid.

The maximum operating voltage is normally fixed by the array designer so as to be on or slightly to the left of the 'knee' of the characteristic at the maximum operating temperature. Note that the output current of a good quality cell is almost constant over the voltage range up to this point, a feature that makes a photovoltaic generator particularly suitable for battery charging.

The maximum output power of a photovoltaic generator under standard conditions of irradiance ($1000\,\text{W/m}^2$) and temperature ($25\,^\circ\text{C}$ or $28\,^\circ\text{C}$) is commonly referred to as its 'peak' rating. Thus, a generator may be said to have a rated power of 100 kW (peak) or 100 kWp. The silicon cell production figures quoted in Section 3.1 were in terms of peak ratings. When comparing photovoltaic generation with conventional electricity generation, it is important to remember that the average output power will be only a fraction of the peak rating, because the irradiance, averaged over 24 hours, is considerably less than 1000 W/m² and solar cells normally operate at temperatures higher than 28 °C. The relationship between average and peak power is illustrated later in Table 3.2.

3.3.4 Effect of Change of Irradiance

Changes of irradiance (Fig. 3.9) affect the short-circuit (generated) current proportionally but have little effect on the open-circuit voltage until low levels are reached, because the latter is logarithmically related to the irradiance. Conversion efficiency is little affected. The operating voltage remains to the left of the knee of the curve, so a battery would continue to be charged at the lower irradiance.

A similar effect is caused as the angle of incidence of a beam of light is increased. The short-circuit current, following the irradiance, varies approximately as the cosine of the angle of incidence. So, if the upper curve in Fig. 3.9 is the performance at normal incidence, the lower one would correspond to an angle of incidence of $\cos^{-1} 0.2$, i.e. about 78.5 °.

In concentrated sunlight, the short-circuit current remains proportional to the irradiance up to extremely high levels, provided the temperature is kept constant. As the open-circuit voltage also increases, albeit logarithmically, one

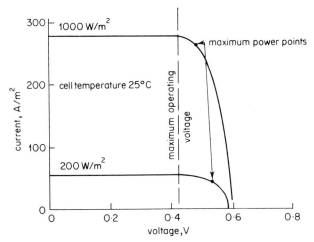

Fig. 3.9. Effect of change of irradiance. (Reproduced by permission of the Institution of Electrical Engineers)

would expect the conversion efficiency to improve and so it does, up to a point. But the series resistance of the cell has a progressively flattening effect on the characteristic and this limits the improvement that can be achieved in practice.

3.3.5 Effect of Temperature Change

An increase in temperature (Fig. 3.10) causes a slight rise in short-circuit current but a sharp fall in open-circuit voltage and maximum power. Typically, the maximum power falls by about 0.5% for every one degree Kelvin rise in temperature, so it is important for the designer to ensure that the cells in a photovoltaic generator run as coolly as possible under operational conditions. Note that the voltage for maximum power also falls with increasing temperature but, provided the operating voltage does not exceed the voltage for maximum power at the highest operating temperature, the output current is little affected by changes of temperature below the maximum.

3.3.6 Performance Limitations

Wolf[11] listed seven factors which limit the performance of solar cells:
(1) *Reflection losses.* Some of the incident radiation is lost through reflection from the surface of the cells and the materials used to protect them from the weather.
(2) *Incomplete absorption.* Photons which have less energy than the energy gap E_g will simply generate heat in the cell or pass right through it. The higher the energy gap, the greater the wastage.

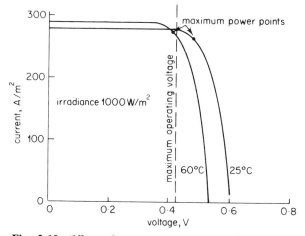

Fig. 3.10. Effect of temperature change. (Reproduced by permission of the Institution of Electrical Engineers)

(3) *Partial utilization of photon energy.* Many of the photons which generate electron-hole pairs have more energy than is needed for this operation. The excess energy is dissipated as heat. In this case, the higher the energy gap, the smaller the wastage. Taking the previous loss factor into account, it is found that an E_g of 0.9 eV gives maximum utilization of the solar spectrum, about 46% of the total incident energy.

(4) *Collection losses.* Only those carriers that reach the junction before recombining are collected and contribute to the output current. The others simply generate heat and do no useful work. Factors affecting the collection efficiency are:

1. The absorption characteristic of the semiconductor, which, with the spectral irradiance distribution of the incident radiation, determines the geometrical distribution of the generated electron-hole pairs in the crystal.

2. The junction depth.

3. The width of the depletion layer.

4. The rate at which electrons and holes combine at the surface (surface recombination velocity).

5. The average distance an electron will travel in the *p*-region and a hole in the *n*-region before recombining (the minority carrier diffusion length).

6. The existence and strength of any inbuilt electric fields, resulting from impurity concentration gradients in the surface and base regions, which help to accelerate carriers towards the junction.

(5) *Voltage factor.* The open-circuit voltage is always less than the energy gap for the following reasons:

1. An open-circuit voltage equal to the potential of the electric field at the junction (the 'barrier height') can be obtained only at extremely high inputs, which can never be reached by photon absorption from unconcentrated sunlight.

2. Because of the effects of doping, the barrier height is always less than E_g.

An increase in E_g will raise the open-circuit voltage but, if it is increased beyond the level for maximum generation of electron-hole pairs, current generation will be reduced. There is therefore an optimum value of E_g for any particular spectral distribution, at which the product of short-circuit current and open-circuit voltage is a maximum. For direct terrestrial sunlight, the theoretical optimum is 1.4 eV.

(6) *Curve factor.* Because the shape of the voltage-current characterisitic is dependent on the characteristic of the junction diode, the maximum power is always less than the product of short-circuit current and open-circuit voltage, even without series resistance. The quality of the junction and hence the 'squareness' of the voltage-current characteristic, improves as E_g is increased.

(7) *Series resistance losses.* Power loss in the cell due to series resistance causes a flattening of the voltage-current characteristic. It can be minimized by good contact grid geometry, good ohmic contacts and low sheet resistance in the surface layer. A measure of the effects of junction quality and series resistance on performance is given by the 'fill factor', which is defined as:

$$\text{Fill factor} = \frac{\text{maximum power}}{\text{short-circuit current} \times \text{open-circuit voltage}}$$

For further information on the operation and performance characteristics of solar cells, the reader is referred to the classical papers on the subject by Prince,[12] Rappaport,[13] and Wysocki.[14]

3.3.7 Present Status

Most present-day commercial silicon solar cells have AM1 conversion efficiencies ranging from 10 to 15% at 25 °C. The theoretical limit under these conditions is about 23%,[12] so there is still considerable scope for improvement, provided this can be done cost effectively.

The best terrestrial cells already have many of the features developed for space solar cells, such as:

(1) Efficient antireflective coatings of evaporated tantalum oxide or titanium oxide.
(2) Shallow junctions (1000–2500 A) for enhanced blue/violet response.[15]
(3) An impurity concentration gradient at the back of the cell ('back surface field'), which accelerates carriers generated deep in the cell towards the junction and thus gives better red response and higher open-circuit voltage.[16]
(4) Moisture-proof contacts of evaporated Ti/Pd/Ag.
(5) Fine grid front contact for high light transmission and low series resistance.

The silicon solar cell is extremely stable, there being no inherent degradation mechanism in the terrestrial environment, although in space it is subject to radiation damage.

3.4 MODULE ENGINEERING

3.4.1 Description

A module, the basic building block of a terrestrial flat-plate photovoltaic solar array, consists of a number of interconnected solar cells, usually but not always in a single series string, encapsulated behind a transparent window. Bolt holes or clamps are provided for fixing it to a supporting framework and it is fitted with either terminals, a pigtail lead, or a plug-and-socket connector for connecting it to other modules or the load.

Figure 3.11 shows a module of current (1981) design. It consists of 33 100 mm diameter silicon cells connected in series and is rated to produce 35 Wp at 155 V and a cell temperature of 25 °C in sunlight at an irradiance of 1000 W/m². The cells are encapsulated in transparent polyvinylbutyral (pvb), between a window of tempered glass and a backing of white acrylic-coated steel foil, the whole being surrounded by a polysulphide rubber edge seal and an anodized aluminium frame. The overall dimensions are 1219 mm long × 305 mm wide × 38 mm thick.

3.4.2 Design Requirements

To offset the high capital cost of solar cells, modules must be capable of

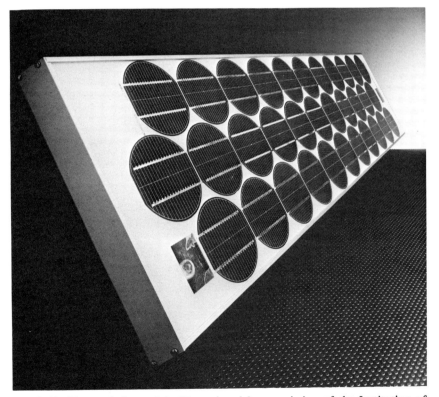

Fig. 3.11. Photovoltaic module (Reproduced by permission of the Institution of Electrical Engineers) (Arco Solar Inc.)

prolonged maintenance-free operation under any conditions they are likely to encounter. The current R & D target is a lifetime of twenty years.

Window and encapsulant materials must be highly transparent to radiation in the waveband 350 nm to 1200 nm and must not be unduly affected by weathering and prolonged exposure to sunlight. The window must have good impact resistance, in particular against the large hailstones encountered in some parts of the world, and a hard, smooth, flat, abrasion-resistant, non-staining surface, which will promote self-cleaning by wind, rain, or spray. It should be free of projections which might provide lodgement for water, dust, and other foreign matter.

The encapsulation system must be highly resistant to the permeation and ingress of gases, vapours, and liquids, as condensation on the cells and circuitry may cause shorts or galvanic corrosion. Failure of the adhesive bonds in the assembly will increase the rate of moisture absorption and chemical reaction at the interfaces and will also increase the cell temperature under operational conditions. Delamination between the window and encapsulant or between the encapsulant and the front surfaces of the cells will increase reflection losses and thus degrade the performance of the module. Materials in the assembly must therefore be compatible and the bonds between them capable of withstanding the extremes of temperature and repeated thermal cycling that will be encountered in practical applications.

To keep operational temperatures as low as possible, the module should be designed to take full advantage of radiative, convective, and conductive cooling and absorb the minimum of unused solar energy.

The solar cells must be well insulated against the high voltages possible when several modules are connected in series. Existing specifications call for modules to withstand the application of 1500 V d.c. for one minute.

Finally, the module must be strong and rigid enough to support the fragile cells before, during, and after installation in the array. It should be capable of accommodating imperfections of the mounting structure, withstand wind-induced vibrations, and take the loads imposed by high winds, snow and ice. Modules should be easy to mount, interconnect, and replace. Mountings, terminals, and plug-and-socket connectors must be non-corrosive.

3.4.3 Encapsulation Systems

At present, most module manufacturers use glass as the window material. It is highly transparent, stable, impervious, easily cleaned, and its impact resistance can be made adequate by tempering or toughening processes or lamination with plastics. But it is rather heavy and not particularly cheap.

Plastics, although lighter and, in some cases, cheaper, are generally less stable and not so resistant to abrasion and contamination. Moreover, they are all to some extent permeable to moisture and gases.[17] Silicone rubber resins, used in some early modules as a window material, are among the most permeable but polyesters and acrylics are reasonably good in this respect. Transparent plastics all suffer some transmission loss after prolonged exposure to ultraviolet radiation, epoxy resins being the worst affected and acrylics, polyesters, and u.v.-stabilized polycarbonate (Lexan) among the least affected.

Glass, aluminium, glass fibre-reinforced polyester sheet, and plastic-coated metal foils have all been used as backing materials, each with its own advantages and disadvantages. A glass backing is moisture-proof and it enables most of the unused radiation to pass through the module, thus reducing operating temperatures. But, again, it is heavy and relatively expensive. Aluminium is strong and light and it facilitates conduction of heat to the back surface but great care must be taken to insulate it adequately from the solar cells. Polyester sheet is light, cheap, and a good electrical insulator but it is not a perfect barrier against moisture. White plastic-coated metal foil, as used in the module shown in Fig. 3.11, combines cheapness and good thermal conduction with good insulation and weather resistance. With circular cells, light reflected from the white surface and re-reflected by the window on to the cells has been found to give a slight gain in performance.

Silicone rubber used to be the most commonly-used encapsulating material. It is highly transparent and is resistant to fatigue failure after repeated heating and cooling cycles. However, it is being superseded by polyvinylbutyral (pvb) and ethylene vinyl acetate (eva), which are cheaper, easier to apply and give a better bond to the window, cells, and backing. To minimize the possibility of moisture ingress in the event of delamination at the edges, many designs of module embody a separate edge seal.

Outdoor exposure tests, field trials and accelerated environmental tests in the laboratory[18,19] have shown that the principal failure modes in photovoltaic modules are: cracked cells, interconnect failures, delamination of the encapsulant, and dielectric breakdown.

More recently, failures in arrays have been attributed to the so-called 'hot-spot' phenomenon.[20] When one element in a series connected string is not matched to the others, because of shadowing or poorly matched or damaged cells, it may dissipate current instead of generating it, thus causing overheating. The extent of the temperature rise will depend on the string voltage, the reverse voltage-current characteristic of the affected cell and the rate at which heat is dissipated from it. Failure can be prevented by connecting shunt diodes across sections of the string at suitable intervals to by-pass the current in the event of an imbalance. Badly matched modules in parallel can be subject to overheating due to current reversal in certain circumstances. This can be prevented by inserting diodes in series with the modules.

Over the past few years, considerable progress has been made towards reliable long-life modules and most manufacturers are now prepared to guarantee their products for five years. Authorities in USA and Europe are preparing specifications of standard design qualification tests,[21] which it is hoped will eventually gain general acceptance in the industry.

3.5 STORAGE

In many applications of photovoltaics, it is necessary to store some of the generated electricity for use during periods of darkness, low insolation, and peak loads. Although many new storage techniques, such as flywheels and the electrolysis of water, are being developed and hydraulic storage may be economic where conditions are favourable, the only practical solution for most small-scale applications at present is the electrochemical battery.

Batteries for photovoltaic systems need to have a high charging efficiency, low open-circuit losses and a long, trouble-free working life under difficult environmental conditions. They must be capable of withstanding daily charge–discharge cycles of varying depth and a superimposed seasonal cycle. No existing battery fulfils all these requirements at low cost but some R & D work in this area is being sponsored by the Commission of the European Communities[22] and the US Department of Energy. In the meantime, lead-acid batteries and, to a lesser extent, nickel-cadmium batteries are being used.

The lead-acid battery is commercially available at a cost of about $60 per kWh for large quantities and has a charging efficiency of 75%. The common vented type, which requires periodic topping-up, does not have good charge retention and cycling characteristics but sealed types with special 'low loss' alloy grids are now available.

Nickel-cadmium batteries have excellent charge retention and cycling properties but their charging efficiency is only 70% and they cost upwards of $300 per kWh at present. They are available in both vented and sealed forms. Vented cells have the lowest open-circuit losses and the best charging characteristics but require occasional topping-up. Sealed batteries can operate in any position and require no maintenance, as the gases generated during overcharge recombine under pressure within the cell.

3.6 SYSTEM DESIGN

A typical stand-alone photovoltaic system (Fig. 3.12) consists of an array of flat-plate modules facing the Equator at a fixed inclination and charging a battery which supplies the load either directly or through power conditioning equipment. A simple electronic charge regulator is included when necessary to protect the battery against overcharge during periods of low usage or after a succession of long sunny days. Solid state diodes are connected between the array and battery to prevent the solar cells loading the battery when they are not illuminated and to protect the battery in the event of a short-circuit

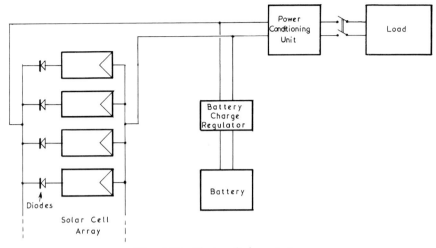

Fig. 3.12. Photovoltaic system

developing in the array. The battery can be substantially reduced in size or even eliminated in applications such as air conditioning, refrigeration, water pumping, and irrigation, where the load cycle can be matched to the solar input or an alternative form of storage, e.g. ice or water reservoir, provided.

The system designer has to work out the cheapest combination of array area, inclination, and storage capacity that will meet his load requirements under the worst conditions at the selected site throughout the demand period. The design steps are essentially as follows:

(1) Determine the maximum operating temperature the solar cells are likely to reach, discounting extremes that may be attained exceptionally on, say, less than 2% of days.

(2) Choose a module or group of modules in series which will give as nearly as possible its maximum power at the working voltage when operating at this maximum temperature. The working voltage will normally be determined by the top-of-charge voltage of the battery, after making due allowance for voltage drop across any protective diodes.

(3) Calculate the mean ampere-hours output of the module or series group for every month of the year from the voltage-current characteristic of the module at the monthly mean operating temperature and the monthly mean global insolation at an inclination equal to the angle of latitude (including an assumed contribution from ground reflection). Reduce these figures by, say, 30% to allow for battery losses, window contamination, unusual weather, etc. to yield net monthly means.

(4) Compare the net monthly mean outputs with the monthly mean load

requirements and calculate the number of parallel modules or series groups required to balance the summer excess against the winter deficit.

(5) Repeat steps (3) and (4) for other inclinations and determine the angle which results in the fewest modules.

(6) From the daily mean charge–discharge cycles, calculate the battery capacity required to meet load requirements during the winter, starting with a fully charged battery in the autumn. Check that this capacity is adequate to carry the load over the longest period of consecutive sunless days likely to be experienced on, say, a 98% probability basis.

(7) Estimate the cost of the complete system, including wiring, support structures, etc.

(8) Repeat steps (6) and (7), assuming more than the minimum number of modules, in order to determine the cheapest combination of array area and storage capacity.

Of course, the optimum will depend on the relative costs of modules and batteries over the projected life of the installation and will be affected by any change in this relativity. At present, the cheapest system is almost always the one with the smallest array but this situation is likely to change in the next decade.

In some applications, there are constraints which may determine the inclination of the array and thus simplify the design procedure. For example, in an installation on a pitched roof, structure costs can be minimized by conforming to the roof line. In cases where a horizontal array is the theoretical optimum, for instance near the Equator, it may be necessary to provide a slight tilt to promote self-cleaning.

As most meteorological stations can at present supply only global insolation data on a horizontal surface, programmes have been devised for computing monthly means of insolation on inclined surfaces.[23,24] Where diffuse insolation data are not available from measurements, they can be estimated.

Most photovoltaic system suppliers have set up computerized design facilities to meet their customers' individual requirements and these are now being extensively used.

Site location does not affect the size of the array as much as might be expected. As a rough guide, Table 3.2 gives the minimum module areas and equivalent peak powers required to yield a mean output of 2.4 kWh per day (i.e. 100 W continuous) in various parts of the world. The calculations are based on Budyko's annual mean global irradiance data (Fig. 3.2), a module conversion efficiency of 10%, an allowance of 25% for battery losses, and a 5% allowance for other losses such as window contamination. The factor relating the size of an array in NW Europe with that of an equivalent installation in the sunniest part of the world, the Red Sea area, is less than 3. This factor would be even smaller if irradiance data on optimally-inclined surfaces

Table 3.2 Approximate array sizes for a mean output of 2.4 kWh/day

Region	Annual mean global irradiance (W/m²)	Minimum module area (m²)	Equivalent peak power (Wp)
NW. Europe	110	12.8	1280
S. France S. Germany Japan N. USA New Zealand	150	9.4	940
Brazil East Indies Greece Italy Spain S. Australia Mid and SE USA	200	7.0	700
California India Middle East Mid-Australia S. Africa Sahara West Indies	250	5.6	560
Red Sea	300	4.7	470

were used instead of horizontal data. The problem in the higher latitudes is not so much the total available solar energy as the marked seasonal variation (Fig. 3.5).

The fixed array is the most common type but it is not the only option. Other possibilities with flat-plate arrays, which may prove more cost effective in certain circumstances, particularly as array sizes increase, are:

(1) *Seasonally-adjusted tilt.* The inclination of the array is manually adjusted at monthly or quarterly intervals to allow for the changing solar elevation at noon.

(2) *Single-axis tracking.* The array is swung about a vertical or polar axis to follow the sun through the day.

(3) *Two-axis tracking.* The array is continuously orientated by a sun-tracking device to keep its active surface normal to the solar beam.

(4) *Augmented irradiance.* Simple planar reflectors are used to boost the useful irradiance on the cells.

In clear conditions, a sun-tracked flat-plate array can collect up to 50% more solar energy than one fixed at the optimum inclination. However, the provision of automatic continuous tracking can greatly complicate what would otherwise be an extremely simple system. Where the necessary labour is available, a better alternative may be manual adjustment, which can be surprisingly effective. For example, a flat-plate array moved to face the sun twice a day in mid-morning and mid-afternoon and given a quarterly tilt adjustment, can collect nearly 95% of the energy collected with full two-axis tracking.

An important aspect of photovoltaic system design is to ensure as far as possible that all energy-consuming components are of the highest efficiency and are well matched to the generator under all conditions of irradiance and load. There is not much point in striving for the utmost in photovoltaic conversion efficiency if an undue proportion of the expensively generated power is dissipated in inefficient appliances. Happily, the importance of the 'total system' approach is being increasingly recognized.

3.7 COST GOALS

In considering the prospects for cost reduction, it is instructive to look at the ambitious programme of photovoltaic research, development and 'commercialization' initiated in the USA in 1975. Figure 3.13 shows the Department of Energy's goals for module price reduction in terms of 1980 dollars, with the prices paid for successive Government-funded 'block buys' indicated by

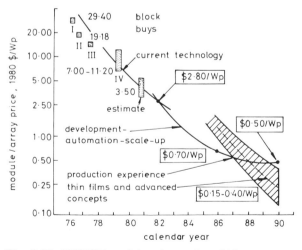

Fig. 3.13. US DOE module price goals and history (1980 dollars). (Reproduced by permission of the Institution of Electrical Engineers)

Table 3.3 US DOE price goals

Year	Module price ($/Wp)	System price ($/Wp)	Prospective market
1982	2.80	6.00–13.00	Small remote loads
1986	0.70	1.60–2.20	Grid-connected residences and intermediate load centres
1990	0.15–0.50	1.10–1.30	Central power stations

shaded rectangles.[25] Specific goals and prospective markets are listed in Table 3.3. The solid line in Fig. 3.13 indicates what it is hoped to achieve with development of the present crystalline silicon technology, coupled with market growth and the automation of production processes. The hatched area shows what may be achieved with thin films and other advanced concepts.

It is now generally accepted that, although silicon module prices may eventually reach a level of $1 or $2/Wp, it will take longer than indicated in the DOE projection. So the time scale is likely to be revised, particularly in view of recent cuts in Federal R, D & D funding. At the present time (1981), modules are being sold commercially in the USA for about $10/Wp, while in Europe prices are somewhat higher. Considerably lower prices have been bid for recent Federal bulk procurements but these prices do not include advertising and marketing costs and probably allow little or no profit margin.

The system price goals in Table 3.3 include 'balance-of-system' items, such as land, fencing, array structure, wiring, power conditioning, energy storage, marketing, transportation, and installation. As modules become cheaper, the cost of these items will assume greater importance. As many balance-of-system costs are area-dependent, an important aspect of cost reduction is to reduce array area by raising conversion efficiency. Most authorities agree that the net array conversion efficiency must be at least 10% if the ultimate price goals are to be achieved.[26] Operating and maintenance costs also figure in the overall life cycle cost of a system, so any reduction in capital cost must not be made at the expense of a disproportionate increase in running cost or a shorter lifetime.

3.8 SILICON COST REDUCTION

3.8.1 Silicon Material

An important part of the current R & D effort is devoted to finding cheaper and more energy-efficient ways of producing the silicon crystal wafer, which at present accounts for about 20% of the module cost.

The present Siemens process for producing the high purity semiconductor-grade polysilicon feedstock for crystal growing consumes about 200 kWh per kg and costs about $63 per kg. Union Carbide, Hemlock, and Battelle are developing improved processes which, on the basis of plants producing 1000 tonnes per year, are expected to cut energy consumption to 90 kWh per kg and meet DOE's cost goal of $14 per kg.[27] Union Carbide plan to have a 100 tonnes per year pilot plant in operation by December, 1981.

Dow Corning[28] and Westinghouse[29] are developing processes for producing low cost 'solar-grade' silicon, which would not be as pure as semiconductor-grade but might be good enough for solar cells.

Efforts to cheapen the present Czochralski crystal growing process are being made by drawing several large ingots from a single crucible, with melt replenishment. One group has succeeded in growing four 120 mm diameter ingots in this way and have made cells of about 10% efficiency with them.[30]

A potentially cheaper alternative to single crystal Czochralski material has recently appeared on the market under the trade names 'Silso' and 'Semix'. This is made by the controlled casting of silicon in square or rectangular crucibles, followed by annealing. The result is a block of multicrystalline or 'semicrystalline' silicon with a grain size of the order of 1 mm. It requires an order of magnitude less energy to produce (about 10 kWh per kg) and, although cells made from it are not yet as efficient as the best single crystal cells, this disadvantage will probably disappear with further development. Square or rectangular cells can be packed closer in the module than circular or semicircular ones, thus increasing module efficiency. In Europe, Wacker Heliotronic, the manufacturers of Silso, have decided to invest heavily in this material and plan to increase their production capacity to 1800 tonnes per year by the end of 1981.[31] Similar plans have been revealed by the Semix Corporation, in co-operation with the US Department of Energy.[32] Solarex, the parent company of Semix, have also decided to build a $20M silicon refining facility to feed their Semix plant. The change to multicrystalline silicon, if successful, will relieve the solar cell industry of its present dependence on single crystal material, for which it has to compete on disadvantageous terms with the more powerful micro-electronics industry.

Another interesting approach is the directional solidification of silicon ingots by the 'heat exchange' method (HEM).[33] This is still under development but Crystal Systems have succeeded in producing 4 kg single crystals of 15 cm × 15 cm cross-section from cheap metallurgical-grade feedstock.

The sawing of ingots into wafers, even with the sophisticated equipment now available, is a time-consuming operation and wastes up to 70% of the crystal. To avoid sawing altogether, many techniques are being developed to produce crystalline silicon in a continuous thin ribbon. This has obvious attractions as part of a fully automated plant. Some of the more important approaches are listed in Table 3.4 and illustrated in Fig. 3.14. Progress is being made but the indications are that it will be some years before ribbons challenge sliced ingots in terms of throughput, reproducibility, and efficiency. Figure

Energy — Present and Future Options

Table 3.4 Silicon ribbon processes

Process	Method	Reference
Self-supported Edge-defined film-fed growth (EFG) (Mobil-Tyco) *and* Capillary action shaping technique (CAST) (IBM)	The ribbon is pulled upwards through a graphite die fed by capillary action from the melt.	34
Inverted Stepanov and Pendant drop growth	The ribbon grows downwards through a non-wetted silica die in the base of the crucible. The melt is pressurized.	35
Dendritic web (Westinghouse)	A thin sheet of silicon is formed between two growing dendrites, which act as shaping guides.	36
Ribbon-to-ribbon (RTR) (Motorola)	A pre-formed ribbon of polycrystalline silicon is passed through a laser beam to enhance the grain size by local recrystallization.	37
Substrate-supported Silicon-on-ceramic (SOC) (Honeywell)	A slightly inclined substrate of aluminium silicate is passed over a narrow trough of molten silicon.	38
Silicon-on-paper (RAD) (LEP, France)	A graphite-coated paper ribbon is pulled downwards through a slit in the base of a graphite crucible of molten silicon.	39

3.15 shows an early Japanese module made with silicon ribbon solar cells. Note the close packing.

3.8.2 Silicon Cell Fabrication

The conventional techniques used in the manufacture of silicon solar cells, such as the thermal diffusion of phosphorus and the vacuum deposition of contacts and antireflective coatings, are batch processes, requiring much handling of the individual wafers. To cut costs, these are gradually being superseded by processes that can be automated. One of the most important developments in this field is junction formation by ion implantation. This is a low energy process and it affects only one side of the wafer, thus eliminating the back junction removal which is necessary with conventional diffusion. It can also be used to implant a back surface field for enhanced open-circuit voltage and red response. In the USA, Spire Corporation[40] have developed implanting equipment for processing 300 wafers per hour and have designs for

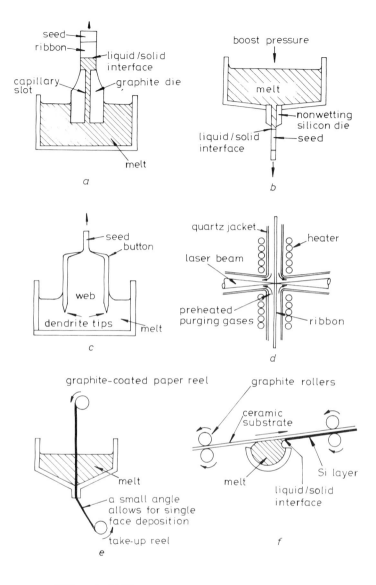

a   EFG and CAST

b   Inverted Stepanov and pendant drop growth

c   Dendritic web

d   Ribbon-to-ribbon

e   Silicon on graphite-coated paper

f   Silicon on ceramic

Fig. 3.14. Silicon ribbon techniques. (Reproduced by permission of the Institution of Electrical Engineers)

Fig. 3.15. Japanese module with silicon ribbon solar cells. (Reproduced by permission of the Institution of Electrical Engineers)

plants with capacities of 10 and 100 MW per year. They claim to have achieved efficiencies of 15% in single crystal and 10% in multicrystalline cells. At present, the cells are thermally annealed after implantation but laser and pulsed electron annealing techniques are being successfully developed.

Diffusion from sprayed or spun-on dopants is another low-cost technique under development but it is not sufficiently reproducible at the present stage. Contact metallization by screen printing,[41] which is adaptable to mass production, is being used already in some factories. There are some doubts as to whether screen-printed contacts will withstand moisture, so an impermeable module package is essential.

Sprayed or spun-on single layer antireflective coatings are being developed with some success. Several of the process sequences under current development are compatible with a module price of $1/Wp (1980 dollars), provided that the expected reductions in silicon material costs are realized.

3.9 CONCENTRATION

The use of concentrated sunlight is an important aspect of cost reduction, as fewer solar cells are necessary for a given output and in some applications the heat from the cooling system can be put to good use. Furthermore, as has already been explained, concentration can increase the conversion efficiency.

Attempts to adapt silicon cells for concentrated sunlight have met with considerable success. Recently, an efficiency of 20% at 50 suns has been claimed[42] and 16% has been reached at 200 suns. The main feature of these cells is a fine front contact grid to reduce internal series resistance.

For higher concentrations, one must turn to gallium arsenide cells, as they are less temperature-dependent and more efficient than silicon. (The power loss per degree Kelvin rise is 0.2 to 0.3%.) Since gallium arsenide, unlike crystalline silicon, is a so-called 'direct bandgap' semiconductor, it has a high absorption coefficient. Because of this, nearly all the electron-hole pairs are generated within a few micrometres of the surface and a long minority carrier life-time is not necessary for good collection efficiency. Figure 3.16 shows the cross-section of a typical modern cell, with the vertical dimensions greatly exaggerated for clarity. The gallium aluminium arsenide window reduces surface recombination and internal series resistance and improves the shunt resistance. The efficiencies of these cells have now reached 25% at 180 suns[43] and 23% at 1000 suns.[44]

Significantly higher efficiencies can theoretically be obtained with multi-junction or cascade structures having layers of semiconductors with pro-gressively diminishing energy gaps. For example, it is theoretically possible to achieve 40% at 1000 suns with a two-junction cascade of gallium aluminium arsenide and gallium indium arsenide, but performances approaching this level have not yet been realized.

Various types of concentrator are being developed and tested in photovoltaic systems. Four of the principal ones are illustrated in Fig. 3.17. The parabolic trough reflector and linear Fresnel lens can be used for concen-trations up to about × 30, while the paraboloidal dish reflector and point-focusing Fresnel lens are necessary for higher concentrations. These devices, of course, are only effective when the sun is shining and, except for low con-centrations, they have to track the sun. The higher the concentration, the more

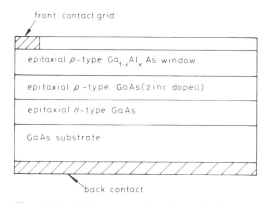

Fig. 3.16. Gallium arsenide solar cell. (Repro-duced by permission of the Institution of Electrical engineers)

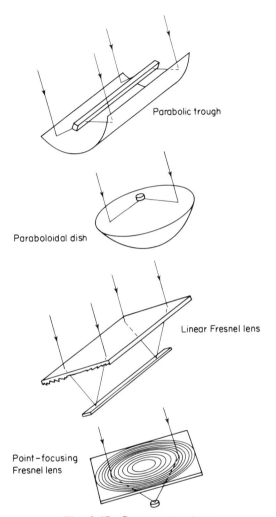

Parabolic trough

Paraboloidal dish

Linear Fresnel lens

Point-focusing
Fresnel lens

Fig. 3.17. Concentrator types

accurate the optical profile and tracking system have to be and the more
sophisticated the cooling system. Thus, the cost of the concentrator rises with
concentration ratio, while the cell area falls and there is an optimum at which
the total cost of the system will be minimum. This optimum, which some
investigators have placed as low as × 30, will tend to diminish as cell prices
fall.

Fresnel lenses can be designed to provide high optical transmission and a
suitable flux distribution for solar cells. They are somewhat limited by off-axis
and chromatic aberrations but can be made cheaply in acrylic and are easy to

clean. Silvered glass reflectors are superior to aluminized ones but cost about twice as much. Concentrator array efficiencies of 11% have been achieved and current installed array prices are running at about $10/Wp.

However, the need to track the sun and the inability to make use of diffuse radiation are serious disadvantages. Concentrator solar cells are likely to remain considerably more expensive per unit area, than cells for flat-plate arrays and there is relatively little scope for cost reduction in concentrators, their support structures and cooling systems. For these reasons, concentrator systems are unlikely to compete with flat-plate systems in the long run, except perhaps in applications where the thermal output can be used to advantage.

3.10 THIN-FILM CELLS

3.10.1 Cadmium Sulphide

The ultimate price goal of $0.15 to $0.50/Wp (Table 3.3) will be achieved only if there is a technological breakthrough to a highly efficient, stable thin-film solar cell. The main contender in this field is the copper sulphide–cadmium sulphide heterojunction, usually referred to as the 'cadmium sulphide' cell. Although invented at about the same time as the silicon cell, this type has developed much more slowly, mainly because of problems with low efficiency, instability, and poor reproducibility. However, there are signs that these difficulties are being overcome. The cadmium sulphide cell is potentially cheap because material costs are low and it can be mass produced.

There are two versions, illustrated in Fig. 3.18. In the frontwall type, the light enters the copper sulphide layer first, while in the backwall the cadmium sulphide is illuminated first. The particular frontwall structure illustrated was developed by the University of Stuttgart[45] and is being manufactured in Germany by Nukem. A columnar layer of cadmium sulphide is vacuum deposited on a metallized glass plate. The copper sulphide barrier layer is then formed by a hot chemical dip process, after which the surface is stabilized by special treatments. The upper contact, a grid of fine gold-plated copper lines, is deposited on another sheet of glass and the cell is completed by pressing the two halves together with a suitable adhesive to form a moisture-proof enclosure. Interconnected cells in small modules are fabricated by leaving masked gaps in the cadmium sulphide layer and substrate metallization and connecting the grid of one cell to the back contact of the next using a conductive paste and pressure. Single cells of this construction, 4200 mm^2 in area and averaging 7.3% in efficiency, have shown no degradation after a 560 day field test on load. The University of Delaware recently achieved an efficiency of over 10% in the laboratory with a frontwall cell incorporating zinc in the cadmium sulphide layer.[46] Frontwall cells of a different construction are in pilot produc-

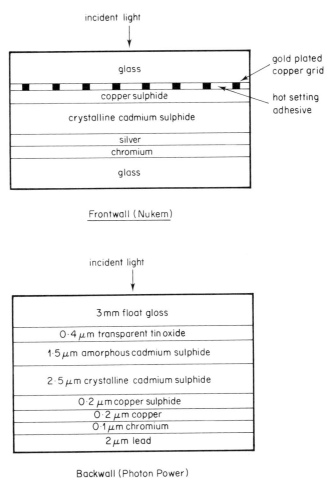

Fig. 3.18. Cadmium sulphide solar cells. (Reproduced by permission of the Institution of Electrical Engineers)

tion in the USA but conversion efficiencies have so far been disappointingly low.

In the backwall cell illustrated in Fig. 3.18, successive thin layers of amorphous and polycrystalline cadmium sulphide are sprayed on to a sheet of tin oxide-coated glass, after which the copper sulphide barrier layer is formed in a hot chemical dip. The cell is completed by evaporating successive layers of copper, chromium, and lead to form the back contact. The conductive, transparent tin oxide forms the front contact. Interconnected cells are formed on a single sheet of glass by suitable masking. Photon Power of El Paso are in pilot production with these cells[47] and hope eventually to reach a conversion efficiency of over 8%. But the performance of current production cells is much lower than this and a number of problems remain to be overcome before this ambition can be achieved.

3.10.2 Amorphous Silicon

The other important thin-film candidate is hydrogenated amorphous silicon, prepared by radio frequency glow discharge decomposition from doped silane, a process pioneered by the University of Dundee.[48] The energy gap of this material is 1.55 eV and therefore near to the optimum of 1.4 eV. Moreover, its optical absorption characteristic is similar to that of gallium arsenide, so only 1 or 2 μm are needed to make a solar cell, instead of the 300 to 400 μm at present used in crystalline silicon devices.

Many research groups are working to develop amorphous silicon solar cells. Although an efficiency of 15% is theoretically possible, values have not yet exceeded 6% in 100 mm^2 cells. At present, nearly all the photocurrent is generated in and collected from the depletion layer at the junction. Little is known about the transport of holes in amorphous silicon but it seems likely that the diffusion length of these carriers is no more than a tenth of a micrometre. This explains the absence of the usual diffusion mechanism which would enable current to be contributed from outside the junction region.

As with other thin-film cells, process control and high series resistance are serious problems, especially when attempts are made to increase the cell area. Nevertheless, RCA have succeeded in making more than 200 100 mm^2 cells with efficiencies of over 5%.[49] A considerable R & D effort into this area of solar cell technology is being applied in Japan, where an efficiency of 3.4% for 10 cm square devices is claimed.[50] Japanese manufacturers have already made prototype modules and plan to develop mass production techniques. Wristwatches and pocket calculators powered by amorphous silicon cells are being marketed.

3.10.3 MIS

Another low cost approach is the metal-insulator-semiconductor or MIS cell, sometimes referred to as the Schottky barrier type. Such cells can theoretically be as efficient as crystalline silicon cells but they are potentially much cheaper to make. The metal can be either in the form of a substrate supporting a thin film of polycrystalline or amorphous semiconductor or a thin transparent layer over a self-supporting semiconductor. The cross-section in Fig. 3.19 shows the essential features of the latter version.

Research on MIS cells is still in the early stages. An efficiency of 13.9% has been reported with a single crystal silicon MIS cell[51] and 17% with single crystal gallium arsenide[52] but single crystals are too expensive and large area

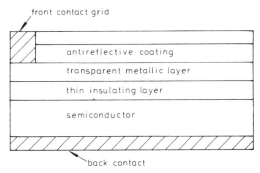

front contact grid

antireflective coating

transparent metallic layer

thin insulating layer

semiconductor

back contact

Fig. 3.19. MIS (Schottky barrier) solar cell

thin films must be used if the full low cost potential is to be realized. The best reported thin film MIS efficiency to date is 4.8% from a 2 mm diameter cell with no antireflective coating.[53]

The control and stability of the thin oxide layer, which is necessary to produce an acceptable open-circuit voltage, pose problems which are proving difficult to overcome.

3.10.4 Other Types

In the continuing search for an advanced, low-cost solar cell, many other types are being studied. For example, heterojunctions embodying II–VI compounds have the potential advantages of direct near-optimum bandgap and ease of fabrication by a variety of deposition methods. Among those being tried are combinations of absorber materials with a bandgap of between 1 and 1.5 eV, such as p-type Si, GaAs, InP, CdTe, or CuInSe$_2$, with a wide bandgap window layer of n-type CdS, ZnCdS, ZnSe, ZnO, or $(In_2O_3)_x(SnO_2)_{1-x}$ (called 'ITO'). However, the practical performances of such structures are still poor and none has yet emerged with any particular promise.

3.11 PRESENT APPLICATIONS

Despite their high initial cost, photovoltaic generators are already commercially competitive in a wide range of low-power stand-alone applications and, as we have seen, the world market is expanding rapidly. Some of the more important applications are discussed below.

3.11.1 Marine Beacons, Minor Navigational Lights, Fog Signals

Marine beacons and minor navigational lights are at present powered by acetylene, pressurized paraffin, or batteries and consume from 30 to 1200 Wh per day. Maintenance of such supplies is, in most cases, difficult and costly. One supplier estimates that batteries for a 75 W light cost about $3000 per year. Because of this, authorities in France, Japan, UK, and USA have been experimenting for some years with solar-powered navigational aids. In the UK, Trinity House has been monitoring experimental installations since 1963. Figure 3.20 shows a solar-powered beacon in Chichester Harbour. The system was installed by the Royal Aircraft Establishment in May, 1977 and has operated satisfactorily ever since.

Fig. 3.20. Solar-powered marine navigation light in Chichester Harbour (Courtesy RAE, Farnborough)

3.11.2 Offshore Oil Rig Lights

Photovoltaic generators have been installed on unmanned offshore oil platforms in the Gulf of Mexico, off Scotland, and off Venezuela to provide power for flashing lights and warnings. The power required is about 7 Wh per day, which can be supplied by a 2 to 4 Wp module.

3.11.3 Remote Aerial Navigation Beacons

Solar cells can be used to advantage to power radio and light beacons in the vicinity of airports, where the cost of running cables or maintaining batteries would be high. One such radio beacon has been operating successfully at St Girons near Bordeaux since 1967. It comprises a 300 Wp array and a 24 V, 200 Ah lead-acid battery. More recently, similar installations have been set up in the Pacific and in Africa. Figure 3.21 shows a French 400 Wp installation. Solar powered light beacons were installed in 1973 on the seven mountain peaks surrounding Medina Airport in Saudi Arabia. Each beacon consumes 144 Wh per day, requiring a 48 Wp array and 400 Ah of 12 V batteries.

Fig. 3.21. 400 Wp array powering an aerial navigation beacon (Courtesy RTC)

Fig. 3.22. Remote automatic weather station, Long Island Sound (Courtesy NASA/Lewis Research Center)

Fig. 3.23. CSIRO data buoy in the Indian Ocean (Courtesy RTC)

3.11.4 Automatic Weather Stations and Other Remote Instrumentation

Solar-powered automatic weather stations can be installed in remote locations such as mountain tops or in mid-ocean, where manned stations would be impracticable. An added advantage of photovoltaic power is that the solar cell can be used as a cheap, reliable, maintenance-free radiometer. A weather station typically consumes 90 Wh per day. Figure 3.22 shows an experimental station installed by NASA's Lewis Research Center in Long Island Sound. There are many other opportunities for powering remote instrumentation, such as seismic sensors and data buoys, which could be met by solar cell devices. As an example, Figure 3.23 shows one of a number of data buoys installed by CSIRO off the west coast of Australia to monitor currents in the Indian Ocean for the rock lobster industry. The data are telemetered to the Nimbus 6 satellite, which passes overhead every 100 minutes. The power requirement is met by a single 8 Wp module.

3.11.5 Remote Telecommunication Links

Developments in solid-state microwave techniques over the past ten years have reduced the power consumption of telecommunication equipment by a factor of 10 to 20. For instance, a repeater capable of carrying one television channel and 1000 telephone channels may consume 25 to 30 W today, compared with

Fig. 3.24. Solar-powered v.h.f. repeater on Islay (Courtesy Lucas BP Solar Systems Ltd)

500 W previously. This has made the introduction of solar-powered networks attractive and many stations have been installed in recent years. An example is the 36 Wp array shown in Fig. 3.24, which was installed by Lucas in 1974 on Islay in the Western Isles to power a v.h.f. repeater for coastguard use. Lead-acid batteries of 500 Ah capacity provide the necessary storage. In 1978, Telecom Australia installed a network of thirteen repeater stations, each with a 792 Wp photovoltaic generator, to provide a 3000-channel microwave link to a large part of Australia. The total cost of the generators, including support structures, wiring, protective devices, installation, etc. was $500 000, which works out at $48.5 per Wp.

3.11.6 TV in Remote Areas

The first solar-powered television set was installed in 1968 by Télédiffusion de France in a school in Goudel, near Niamey, Niger. The operation was so successful that, by 1977, 125 schools had been similarly equipped. More recently, the government of Niger initiated a programme to bring communal TV to the villages. By the end of 1980, 225 solar-powered centres had been set up and a further 250 were under construction. Figure 3.25 shows a typical school installation.

The solid-state black-and-white receivers consume 32 W and are used on average about 45 hours per week. The power system consists of a solar array of 29 to 52 Wp, depending on location, a 40 Ah battery and an electronic

Fig. 3.25. Solar array powering educational TV in Niger (Courtesy RTC)

charge regulator. The modules carry a five-year guarantee. Experience has shown that they require no maintenance apart from cleaning three or four times a year. The battery electrolyte is checked annually. For the village centres, an improved 61 cm receiver consuming less than 20 W is powered by a 33 Wp module with 140 Ah battery. Automatic regulation and protection systems for the battery are incorporated in the television set. The total cost of the solar generator, including support, wiring, battery, packing, transport, installation and commissioning, is 7700 FF or about $54 per Wp.[54]

3.11.7 Highway Traffic Control Warning Lights and Breakdown Telephones

For these applications, as in others, it is often cheaper to use solar cells than to run cables or maintain batteries. Loads are quite small and the module can be mounted on a pole to protect it from accidental damage and vandalism. An example of the many systems already in operation in Africa, Europe, and USA is an experimental 100 Wp system near Phoenix, Arizona, which warns motorists of impending sand and dust storms. If the experiment is successful, the State plans to replace the existing propane system completely, at an estimated saving of $12 000 per year.

3.11.8 Cathodic Protection

Cathodic protection systems provide essential corrosion control for wellheads, bridges, pipelines, and transmission cables. A small direct current is impressed on the structure or at intervals along pipelines and cables, sufficient to offset any current occurring naturally due to electrochemical reaction between the structure and the ground. In a conventional system, long lengths of cable have to be run from mains-fed transformer/rectifier units or diesel generators and high voltage has to be used to offset line drop, with a consequent waste of energy. Distributed photovoltaic sources eliminate this waste and the difficulties of maintaining and fuelling diesel generators in remote areas. In recent years, many solar-powered cathodic protection systems have been installed. One such, installed by Lucas BP Solar Systems, protects some 40 miles of oil pipeline in Qatar, Persian Gulf. Comprising a 300 Wp array and 10 kWh of battery storage, it delivers 60 W continuously.

3.11.9 Alarm and Monitoring Sytems

Their reliability and independence from mains supply make photovoltaics attractive for alarm and monitoring applications such as railway signals, alarm systems, fire hazard warnings, and electric fences. The power demand for railway signalling depends on the duty cycle but a typical American system to indicate train location over a two-mile block has a 5% duty cycle and requires 3.8 Wh per day at 2 V.

Solar-powered radio alarm systems can be used to give warning of bush or forest fires or floods and telemeter coded messages to indicate the status of pipeline valves, irrigation systems, pressure levels in compressor stations, etc. A network of solar-powered stations for the detection of bush fires has been set up in Australia. If a fire breaks out, its location is radioed to a central station on Mount Tennant, which initiates the necessary control action. Solar-powered environmental monitoring and supervisory control systems have also been installed in Ethiopia, Fiji, New Guinea, Italy, and USA.

3.11.10 Battery Charge Maintenance

Solar cells can be used to maintain batteries on lifeboats, rafts, and other emergency equipment in a state of readiness. Battery charging in infrequently used leisure equipment like sailing boats and caravans is another possible application. Several entries in the 1976 single-handed transatlantic sailing race were equipped with solar cells. Figure 3.26 shows one of these installations, in which glass-sandwich modules were employed. Note the transmission of unused light between the circular cells.

3.11.11 Dry Battery Replacement

Power from primary cells or so-called 'dry' batteries is expensive, costing from $40 to $60 per kWh, so they are likely to be replaced by solar cells in many

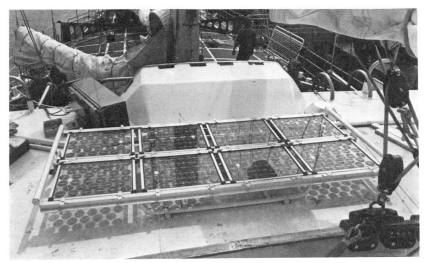

Fig. 3.26. Solar array on a sailing boat (Courtesy RTC)

applications. Solar-powered pocket calculators, watches, and transistor radios are already on the market.

3.11.12 Military Applications

There are many military requirements for mobile and remote electrical power, which are at present met by batteries, hand-cranked generators, and diesel generators. These power sources present logistic and operational problems which could be eased by the introduction of photovoltaics.

In 1976, the US Department of Defense initiated a programme for assessing photovoltaics for a range of military uses, including remote instrumentation, radio relays, telephone communication centres, water purification, and remote radar. One example of a small military photovoltaic generator is a lightweight 8 Wp array designed for use in a manpack radio haversack. The foldable array has its own stand and built-in lightmeter to facilitate optimum orientation when charging a 12 V nickel-cadmium battery. It weighs 1.8 kg and, when folded, measures 280 mm × 292 mm × 25 mm. At the other end of the military scale, the largest system installed to date is the 60 kWp array at the US Air Force Radar Station on Mount Laguna, San Diego, which was commissioned in 1979.

3.12 FUTURE PROSPECTS

3.12.1 Water Pumping and Purification

Perhaps the most important use to which photovoltaics will be put in the 1980s is in the provision of pumped water for agriculture and for human and animal consumption, particularly in the developing countries. In many parts of the world, water has to be laboriously drawn or pumped by hand and carried long distances to the people's dwellings. In some places, water is available but it is too brackish for human consumption. Life in these arid places would be transformed if potable water could be made available at reasonable cost. For example, a 20 m well in the Sahara desert could provide 20 m^3 of water per day, enough to sustain some hundreds of people and as many head of cattle.

Some 250 millions of the world's population live on less than 500 000 km^2 of prime agricultural land in the valleys and deltas of large rivers. Most of these areas experience extended dry seasons during which crop production is impossible without irrigation. With proper land management and water resources, current production of food could be at least doubled.

Water pumping, disinfection, and desalination are attractive applications for photovoltaics because, with an adequate reservoir, pumping can be restricted to the daylight hours and no electrical storage is required apart, perhaps, from a small buffer battery. There are at present some 75 photovoltaic water pumps in operation around the world, with array sizes ranging from 220 Wp (Islamabad, Pakistan) to 27 kWp (Montpellier, France). Experience has shown them to be simpler, more reliable, and easier to maintain than other types, including hand and foot pumps, solar thermal, and wind pumps. In Mali, the 26 photovoltaic pumps installed since 1977 are now serving 4500 people and saving 1000 litres of gas oil every month.[55]

Figure 3.27, for example, shows a small photovoltaic irrigation pump for the Third World farmer, which was designed by MIT Lincoln Laboratory[56] and developed and marketed by Solar Electric International of Bedford, Mass. It consists of a 220 Wp portable array powering a 65 V permanent magnet brushless d.c. motor direct coupled to an immersed centrifugal pump. There is no battery but an electronic maximum power tracker interposed between the array and motor ensures early start-up in the morning and good matching under all operational conditions. In bright sunlight, the pump will deliver 2.5 l/s (40 gal/min) against a 5 m head. One advantage of a portable array is that it can be orientated manually to boost the output at times of maximum demand. Another is that it can be moved from field to field and taken in at night for protection.

An economic analysis[57] comparing the costs of photovoltaic and diesel pumps in rural India showed that the former becomes cost effective for drink-

Fig. 3.27. Portable photovoltaic water pump (Courtesy Solar Electric International Inc.)

ing water supplies at a module cost of $6/Wp and for irrigation at $3.50/Wp (1978 dollars).

In 1979, the World Bank, acting for the United Nations Development Programme, launched a programme to assess small-scale solar water pumps, photovoltaic and thermal, for irrigation in developing countries. Field tests were carried out with local participation on a number of selected systems in Mali, the Sudan and the Philippines. These were complemented by laboratory tests on components and sub-systems. A design study was also conducted to compare the cost effectiveness of various systems and determine how this is affected by changes in various parameters. The ultimate aim of this programme is to develop a robust, reliable, easily maintained system for the small farmer, capable of providing a flow of at least 1 l/s against heads of up to 6 m at an overall cost of no more than $0.05 per m^3 (1979 dollars). This corresponds to a module price of about $2.80/Wp.

An example of a larger irrigation scheme is the MIT–Lincoln 25 kWp system at Mead, Nebraska,[58] which has been in use since July, 1977 irrigating a 32 hectare field of corn with a flow of 63.3 l/s. Power from the photovoltaic array is also used for crop drying and the manufacture of nitrogenous fertilizer.

3.12.2 Rural Electrification

About 2000 million people in the developing countries, some 80% of their total population, live in rural areas without electricity. Even in industrialized countries, there are many isolated communities, for example in the Western Isles of Scotland and the Mediterranean, who either have no electricity or are supplied by expensive diesel or gasoline generators.

Photovoltaics, with other forms of solar energy conversion, offer a means of tackling this energy supply problem without nuclear proliferation, environmental damage or undue depletion of gas and oil reserves. On-site solar generators, serving individual villages, could be introduced gradually without the detailed planning and long lead times associated with networks of large central power stations. The provision of solar electricity on a quite modest scale might help to halt the drift to the towns and encourage skilled people to move to the villages, with consequent benefit to the economy. For example, a 3 kWp array, occupying an area of 30 to 40 m², could provide basic lighting, refrigeration, pumped water, and TV for fifty African houses. Wolfe[59] maintains that, even at present prices, photovoltaic systems can be cheaper in the long run than alternatives for average loads smaller than 200 to 2000 W, depending on local conditions. The 'breakeven' load will, of course, increase as solar cell prices fall. Solar powered refrigerators and high-efficiency lighting units are already being marketed for use in remote areas.

Details of some of the pilot rural electrification schemes which are already in operation or are to be built are given in Table 3.5. The objectives of these schemes are to gain experience in system design, test the hardware under operational conditions, gauge the attitudes of the intended beneficiaries towards the new technology, and demonstrate photovoltaics in action to the public at large.

The Schuchuli and Tangaye projects, sponsored by the US Agency for International Development, and the Ladakh project, sponsored by the Indian Department of Science and Technology, are the only ones in the list providing a d.c. supply. Although this eliminates inverter losses, it means that special d.c. appliances may have to be developed and these are likely to be more expensive than equivalent a.c. appliances which enjoy a larger market and are more readily available.

The small system at Choglamsar, Leh, Ladakh is of particular interest, as it provides electric lighting for a hospital and community centre at the SOS Childrens' Village and was designed and constructed entirely in India by Central Electronics Ltd. Figure 3.28 shows one of the two arrays supplying sixteen 20 W fluorescent lamps in the hospital. The arrays are manually adjusted three times a day about the polar axis and once a month about the horizontal axis to increase the solar input.

The 350 kWp system that is to provide power for two villages about 50 km north-west of Riyadh, Saudi Arabia under the joint US–Saudi SOLERAS

Table 3.5 Rural electrification projects

Location	Array type	Peak power (kWp)	Battery storage (kWh)	System voltage	Initial operation
Schuchuli, Arizona	Flat-plate, fixed	3.5	286	120 V d.c.	Dec. 1978
Tangaye, Upper Volta	Flat-plate, fixed	1.8	65	120 V d.c.	Mar. 1979
Choglamsar, Leh, Ladakh, India	Flat-plate, fixed manually orientated	0.54	11.5	48 V d.c.	Oct. 1980
Saudi Arabia	Concentrator	350	1100	480/277 V a.c.	1981
Rondulinu, Corsica	Flat-plate, fixed	44	432	380/220 V a.c.	June 1983
French Guiana	Flat-plate, fixed	35	480	380/220 V a.c.	June 1983
Giglio Is., Italy	Flat-plate, fixed	43	48	380/220 V a.c.	June 1983
Alicudi Is., Italy	Flat-plate, fixed	76	500	380/220 V a.c.	June 1983
Tremiti Is., Italy	Flat-plate, fixed	65	500	380/220 V a.c.	June 1983
Aghia Roumeli Canyon, Crete	Flat-plate, fixed	50	480	380/220 V a.c.	June 1983
Kythnos Is., Greece	Flat-plate, fixed	100	600	380/220 V a.c.	June 1983

Fig. 3.28. Array supplying lighting for hospital in Ladakh, India (Courtesy Central Electronics Ltd., India)

project is the only one to employ concentration. The generator consists of 160 concentrator arrays with point-focusing acrylic Fresnel lenses. Each array is computer-controlled to track the sun.

The other projects listed in Table 3.5 are all part of the Commission of the European Communities' Photovoltaic Pilot Project Programme, which was launched in 1980. The plants have been designed by different consortia for locations ranging from Greece to French Guiana but it is significant that all of the designs are for fixed flat-plate arrays and ac systems. Completion of these plants is planned for June, 1983, after which they will be carefully monitored. The installation at St Nicola in the Tremiti Islands is special in that it will be used to provide the islands with drinking water, which at present has to be imported by ship. Seawater will be desalinated, using the reverse osmosis process, which consumes $10 \, \text{kWh/m}^3$. During the winter, when there are fewer tourists, excess power from the photovoltaic generator will be used for lighting.

3.12.3 Other Stand-alone Applications

Two other projects in the Commission of the European Communities' programme illustrate the fact that stand-alone photovoltaic generators can be used for purposes other than water pumping and rural electrification. The first of these is a 30 kWp flat-plate generator which will be installed on the roof of a semiconductor factory at Hoboken, near Antwerp, Belgium to produce hydrogen by electrolysis. The target is to produce $1\,m^3$ of hydrogen at normal temperature and pressure for a consumption of 3.5 kWh. The maximum production rate is expected to be $8.57\,m^3/h$ and the hydrogen will be stored at low pressure in a $20\,m^3$ tank.

In the second project, a 50 kWp array will be erected on the roof of a technical building at Nice Airport to supply dc current for monitoring and controlling the allocation of electricity to substations on the airport.

3.12.4 Grid-connected On-site Plants

The US Department of Energy has predicted (Table 3.3) that, if and when photovoltaic system prices fall to $1.6 to $2.2/Wp, localized roof-top arrays, generating power where it is consumed and using the grid as a buffer store, will begin to be used on a significant scale for residences and what they term 'intermediate load centres', such as hospitals, schools, shopping centres, offices, and factories. Even in the UK, a $50\,m^2$ roof array would generate an average of nearly 10 kWh per day, more than enough for the average household. Modules can be easily installed on existing roofs of suitable orientation and, in new houses, they can be designed to take the place of roofing tiles, thus reducing the overall cost. Each consumer would have his own power conditioning unit to enable the array to operate always at its maximum efficiency and convert the dc to ac of a waveform suitable for synchronous connection to the grid. The grid interface raises non-technical as well as technical problems, which are being studied, notably in USA and Italy.

Table 3.6[60] shows the predicted conventional and on-site photovoltaic energy prices in 1986 for residential and selected intermediate load centres in three American towns, assuming optimistically that DOE's system price goal of $1.6/Wp is met by that date. The conventional energy prices are based on an assumed annual escalation of 3% above inflation.

Table 3.6 Predicted conventional and on-site p.v. energy prices in 1986

Load	Location	Predicted conventional energy price (c/kWh)	Photovoltaic energy price if DOE goal is met (c/kWh)
Residential	Phoenix	5.7	5.2
	Miami	5.5	6.9
	Boston	9.4	8.7
Selected intermediate	Phoenix	6.4	5.5
	Miami	7.0	7.3
	Boston	8.0	9.2

On the assumptions made, the figures suggest that on-site photovoltaics would be able to compete over much of the USA by 1986. Interestingly, although Boston is less sunny than Miami, it is more favourable for photovoltaics because the local electricity tariffs are higher. It has been independently predicted[59] that on-site solar power will become cheaper than mains electricity in the United Kingdom in the 1990s.

In the USA, the market for domestic roof-top arrays is being stimulated by generous Federal and local tax incentives and by recent legislation which requires utility companies to pay private owners of small photovoltaic arrays and other types of generator for excess power fed into their grids.

Prototype residential systems for heating, cooling, and electrical power are being built and evaluated in the USA, following a series of conceptual design studies sponsored by the Department of Energy. For example, a complex of nine such houses is being constructed in the south-west. These all have roof-mounted arrays of various types with peak powers ranging from 4.5 to 6.7 kWp. However, there are signs that, with the present tax incentives, there is already a commercial market for solar photovoltaic houses in the more affluent parts of the United States. One New England builder has already sold one such house and has orders for several more.

Plans are also in hand for a number of prototype systems for intermediate load applications in USA and Europe, the latter supported by the Commission of the European Communities under their Pilot Projects Programme. Table 3.7 lists some of these projects. Figure 3.29 shows a model of the 300 kWp array which is to be installed on Pellworm Island in the North Sea to supply power to a vacation centre. When completed, it will be the largest flat-plate photovoltaic generator in the world.

An interesting project in the intermediate load category, not listed in Table 3.7, is the 'Solar Breeder' facility, which the Solarex Corporation plan to build in the Washington DC area. Figure 3.30 shows the architect's proposed design. A 200 kWp array of solar cells on the south-facing roof will provide all the power required to operate the factory and produce more solar cells. Dr Lindmayer, President of Solarex, estimates that after a year's operation the cells in the array will have paid for themselves and from then on the factory will enjoy free electricity. The structure is scheduled to be completed by June 1982.

3.12.5 Central Power Stations

The concept of large central photovoltaic power stations or 'solar farms' feeding power into the grid has been the subject of a number of design studies. The balance-of-system costs of such stations are likely to be higher than in the case of on-site generation, because the land has to be acquired, the site made

Fig. 3.29. 300 Wp array to be installed on Pellworm Island (Courtesy CEC)

Table 3.7 Grid-connected intermediate load centre prototypes

Location	Application	Array type	Peak power (kWp)	Initial operation
USA				
Kauai, Hawaii	Hospital hot water and electricity	Concentrator, × 100 paraboloid	60	Aug. 1981
Phoenix, AZ	Airport building electricity	Concentrator, × 70 cassegrain	225	Nov. 1981
Albuquerque, NM	Office building electricity and winter heating	Concentrator, × 24 para. trough	50	Feb. 1981
Dallas, Texas	Airport electricity	Concentrator, × 25 lin. Fresnel	27	July 1981
Orlando, FL	Electricity and cooling for 'Sea World' Park	Concentrator, × 25 para. trough	110	Dec. 1981
Lovington, NM	Shopping centre electricity	Flat-plate, fixed	100	Jan. 1981
El Paso, NM	Generating plant computer control	Flat-plate, fixed	20	Nov. 1980
Oklahoma City	Science and Art Centre electricity	Flat-plate, fixed with boosters	135	May, 1981
Beverly, MA	High school electricity	Flat-plate, fixed	100	Feb. 1981
Europe				
Chevetogne, Namur, Belgium	Auxiliary power for solar-heated swimming pool	Flat-plate, fixed	63	June 1983
Pellworm Island, Germany	Electricity for a vacation centre	Flat-plate, fixed	300	June 1983
Mont Bouquet, France	Power supply for FM transmitter	Flat-plate, fixed	50	June 1983
Fota Island, Cork, Ireland	Electricity for dairy farm	Flat-plate, fixed	50	June 1983
Terschelling Is., Netherlands	Power supply to marine training school	Flat-plate, fixed	48	June 1983
Marchwood, Southampton, UK	Power supply to grid	Flat-plate, fixed	80	June 1983

secure, and transmission lines laid to the grid. The supporting structures cannot be used for a dual purpose, as they can be in pitched roof installations. Moreover, calculations of the cost of integrating large central photovoltaic power stations into a grid network must take into account the back-up conventional generating capacity that must be provided if grid reliability is not to be adversely affected. When these factors are taken into consideration, it turns out that the module price must fall to between 15 and 50 c/Wp (1980 dollars) to compete with coal, oil, or nuclear stations. As we have already seen, a breakthrough to a highly efficient thin-film cell is necessary before this can happen.

Fig. 3.30. 'Solar Breeder' concept (Courtesy Solarex Corp.)

142

3.12.6 Solar Power Satellites

An extension of the central power station idea is the solar power satellite (s.p.s.), a concept first proposed by Glaser[61] in 1968 and expounded by him at the UNESCO Conference 'The Sun in the Service of Mankind' in 1973. He envisaged a system of power satellites in geostationary orbit, where an irradiance of $1350\,W/m^2$, nearly 40% higher than the brightest terrestrial sunlight, is continuously available, except for short periods of eclipse at the vernal and autumnal equinoxes, and the environment is relatively benign.

In 1977, the US Department of Energy, in collaboration with the National Aeronautics and Space Administration, initiated a concept development and evaluation programme, which comprised both in-house studies and contracts placed with a large number of industrial and academic institutions. These studies covered not only the technical aspects but also the economic, environmental and social issues involved. Comparisons were also made with other future energy sources which may become available early in the next century. The 5 GW Reference System used as a basis for these studies consisted of a satellite with a $55\,km^2$ solar array and a microwave antenna transmitting power to one or more receiving stations or 'rectennas' on the ground, each covering an area of 100 to $200\,km^2$.

More recently, the Department of Industry in UK funded a joint study by the British Aerospace Dynamics Group, Marconi Space and Defence Systems, and ERA Technology Ltd of the implications and potential of an s.p.s. project for British industry.

All studies so far made agree on the technical feasibility of the concept, though many serious problems remain to be overcome and assumptions confirmed. Among the problems is that of finding suitable sites for receiving stations in heavily populated countries, integrating a 5 GW station into the grid and providing the necessary short- and long-term back-up. Re-usable 'Shuttle' launchers much bigger than the present model would be required to transport material for the satellites into geostationary orbit.

The future of the project awaits a decision by the US Government following the Final Review and Symposium, which was held at Lincoln, Nebraska in April 1980.

3.13 SUMMARY AND CONCLUSIONS

Although still expensive, photovoltaic generators have become progressively cheaper in recent years and are already cost effective in a wide range of low power stand-alone applications. These applications are being exploited commercially and the world market is growing rapidly. The potential economic and social benefits of this technology are such that governments and international agencies are sponsoring extensive efforts to encourage market growth and cut costs still further. Cheaper silicon is being developed and production capacity increased to ensure adequate supplies for an expanding solar cell industry. Cell and module fabrication is being gradually automated. It is anticipated that, in the next decade, silicon module prices will fall to $1 to $2/Wp, at which level photovoltaic generators will be cheaper to run than diesel generators in most places.

If this happens, solar electricity is likely to be adopted for water pumping, disinfection, desalination, irrigation, and rural electrification, especially in developing countries. At a later stage, grid-connected on-site photovoltaic generation may be increasingly used for dwellings, hospitals, schools, shopping centres, offices, and factories. Many pilot plants are being set up to develop system design and test the hardware in readiness for the future.

But photovoltaic generation is unlikely to have any significant impact on the large-scale generation of electricity in central power stations unless a highly efficient, stable thin-film solar cell can be developed and mass-produced. Considerable progress has been made in this direction with cadmium sulphide and amorphous silicon.

3.14 ACKNOWLEDGEMENTS

The author is indebted to the Institution of Electrical Engineers for permission to draw on information from his IEE Review 'Solar Cells', which was published in *IEE Proceedings,* Vol.127, Pt.A, No.8 in November 1980. He would also like to thank the Meteorological Office, Bracknell for insolation data and the Commission of the European Communities, the Royal Aircraft Establishment, NASA/Lewis Research Center, Cleveland, Ohio, the US Department of Energy, Arco Solar Inc., Chatsworth, CA, Central Electronics Ltd, Sahibabad, India, La Radiotechnique Compelec (RTC), Paris, Lucas BP Solar Systems Ltd, Haddenham, Bucks, Solar Electric International Inc., Bedford, MA, and Solarex Corporation, Rockville, MD for information and illustrations.

3.15 REFERENCES

1. Becquerel, E. (1839). On electric effects under the influence of solar radiation, *Compt. Rend.,* **9**, 561.
2. Adams, W. G., and Day, R. E. (1877). The action of light on selenium, *Proc. R. Soc.* London ser. A, 25, p. 113.
3. Lange, B. (1930). New photoelectric cell, *Zeit. Phys.,* **31**, 139.
4. Grondahl, L. O. (1933). The copper-cuprous oxide rectifier and electric cell, *Rev. Mod. Phys.,* **5**, 141.
5. Schottky, W. (1930). Cuprous oxide photoelectric cells, *Zeit. Phys.,* **31**, 913.
6. Chapin, D. M., Fuller, C. S., and Pearson, G. L. (1954). A new silicon $p–n$

junction photocell for converting solar radiation into electrical power, *J. Appl. Phys.*, **25**, 676–7.
7. Reynolds, D. C., Leies, G., Antes, L. L., and Marburger, R. E. (1954). Photovoltaic effect in cadmium sulfide, *Phys. Rev.*, **96**, 533–4.
8. Schuster, G. (1980). The future of photovoltaics in Europe, *Proc. 3rd EC Photovoltaic Solar Energy Conference*, pp. 4–9.
9. Budyko, M. I. (1958). The heat balance of the earth's surface, *English Translation by N. Steponova, US Department of Commerce Weather Bureau, Washington DC.*
10. Ougham, A. J. (1978). *Private communication.*
11. Wolf, M. (1960). Limitations and possibilities for improvement of photovoltaic solar energy converters, Part 1, Considerations for Earth's surface operation, *Proc. IRE*, **48**, 1246–63.
12. Prince, M. B. (1955). Silicon solar energy converters, *J. Appl. Phys.* **26**, 534–40.
13. Rappaport, P. (1959). The photovoltaic effect and its utilization, *RCA Rev.*, **20**, 373–97.
14. Wysocki, J. J., and Rappaport, P. (1960). Effect of temperature on photovoltaic solar energy conversion, *J. Appl. Phys.*, **31**, 571–8.
15. Lindmayer, J., and Allison, J. (1972). The violet cell: an improved silicon solar cell, *Proc. 9th IEEE Photovoltaic Specialists' Conference*, pp. 83–4.
16. Wolf, M. (1963). Drift fields in photovoltaic solar energy converter cells, *Proc. IEEE*, **51**, 674–93.
17. Carroll, W., Cuddihy, E., and Salama, N. (1976). Material and design considerations of encapsulants for photovoltaic arrays in terrestrial applications, *Proc. 12th IEEE Photovoltaic Specialists' Conference*, pp. 332–9.
18. Anagnostou, E., and Forestieri, A. F. (1978). Endurance testing of first generation (Block 1) commercial solar cell modules, *Proc. 13th IEEE Photovoltaic Specialists' Conference*, pp. 843–6.
19. Shumka, A., and Stern, K. H. (1978). Some failure modes and analysis techniques for terrestrial solar|cell|modules, *Proc. 13th|IEEE Photovoltaic Specialists' Conference*, pp. 824–34.
20. Larue, J. C., and du Trieu, E. (1980). Effect of partial shadowing on solar panels—hot spot or breakdown?, *Proc. 3rd EC Photovoltaic Solar Energy Conference*, pp. 490–5.
21. Hoffman, A. R., and Ross, R. G. (1978). Environmental qualification testing of terrestrial solar cell modules, *Proc. 13th IEEE Photovoltaic Specialists' Conference*, pp. 835–42.
22. Tofield, B. C., Dell, R. M., and Jensen, J. (1978). Advanced batteries, *Nature*, **276** (5685), pp. 217–20.
23. Kondratyev, K. Y., and Fedorova, M. P. (1977). Radiation regime of inclined surfaces, *Technical Note No. 152 (WMO No. 467)*, published by the World Meteorological Organization, Geneva.
24. Rodgers, G. G., Page, J. K., and Souster, C. G. (1979). Mathematical models for estimating the irradiance falling on inclined surfaces for clear, overcast and average conditions, *Proc. UK-ISES Conference C18 on Meteorology for Solar Energy Applications*, pp. 48–62.
25. Forney, R. G. (1979). Photovoltaics in the USA—a progress report, *Proc. UK-ISES Conference C21 on Photovoltaic Solar Energy Conversion*, pp. 81–91.
26. Durand, H. L. (1978). Present status and future prospects of silicon solar cell arrays and systems, *Discussion Meeting on Solar Energy, Royal Society, London, 15 November, 1978.*

27. Lutwack, R. (1980). A US view of silicon production processes, *Proc. 3rd EC Photovoltaic Solar Energy Conference*, pp. 220–7.
28. Hunt, L. P., and Dosaj, V. D. (1979). Progress of the Dow Corning process for solar grade silicon, *Proc. 2nd. EC Photovoltaic Solar Energy Conference*, pp. 98–105.
29. Reed, W. H., Meyer, T. N., Fey, M. G., Harvey, F. J., and Arcella, F. G. Development of a process for high capacity arc heater production of silicon, *Proc 13th IEEE Photovoltaic Specialists' Conference*, pp. 370–5.
30. Kachare, A. H., Uno, F. M., Miyahira, T., and Lane, R. L. (1980). Performance of silicon solar cells fabricated from multiple Czochralski ingots grown by using a single crucible, *Proc. 14th IEEE Photovoltaic Specialists' Conference*, pp. 327–31.
31. Freiesleben, W. (1980). The solar material market—projections, needs and commitments, *Proc. 3rd EC Photovoltaic Solar Energy Conference*, pp. 166–70.
32. Lindmayer, J. (1980). Industrialization of photovoltaics, *Proc. 3rd EC Photovoltaic Solar Energy Conference*, pp. 178–85.
33. Schmid, F., Basaran, M., and Khattak, C. P. (1980). Directional solidification of mg silicon by heat exchanger method (HEM) for photovoltaic applications, *Proc. 3rd EC Photovoltaic Solar Energy Conference*, pp. 252–6.
34. Mackintosh, B., Kalejs, J. P., Ho, C. T., and Wald, F. V. (1980). Multiple EFG silicon ribbon technology as the basis for manufacturing low cost terrestrial solar cells, *Proc. 3rd EC Photovoltaic Solar Energy Conference*, pp. 553–7.
35. Kim, K. M. (1977). Silicon ribbon growth by the inverted Stepanov technique, *DOE/JPL Final Report 954465*.
36. Duncan, C. S., Seidensticker, R. G., McHugh, J. P., Hopkins, R. H., Heimlich, M. E., Driggers, J. M., and Hill, F. E. (1980). Development of processes for the production of low cost silicon dendritic web for solar cells, *Proc. 14th IEEE Photovoltaic Specialists' Conference*, pp. 25–30.
37. Gurtler, R. W., Baghdadi, A., Legge, R. N., and Ellis, R. J. (1979). Potential for improved silicon ribbon growth through thermal environment control, *Proc. 2nd EC Photovoltaic Solar Energy Conference*, pp. 145–52.
38. Zook, J. D. (1980). Evaluation of silicon-on-ceramic material for low cost solar cells, *Proc. 3rd EC Photovoltaic Solar Energy Conference*, pp. 569–73.
39. Belouet, C., Belin, C., Schneider, J., and Paulin, J. (1980). Progress in the growth and performance of RAD silicon sheets, *Proc. 3rd. EC Photovoltaic Solar Energy Conference*, pp. 558–62.
40. Kirkpatrick, (1980). Low-cost ion implantation and annealing technology for solar cells, *Proc. 14th IEEE Photovoltaic Specialists' Conference*, pp. 820–4.
41. Michel, J., and Baudry, H. (1980). Reliable screen-printed contacts on silicon solar cells, *Proc. 3rd EC Photovoltaic Solar Energy Conference*, pp. 679–83.
42. Weaver, H. T., and Nasby, R. D. (1980). High efficiency silicon solar cells, *Proc. 3rd EC Photovoltaic Solar Energy Conference*, p. 1098.
43. Sahai, R., Edwall, D. D., and Harris, J. S. Jr. (1978). High efficiency AlGaAs/GaAs concentrator solar cells, *Proc. 13th IEEE Photovoltaic Specialists' Conference*, pp. 946–52.
44. Van der Plas, H. A., James, L. W., Moon, R. L., and Nelson, N. J. (1978). Performance of AlGaAs/GaAs terrestrial concentrator solar cells, *Proc. 13th IEEE Photovoltaic Specialists' Conference*, pp. 934–40.
45. Arndt, W., Bilger, G., Pfisterer, F., Schock, H. W., Woerner, J., and Bloss, W. H. (1980). Integrated Cu_xS–CdS thin film solar cell panels with higher output voltages, *Proc. 3rd. EC Photovoltaic Solar Energy Conference*, pp. 798–802.

46. Hall, R. B., Birkmire, R. W., Phillips, J. E., and Meakin, J. D. (1980). Thin-film polycrystalline (CdZn)S/Cu$_2$S solar cells of 10% conversion efficiency, *Proc. 3rd EC Photovoltaic Solar Energy Conference*, pp. 1094–6.
47. Roderick, G. A. (1980). Sprayed cadmium sulphide solar cells, *Proc. 3rd EC Photovoltaic Solar Energy Conference*, pp. 327–34.
48. Spear, W. E., and Le Comber, P. G. (1975). *Solid State Communications*, **17**, 1193–6.
49. Carlson, D. E. (1980). The status of amorphous silicon solar cells, *Proc. 3rd EC Photovoltaic Solar Energy Conference*, pp. 294–301.
50. Kuwano, Y., and Ohnishi, M. (1980). Development of amorphous Si solar cells in Japan, *Proc. 3rd EC Photovoltaic Solar Energy Conference*, pp. 309–16.
51. Green, M. A., Godfrey, R. B., and Davies, L. W. (1978). The short wavelength response of single- and poly-crystalline MIS solar cells, *Proc. 13th IEEE Photovoltaic Specialists' Conference*, pp. 651–5.
52. Stirn, R. J., Yeh, Y. C. M., Wang, E. Y., Ernest, F. P., and Wu, C. J. (1977). Recent improvements in AMOS solar cells, *Tech. Dig. 1977 IEEE IEDM, Washington DC*, pp. 48–50.
53. Yeh, Y. C. M., Ernest, F. P., and Stirn, R. J. (1978). Progress towards high efficiency polycrystalline thin-film GaAs AMOS solar cells, *Proc. 13th IEEE Photovoltaic Specialists' Conference*, pp. 966–71.
54. Polgar, S. (1980). Centres de reception communautaire de television alimentes par generateurs solaires au Niger, *Proc. 3rd EC Photovoltaic Solar Energy Conference*, pp. 1018–1022.
55. Verspieren, B. (1980). The application of photovoltaics to water pumping and irrigation in Africa, *Proc. 3rd EC Photovoltaic Solar Energy Conference*, pp. 439–45.
56. Matlin, R. W. (1979). Design optimization and performance characteristics of a photovoltaic micro-irrigation system for use in developing countries, *MIT-Lincoln Laboratory Report COO/4094–33*.
57. Sangal, S. K., Gnanainder, Y. A., and Bhaskar, E. V. (1978). An experimental study of solar photovoltaic water pumping system for rural India, *Proc. 13th IEEE Photovoltaic Specialists' Conference*, pp. 1278–82.
58. Matlin, R. W., Romaine, W. R., and Fischbach, P. E. (1977). 25 kilowatt photovoltaic powered irrigation and grain drying experiment, *MIT-Lincoln Laboratory Report COO/4094–2*.
59. Wolfe, P. R. (1981). Photovoltaic systems for power generation, *Proc. Third International Conference on Future Energy Concepts, IEE Conf. Pub.* No. 192, pp. 165–8.
60. Clorfeine, A. S. (1980). Economic feasibility of photovoltaic energy systems, *Proc. 14th IEEE Photovoltaic Specialists' Conference*, pp. 986–9.
61. Glaser, P. E. (1973). Space solar power, *UNESCO Conference 'The Sun in the Service of Mankind', Paris, July, 1973*.

Energy—Present and Future Options, Volume 2
Edited by D. Merrick
© 1984 John Wiley & Sons Ltd

J. T. McMULLAN
Energy Study Group
School of Physical Sciences
The New University of Ulster
Northern Ireland

4

The Heat Pump

4.1 INTRODUCTION

The principal feature of the heat pump which distinguishes it from all other types of heat supplying equipment is that it absorbs energy from a heat source at a low temperature and delivers energy to a heat sink at a higher one. In order to achieve this transfer, energy also has to be supplied to drive the process. The net result is that the heat delivered to the sink includes both the heat absorbed from the source and the heat equivalent of the driving energy. Thus, in contrast with fuel-burning equipment, the heat pump delivers more energy than was used to drive it. This suggests that the heat pump might have an important role to play in reducing national energy consumption, and it is on this aspect that we will concentrate in this chapter, although certain technical considerations are unavoidable.

The two areas where the heat pump has an obvious role to play are in space heating and in heat recovery, and it is here that the greatest contribution can be made in the coming transitional period for energy use patterns. The only reason why space heating is required to heat a building is that the inside is at a higher temperature than the outside and continually loses heat to the environment. Thus, the heat supplied by a conventional heating system produces an environmental heat load which is equal to the heat demand of the building.

By contrast, the heat pump is able to meet the heat demand with a reduced energy input, and with a reduced environmental heat load. It achieves this by using the environment itself as the source of low grade heat, ensuring that the net environmental load is only the mechanical energy supplied to the heat pump and not the total heat loss. Thus, a fuel-fired heat pump will cause a lower environmental heat load than a conventional fossil fuel heating system and an electric heat pump will cause a lower heat burden than an electric heating system. Comparison of electrical and fuel-fired systems is complicated by the existence of the power station with its 30% generation efficiency. That the environmental burden is reduced is ultimately true whether the heat source is the outside air, the ground, a river, or the sea, and it emphasizes the basic benefit that the heat pump has to offer—heat recovery.

The same principle applies in industrial heat recovery, with the difference that the heat source is an industrial waste effluent and the heat sink is represented by the demand for heat at a higher temperature, for example as process heat or as space heating to the factory buildings. There are a large number of possibilities for this type of heat recovery, and the range can be extended to include the recovery of latent heat in drying processes, or to heat recovery in shops and swimming pools. The intuitive attractiveness of such projects lies partly in the fact that each is fairly large in scale so that the amount of heat recovered is usually substantial. Further, industry is sensitive to the importance of energy conservation so that overt industrial heat recovery applications offer attractive opportunities for heat pump applications. Unfortunately, however, process heat in manufacturing industry typically accounts for only about 8% of national energy demand, while space heating absorbs about 38%. It is clear, therefore, that this is where the largest potential for heat pumps lies—at least in principle. In practice, the industrial sector is easier to tackle because of the larger size of the projects, and because widespread application in the space heating sector would require the combined effort and cooperation of a large number of individuals. This is equally true for any activity involving the private consumer and is not likely to be possible without considerable government encouragement and pressure.

It is a common misconception that, since refrigeration has been available for at least 150 years, and since the heat pump is 'nothing but a refrigerator', no research or development work on heat pumps is required. Despite the similarities in the thermodynamic cycles, and the fact that most heat pump hardware is bought from refrigeration manufacturers, heat pumps differ from refrigerators in several important respects. They are subject to much harsher operating conditions, to an exceptionally large range of load factors, to control problems that do not arise in refrigeration practice and to demands on efficiency that have never been faced by the refrigeration industry.

The reasons for the difference in emphasis on efficiency are not hard to find. Until recently the main concern of the refrigeration industry lay, quite properly, in reliability rather than in efficiency. Refrigeration plant operators would be more upset if their equipment broke down than if it were slightly more expensive to run. In effect, people are willing to pay for cooling because there is no competing technology. For heating applications, however, the situation is different. Oil, coal, gas, and wood-fired boilers provide competition and the high capital cost of the heat pump must be offset by operating cost savings that make it attractive over a realistic accounting period. Reliability, therefore, becomes an insufficient criterion for judging the performance of the equipment and efficiency takes on an importance not present in refrigeration.

Heat pumps currently fall into two general categories, Rankine cycle machines which require a mechanically driven compressor to effect the transfer of the heat from the low temperature source to the high temperature sink, and absorption cycle units which depend on a supply of heat for their operation. The Rankine cycle machine is the most common type and it can be driven either by an electric motor or by an internal combustion engine. The electric motor is the most common source of motive power, although diesel engine units are used in the larger equipment. The present emphasis on electric drives is likely to be reinforced in the long term if energy supplies become more dependent on alternative energy sources such as wind power, solar power, wave power, or nuclear power. All of these will require increased electrical energy distribution if they are to be fully exploited. Consequently, electrically driven heat pumps may be expected to become increasingly attractive as such a transition progresses.

Historically, the heat pump is credited to Lord Kelvin for a design produced in 1852, although refrigeration can be traced back rather further. The first reliably recorded reference is to William Cullen who produced refrigeration by evaporation of ether into a partial vacuum at the University of Glasgow in 1748, while the vapour compression cycle can be dated to 1834 and a design by Jacob Perkin. Ice factories were working by 1850, and refrigeration plant was installed in breweries by 1851. If one is in the proper frame of mind, there are fascinating references in *The Voyage of Wulfstan,* translated by Alfred the Great about AD 900 from a fifth century original, and in Muirchu's seventh-century *Life of St Patrick* which can be interpreted as reporting very early instances of man-made refrigeration. This is not necessarily the stuff of fairy tales as the basic materials required for the absorption refrigerator, ammonia or sulphuric acid and water, were all well known. This has been reported more extensively elsewhere[1] and so will not be discussed further here.

The heat pump *per se* appeared when Haldane built the first practical machine in about 1930, and used it to heat his home in Scotland. He used the atmosphere as the heat source and backed it up with the local water supply when atmospheric conditions were not favourable. There was little interest in the heat pump until the 1950s when extensive studies were undertaken and a number of developments made. However, the rapid decline in the real price of oil made energy conservation much less attractive and interest waned.

With the expansion of the refrigeration industry, many of the early operating difficulties were overcome, but it was not until the rise in oil prices of 1973–74 that interest revived. Since then, there has been a steady increase in interest, first in heat pump development and subsequently in application areas, with the result that it is now possible to foresee a healthy heat pump industry in the coming decades.

The main difficulty that the heat pump faces in breaking into the market is its high capital cost, and here several distinct market areas can be identified. The domestic market is the one in which this high first cost is probably most important. It is difficult to convince a private home owner that he should invest in a capitally intensive piece of equipment when he can apparently obtain the same or greater heating effect at less than half the price by using a conventional fuel boiler. The fact that there is a realistic payback period (except against natural gas which presently still enjoys a uniquely advantageous price when compared with all other fuels, but which is available to only about half of the population) counts as naught because private individuals do not cost their personal expenditure in the same way as companies. They tend not to cost the operation of the system over a period of time, but to regard the capital cost and the running cost as separate entities. The result is that a system which is either expensive to install but cheap to operate or the reverse is likely to be rejected in favour of one which is middle of the road on both counts. This factor would make heat pump installation more attractive in new houses than in the existing housing stock, but the argument is complicated by yet another factor. Private individuals are influenced by publicity and advertising and are likely to react to a 'perceived' cost rather than the actual one. Thus, for a period in the late 1970s, large numbers of householders were replacing existing off-peak electrical heating systems with new boiler systems, many oil-fired, using the high cost of electricity as the justification, but ignoring the £1000 or so (plus interest charges) incurred as capital expenditure.

For the domestic sector, therefore, the development of heat pumps must include the reduction of the high first cost to make the system initially more attractive, and improvements in the amount of heat transferred per kilowatt of power supplied.

Also important are the systems aspects of domestic heat pump installations. In the United States, where there is a requirement for cooling as well as heating, the heat pump can act as an air-conditioner in the summer. As a result, the equipment is used throughout the year and a large part of the capital cost is attributed to the cooling function for which there is no alternative technique available. In Northern Europe, by contrast, air conditioning is not a necessity and the installation has to be justified on its performance during the heating season alone—a more difficult task.

Industrial applications fall into a number of categories: space heating as in the domestic sector and with similar problems and scope; heat recovery from cooling plant, from process heat or from waste heat; drying; and heat recycling. A feature of industrial applications is that each is essentially a one-off installation and requires individual examination, planning, and justification. These possibilities will be considered later, although not in the detail that is to be found in McMullan and Morgan[1] on the fundamental or theoretical aspects, and Macmichael and Reay[2] or Heap[3] on the applications and practical side. These three books will provide readers with an extensive treatment of the subject and a large bibliography.

4.2 FUNDAMENTAL PRINCIPLES

The heat pump or refrigerator is simply a heat engine operating in reverse, as illustrated in Fig. 4.1. In a heat engine, heat is supplied from a high temperature source to provide mechanical power, with the waste heat being rejected at a low temperature sink, whereas in a heat pump, mechanical power is supplied to effect the transfer of heat from a low temperature source to a high temperature sink, i.e. against the temperature gradient.

For the engine we can write

$$Q_h = Q_c + W$$

where heat is represented by Q and mechanical work by W, and the subscripts represent hot and cold respectively. The corresponding temperatures can be taken as T_h and T_c.

The efficiency of the heat engine can be defined as

$$\eta = W/Q_h = (Q_h - Q_c)/Q_h$$

In the case of the heat pump, the first equation still applies, but Q_c now represents the heat taken from the source, while Q_h represents that rejected at

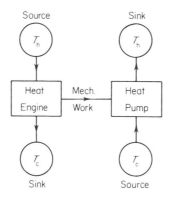

Fig. 4.1. Comparison of heat engine and heat pump cycles

the high temperature sink. *W* also represents mechanical work supplied *to* the machine rather than produced *by* it.

It we now try to define an efficiency for the heat pump, we run into certain difficulties. The obvious expression is the amount of heat absorbed from the cold source per unit of mechanical work required to transfer it to the sink. This has the advantage of being consistent with the earlier definition used in heat engine practice, but has the disadvantage that it does not provide the information of interest to the consumer, i.e. the heat supplied. It is suitable, however, for refrigeration, where cooling is the primary objective. Thus, the logical step is to define the efficiency as Q_h/W, that is as the heat *rejected,* at the high temperature sink per unit of mechanical work supplied. The unsatisfactory aspect of this definition is that the efficiency must have a value greater than unity (since $Q_h = Q_c + W$) and therefore runs contrary to conventional ideas of the definition of an efficiency. To overcome this, we redefine the quantity as a *coefficient of performance* (COP) and define separate coefficients COP_h and COP_c to describe the heating and cooling performance of the machine. These are given respectively by

$$COP_h = Q_h/W = Q_h/(Q_h - Q_c)$$
$$COP_c = Q_c/W = Q_c/(Q_h - Q_c)$$

The tempting corollary $COP_h = COP_c + 1$ is valid only in the ideal case when there are no heat losses from the system.

The maximum theoretical efficiency of a heat engine is the Carnot cycle efficiency and corresponds to a reversible heat engine. In this case,

$$\eta = (T_h - T_c)/T_h.$$

Correspondingly, the coefficients of performance for a reversed Carnot cycle machine are

$$COP_h = T_h/(T_h - T_c)$$
$$COP_c = T_c/(T_h - T_c).$$

These two equations are important as they represent the highest performance that can be achieved by an ideal heat pump operating at prescribed source and sink temperatures. They can never be achieved in practice, because real heat exchangers require a finite temperature difference for their successful operation, the heat transfer is not isothermal and the compression and expansion are not adiabatic. Thus, the real cycle is not reversible, there is a consequent increase in the entropy of the system, and a corresponding increase in the energy input and decrease in the coefficient of performance. These effects can be overcome to a greater or lesser extent by various design and construction techniques, but their effect is to reduce the performance that can be achieved, and to force a compromise to be reached between the various constraints, including economic considerations.

The Carnot expression given above would suggest that a coefficient of performance of about 19.5 is possible for extracting heat from the atmosphere at 5 °C and delivering it to heat a house at 20 °C. If we accept that this is unrealistic and allow the heat exchangers on the high temperature side to operate at 50 °C and on the low temperature side at 0 °C, then the COP_h is still highly attractive at 6.5. In fact, even this value is not achievable except under unusual circumstances, and values for air source units generally lie between 2 and 3. Some manufacturers have claimed values significantly higher than 3 for domestic equipment, but by and large these have proven not to stand up to experimental confirmation.

All practical heat pumps operate, as shown in Fig. 4.2, by boiling a fluid (the refrigerant) at a low pressure and temperature so effecting the transfer of the latent heat from the source at as near to a constant temperature as possible. The resultant vapour is then compressed, raising its temperature above that of the heat sink so that heat can be rejected by condensing the high temperature vapour back to a liquid again. This high temperature liquid is then expanded to the initial low pressure and temperature state, usually through a nozzle or capillary.

As has been said, all heat pumps operate on this basic cycle but there are currently two main practical arrangements: the vapour compression cycle and the absorption cycle. In vapour compression machines, a compressor alone is used to heat and compress the vapour. In absorption cycle machines, however, the cool vapour is dissolved in a non-volatile liquid which is compressed using a pump. The refrigerant is then driven out of the liquid absorber by supplying heat from any convenient source. The similarities between the two systems are shown in Fig. 4.3, the primary difference being in how the energy is supplied. In the vapour compression cycle all of the energy is supplied as mechanical work to the compressor. The power requirement for compressing the liquid in the absorption cycle is small by comparison, and most of the energy is provided as heat to separate the two fluids in the generator. As a result, the

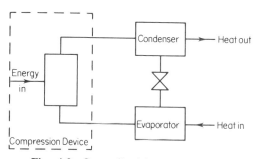

Fig. 4.2. Generalized heat pump cycle

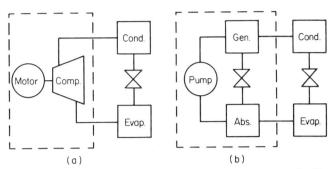

Fig. 4.3. Comparison of (a) vapour compression and (b) absorption cycles

absorption cycle machine can operate mainly on natural gas, coal, wood, etc., according to choice and availability.

Despite this obvious advantage, the absorption cycle heat pump is not commercially available. There are a number of problems that are not as easily overcome for heat pump applications as for refrigeration, and the technology is still at the laboratory stage. Thus, for the short term, the vapour compression heat pump will be the dominant type.

4.2.1 The Vapour Compression Cycle

In view of the importance of this cycle, it is worth quickly reviewing its operation. The refrigerant is a condensible vapour, and the heat exchanger conditions are chosen so that liquid refrigerant is boiled in the low-temperature heat exchanger (*evaporator*), gaining latent heat of evaporation. It is recondensed in the high temperature heat exchanger (*condenser*), releasing the latent heat of evaporation together with any heat that may have been added during the pumping process. A schematic diagram is shown in Fig. 4.4, together with the performance of an ideal cycle illustrated by a pressure-enthalpy chart.

On this diagram, a quantity of heat $h_1 - h_4$ is absorbed from the low temperature heat source by the boiling refrigerant; the work of compression adds $h_2 - h_1$, and the total $h_2 - h_4$ is rejected at the condenser. The high-pressure, high-temperature liquid then is passed through the expansion valve to reduce both temperature and pressure before it is readmitted to the evaporator.

One feature which is apparent even in this idealized scheme is that the vapour discharged from the compressor will be considerably superheated. This is evidenced by the shape of the constant temperature lines as indicated in the *P–h* diagram, and gives the first indication that the Carnot approximation cannot be used. The refrigerant at the compressor discharge port must be at

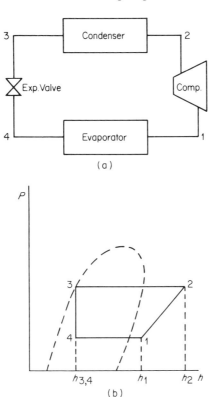

Fig. 4.4. (a) Schematic diagram of vapour compression heat pump. (b) Corresponding cycle on pressure-enthalpy chart

a higher temperature than that corresponding to the condensing temperature and, as a consequence, a loss in efficiency due to irreversibilities in the cycle has been introduced. This is only the first of the irreversibility losses that must be accounted for.

The COP_h, i.e. the heat rejected per unit of mechanical work supplied, is given by

$$COP_h = (h_2 - h_4)/(h_2 - h_1).$$

If the compression is taken as being isentropic then the cycle is a reversed Rankine cycle and represents the best that can be achieved under the operating conditions specified by the particular evaporating and condensing temperatures and using the refrigerant to which the particular diagram applies. Any practical unit will perform less well than this.

4.3 SOURCES OF HEAT

A heat pump is at its most efficient when the heat source and the heat sink are as nearly as possible at the same temperature. Thus, it is worthwhile looking for the warmest possible source of heat. For domestic applications, the air, or ground water, or a river, or even the ground itself, can act as possible heat sources. Other considerations may dictate whether or not a particular source is suitable. For example, it may not be possible to use the air as a heat source for a large number of individual heat pumps in a high density housing estate because of noise, or it may not be practical to bury heat exchanger coils in the ground to serve a high rise apartment building. Equally, it is worth remembering that the heat source should be at a temperature less than that needed for its direct utilization, otherwise there is no need for the heat pump at all.

Apart from these natural sources of heat, there are a number of others which make the application of heat pumps interesting for industrial applications. The principal case is where there is an existing cooling load as in the food processing industry. Here, it makes sense to use the heat rejected to perform some useful task such as office heating or preheating boiler water. A related example is where there is an environmental requirement to lower the temperature of industrial effluent. The usual method is to use a heat exchanger to remove the heat before discharging the effluent. The introduction of a heat pump would allow the heat extracted to be put to some useful purpose. Incidentally, it is worth noting that current methods do not reduce the environmental heat burden since they merely reduce the temperature of the rejected heat, and not its quantity.

The principal advantage of using the air as a source of heat is that it is readily available to any potential user. This is not true of either river or lake water, or of the ground, although all of these have the advantage that their temperature is more stable throughout the heating season. The disadvantage of air as the heat source is that its properties are constantly varying, and it is therefore somewhat more difficult to design for.

This is well illustrated by looking at the expression for the COP of a Carnot cycle heat pump. Here, an air temperature of $0\,^{\circ}C$ (corresponding to an evaporating temperature of about $-5\,^{\circ}C$) could result in a COP of 6.5, while a rise in air temperature to $10\,^{\circ}C$ might raise the COP to 7.3. Thus, the theoretical (Carnot cycle) performance of the heat pump improves as the heat demand reduces.

This situation is bad enough, but is compounded by the way in which the actual capacity of the machine changes. As the evaporating temperature falls, the density of the refrigerant vapour at the compressor inlet also falls. Since a compressor has a fixed volumetric displacement, the mass flow through it decreases and the compressor power consequently falls. This combines with the reduced COP to produce an even lower heat output than the COP variation alone would suggest. For example, a unit designed to produce 9 kW of heat at an outside air temperature of 0 °C, would typically produce about 12 kW at an air temperature of 10 °C.

There is then the further complication that, for the same temperature variation, the heat demand of the house will have dropped from the design value of 9 kW to only 4.5 kW. Under these conditions, the utilization of the heat pump falls to 4.5/12, or about 37% of its former value. This emphasizes the desirability of being able to modulate the capacity of the heat pump to match the heat demand more closely.

The other problem that arises, particularly in the case of air source heat pumps, is the formation of frost on the evaporator coils under certain atmospheric conditions. If frost is not removed periodically, then it will build up to form a solid block of ice which reduces heat extracted from the atmosphere by the evaporator, and thus detracts from the performance of the unit. This is the much publicized *defrost problem* to which we will return later.

Notwithstanding these problems, it is likely that air will provide the principal source of heat for domestic heat pump units both because of its availability, and also because more site work is required with the other sources. Only an air source unit can be completely prefabricated at the factory and transported to the building site as a sealed unit. This has enormous benefits in both time and financial terms.

4.4 HEAT PUMP EQUIPMENT

All heat pumps have three basic components, the heat exchangers, the compressor, and the expansion valve. While there may be variations, such as the absorption cycle, the basic cycle is unaffected as was shown earlier.

The choice of heat exchanger is dictated by the nature of the heat source and the heat distribution system. If water is the heat source, the evaporator will probably be of the shell and tube type, with the refrigerant in the tubes and the water flowing through the external shell. With ground source machines, it is possible, although unlikely, to have a single refrigerant coil laid in the ground. More likely, a separate heat transfer fluid will flow through the ground coils and will then transfer heat to the refrigerant in a shell and tube condenser of the same type as above. By contrast, an air source unit will almost certainly utilize a finned tube evaporator of the type shown in Fig. 4.5. This provides the large air-side surface area needed to collect the desired heat input from the atmosphere. The effective surface area is increased even further by the use of fans to move the air across the fins.

In industrial applications, other designs, such as plate heat exchangers, may be favoured because of factors related to the production process.

The analysis of heat pump heat exchanger performance is complicated, partly because heat transfer calculations are inherently difficult, and partly because

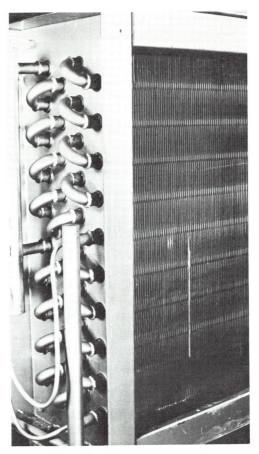

Fig. 4.5. Air source evaporator

the heat transfer media are not simple fluids. On the heat source side, the analysis for water source units is fairly straightforward, but if the heat source is the atmosphere, then the calculations are more complicated. This is mainly because the atmosphere contains a proportion of water vapour which may condense out or even freeze under certain conditions. Analyses proposed for air-side heat transfer must therefore be able to cope with these factors. It is worth noting that most such analyses are designed to deal with the problem of a small amount of uncondensable gas impurity in an otherwise condensable medium. Air is exactly the reverse—a small amount of condensable vapour in an otherwise uncondensable gas. Thus, the common models may be expected to show some limitations, as indeed, they do. The influence of the moisture in the air is an important factor because, in the UK climate, up to 15% of the heat collected at the evaporator may arise from the latent heat of the condensing water vapour.

On the refrigerant side, since the fluid changes phase from liquid to vapour or vice versa, there are the conventional problems of dealing with a fluid which is in different phyical states at different parts of the heat exchanger. In addition to this, however, is the extra, and often overlooked, difficulty that there is a quantity of lubricating oil circulating around the system with the refrigerant. This can represent typically between 1 and 12% of the total volume flow depending on the design of the compressor. The implications for the performance of the heat pump are serious as lubricating oil has a high solubility for most refrigerants. Further, since the oil is non-volatile, it effectively removes the dissolved refrigerant from the heat transfer process and reduces the capacity of the unit. In effect, it changes the heat transfer problem from a complicated, but fairly well understood two-phase problem, to one involving two liquids, three phases, and saturation temperatures that change throughout the heat exchanger as the quality of the refrigerant changes; that is, to one which is not at all well understood.

The heart of the heat pump is the compressor, which drives the refrigerant from its low-temperature, low-pressure state to the high pressures and temperatures which are required at the condenser. It can be one of four types: reciprocating, rotary sliding vane, centrifugal, or rotating screw. The differences are fairly evident from the names. The reciprocal compressor uses pistons sliding up and down in conventional cyclinders. The rotary sliding vane type uses a roughly circular compression chamber with an asymmetrically placed rotor which incoporates a number of radial vanes which are free to slide in and out; centrifugal forces keep the vanes tightly pressed against the inside of the compression chamber and the geometry is arranged so that the largest volume cell between consecutive vanes is at the suction inlet to the compressor while the smallest cell—of as near to zero volume as possible—is at the discharge port. The centrifugal compressor uses an axial turbine type of construction in which the suction vapour is drawn in along the axis and is then expelled radially by the quickly rotating blades of the compressor. The rotating screw compressor consists of a pair of screws meshed in such a way that the vapour is compressed in the channels of the female helix by the semicircular lobes of the opposing male helix. By and large, the last two types are used in large scale applications, while the first two are used for smaller applications.

The most widely used compressor is undoubtedly the reciprocating type. This can exist as a hermetic unit in which the compressor and the motor are integrally constructed and contained within a hermetically sealed container; as a semi-hermetic unit in which the compressor and the motor are separable, but are assembled within a single shell with removable covers sealed using gaskets;

or as an open unit in which the compressor and motor are assembled separately and the mechanical linkage is through belts or a drive shaft and coupling.

The hermetic type is the one commonly seen in domestic refrigerators. Considerable care is taken to reduce noise levels from the compressor by the use of labyrinth vapour channels on the suction and discharge ports, and by suspending the compressor-motor assembly on springs inside the casing. The main disadvantages of this type of compressor for heat pump applications are the reduced efficiency because of the anti-noise precautions, and the fact that repair of the compressor in the event of failure is virtually impossible—replacement being the only practical solution in field operation.

Semi-hermetic compressors are most commonly seen in commercial and industrial applications. They are much more robust in construction than the hermetic type, being produced for more rigorous operating conditions. They have a higher adiabatic efficiency, but are somewhat noisier as the noise limiting measures taken in hermetic units are usually omitted. Their attraction lies in their higher efficiencies and in the fact that they can be repaired on-site in the event of failure. Open compressors have basically the same characteristics as semi-hermetic units, but have the added advantage that the compressor speed can be changed by changing the drive ratio. This allows greater freedom in matching the compressor performance to the requirements of the specific application.

The other essential component is the expansion device which reduces the pressure (and consequently the temperature) of the liquid refrigerant so that it can boil in the evaporator, thus completing the cycle. There are two possible approaches: to use an expander, such as a turbine, which produces mechanical work from the expansion process, or simply to allow the fluid to expand through an orifice. The first approach is used in air cycle air conditioners (such as those in aircraft), because the amount of cooling that is induced by expanding a gaseous (i.e. non-condensable) refrigerant through a nozzle is small. For conventional refrigerants, however, a large temperature drop occurs on expansion and the efficiency of conversion to mechanical energy has been so small that there has been no incentive to use mechanical expanders. Expansion through a nozzle has therefore been the normal practice.

The nozzle either takes the form of a long length of capillary tubing, or of a valve with a controllable orifice size. The length of capillary is selected on the basis of the most likely conditions that the system is liable to encounter, and this approach is used in domestic refrigerators. For larger plant, or in applications where the operation range may be large (for example, heat pumps), an expansion valve is used. The expansion valve may have a fixed orifice size if the operating conditions are fairly static but, more often a *thermostatic expansion valve* is required. This is as shown in Fig. 4.6 and serves two important purposes. The valve both regulates the refrigerant temperature and pressure, and also determines the flow of refrigerant so that the evaporator

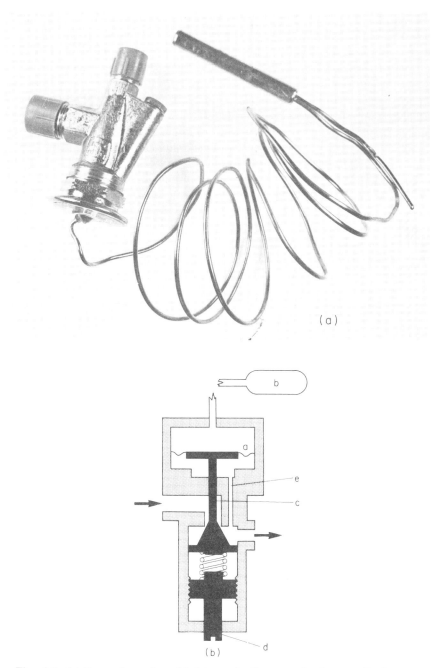

Fig. 4.6. (a) Expansion valve. (b) Schematic diagram of valve. a—diaphragm, b—bulb, c—needle, d—superheat adjustment, e—pressure equalizing tube

receives just that amount of refrigerant that can be evaporated. It achieves this by using the vapour pressure in the temperature sensing bulb to displace a diaphragm in the valve body, and so to determine the position of the needle in a variable cross-section orifice. If the temperature is too high then the needle is pushed out of the orifice, increasing the refrigerant flow. If it is too low, then the needle is drawn into the orifice, decreasing the flow. The bulb is attached to the suction pipe between the evaporator and the compressor and, as the system pressure is admitted to the other side of the diaphragm, the temperature difference between the saturated vapour and the superheated vapour emerging from the evaporator is measured.

In order to stabilize the system, a spring is also applied to the bottom of the diaphragm and so a predetermined amount of superheating is required before the valve reaches its equilibrium setting. This is illustrated in Fig. 4.6(b) which shows a so-called *internally equalized* valve. Such an arrangement is suitable for small evaporators where the pressure drop is small. For larger evaporators, however, where the pressure drop is such that the saturation conditions at the outlet from the evaporator are significantly different from those at the inlet, then an *externally equalized* valve is used. In this system, the internal connection shown in Fig. 4.6(b) is replaced by an external length of pipe which is connected to the suction pipe at the exit from the evaporator. The preset degree of superheating is usually about $5\,^{\circ}$C. This is sufficient to ensure that liquid refrigerant is not allowed to be sucked into the compressor, where it could cause damage since the liquid is virtually incompressible.

It is possible to control the flow of refrigerant by using float valves which determine the level of the liquid refrigerant in the evaporator. This obviously achieves the same result as the amount of refrigerant admitted is the same as that boiled off. However, float valves are not as common as expansion valves for heat pump applications, and so will not be discussed further.

4.5 REFRIGERANTS

The refrigerant is the working fluid of the heat pump. Heat is extracted from the cold source by evaporating the refrigerant, and is rejected at the hot sink by condensing it. The choice of refrigerant is determined mainly by its thermodynamic properties, although it is also affected by a number of other factors.

The ideal refrigerant would be one which has the following properties:

(1)   A high vapour pressure at the evaporating temperature but a low vapour pressure at the condensing temperature. This would result in a low compression ratio.

(2)   A high latent heat of evaporation so that a large quantity of heat is transferred per unit mass of circulating refrigerant. This would mean that the amount of refrigerant required is small.

(3) A low viscosity and surface tension so that the losses in the system are reduced. There is, however, the penalty that droplet formation in the condenser is inhibited in this case, and some compromise must therefore be reached.
(4) Because of the possibility of leaks from the system, the refrigerant should not be explosive, inflammable, poisonous, or detrimental to the environment.
(5) Cheap and readily available.

No such ideal material exists.

The questions of toxicity and environmental impact are particularly vexed problems for the refrigeration and heat pump industries. Safety has always been considered of importance but, during the 1970s, these considerations took on a new significance. Indeed, only fairly recently, the maximum permissible concentration in air of R21, one of the common refrigerants, was changed from 1000 p.p.m. to only 10 p.p.m. This was because it had become apparent that R21 has long term effects on the liver and is also apparently a foetal toxin. The other aspect of this consideration, potential environmental damage, is exemplified by the continuing debate on the possible effects that some of the halogenated hydrocarbon refrigerants might have on the atmospheric ozone layer, and on whether or not this is of importance for the radiation balance at the earth's surface. Thus, there is continuing pressure to ensure that, as far as possible, the refrigerant gases in common use are 'safe'.

There is a large number of available refrigerant materials, including inorganic compounds such as ammonia, carbon dioxide, and sulpur dioxide, saturated aliphatic hydrocarbons such as methane and ethane, and halogenated hydrocarbons—the most important group of modern refrigerants. The most common of these are derived from methane and ethane by systematically replacing the hydrogen atoms by chlorine and/or fluorine. The relationships between the methane based derivatives are shown in Fig. 4.7. The numbers in the figure represent the conventional notation for these derivatives. The first digit is one less than the number of carbon atoms and is omitted if it is zero. The second is one more than the number of unsubstituted hydrogen atoms, and the third is the number of fluorine atoms. Thus, methane is R50, ethane is R170, carbon tetrachloride is R10, carbon tetrafluoride is R14, and R12, one of the most common refrigerants, is CCl_2F_2. R12 has a boiling point at atmospheric pressure of $-29.8\,^\circ C$, a critical temperature of $111.5\,^\circ C$, a saturation pressure at $0\,^\circ C$ of 3.09 bar and at $50\,^\circ C$ of 12.12 bar, and a latent heat of vaporization of 154.77 kJ/kg at $0\,^\circ C$.

Given that, for most applications, the refrigerant will be one of these halogenated hydrocarbons, there is still a fairly wide choice available. The final selection depends upon the application and, in practice, the number of possibilities considered is small. This is partly because of toxicity and other

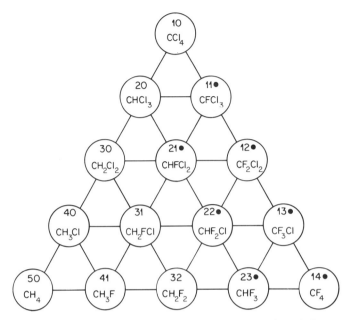

Fig. 4.7. Halogenated hydrocarbons. Relationships of the methane based series. ● Denotes the common refrigerants

problems, but primarily because price and availability are extremely important. Thus, the common refrigerant materials R12 and R22 tend to be preferred unless overriding reasons force a different choice.

One of these considerations may be the operating temperature range proposed for a given installation. If the condensing temperature is high, say around 100 °C, then the high condenser pressure (about 30 bar in the case of R12) may force the selection of an alternative refrigerant, such as R114, which has a lower vapour pressure. In this case, the lower density means that higher volumes must be circulated for the same degree of refrigeration, and so physically larger compressors are required. Once again, a compromise must be struck between the competing requirements for low pressures, low volume flows, and physically small equipment.

4.6 APPLICATIONS

The first heat pump application that generally springs to mind is space heating. This is the area in which there is the greatest potential for energy saving using heat pumps since almost 40% of national energy consumption in a European or American climate is typically devoted to space heating. Typically about 30% of national energy demand is accounted for by domestic and commercial space heating, industrial process heat accounting for less than 10% of the total. This gives energy conservation a rather different perspective from the common or intuitive view in which the large scale energy consumption of industry is perceived as being the primary target, and the small scale consumption of a single dwelling is seen as being negligible in comparison. This is true as long as ICI is compared with 15, Consumer Rd, Manchester. If the basis of comparison is extended to include *all* dwellings in Manchester, the picture changes dramatically, and it becomes apparent why a significant penetration of heat pumps into the domestic market has important implications for national energy consumption.

4.6.1 Space Heating

There are three important differences between domestic and industrial or commercial space heating applications.

First, there is the obvious difference of scale. Domestic equipment is smaller and usually incurs larger efficiency losses. It is also worth noting that domestic equipment is not as well maintained as industrial or commercial plant and that this has implications for the design and operation of domestic systems.

Secondly, the opportunities for heat recovery are reduced in domestic systems. For example, in a large department store, the necessity of maintaining air quality, both to remove body odour and excess body heat during busy periods, ensures that extensive ventilation equipment must be incorporated, with air extraction rates significantly greater than those needed in one-family dwellings. Thus, large quantities of air are being expelled from the building, and the replacement air has to be heated to the required temperature. Exhaust air heat exchangers or heat pumps provide the means of recovering some of the heat from the outgoing air and transferring it to the incoming airstream. The heat pump provides the only way of transferring the heat at the desired delivery temperature; heat exchangers act only as preheaters in these applications and their output still requires further heating. Incidentally, since the outgoing air from such a department store will frequently be heavily moisture-laden, the heat pump, with its cold evaporator, provides a way of recovering the latent heat of condensation of the exhaust water vapour and returning it to the incoming air stream as sensible heat. Most other methods of latent heat recovery involve recycling the actual water and so do little to assist with dehumidification.

Thirdly, in domestic applications, the likelihood that there is a readily available heat source at some elevated temperature is generally small. Thus, locally available natural heat sources must be used. In many industrial applications, there is a warm effluent stream available which can be tapped to reduce both the energy cost of heating the office or storage buildings and the temperature of the rejected effluent. Even if the extent of this cooling is small, it still represents a twofold benefit which is not available to the domestic market.

Domestic units must therefore use air, soil, or water as their heat source. Certainly, in many applications, ground water or a lake or stream will be available and would then represent an attractive option, but for most potential users the air represents the best (or only) choice. Not many tenth floor apartment occupants have access to a passing stream. The consequence of this is that the problems (referred to earlier) of air as a heat source must be tackled and solved to the point where the customer does not notice that they exist. That is, the difficulties associated with part-load operation, the removal of frost, fan noise from the evaporator, capacity limitations in cold weather, the frequent restriction of motor sizes on single phase electricity supplies, the reliability of small fossil fuelled engines over prolonged periods of use, etc., must be overcome to the extent that they are transparent to the user before heat pumps can be expected to take a large slice of the domestic heating market.

The problem of the restriction of the capacity of a heat pump because of the maximum size of the electric motor that can be used on single phase mains supplies is important and can be tackled in two ways.

The first is to develop a heat pump which does not rely on an electric motor as its prime mover. One possibility is to use an internal combustion engine of some kind to drive the compressor (and incidentally to permit recovery of the waste heat from the engine); the other is to develop an absorption cycle heat pump which also uses fossil fuel of some type as the energy input. Unfortunately, neither of these possibilities is presently viable, although considerable R & D effort is being expended in both directions. Suitable absorption cycle units do not exist, and a primary problem with any domestic sized engine driven unit is that of engine maintenance. It must always be borne in mind that normal heat pump duty of, say, a 50% duty cycle for six months of the year is equivalent to an annual usage of about 75 000 miles, and that this is being expected from an engine of about the capacity normally used for a lawnmower. The argument that the running conditions of the engine are constant, so making its life easier, are invalid because one of the advantages of engine drives is that they are much easier to control than electric motors. Their operational duty is therefore likely to involve load following and, consequently, changes in speed.

The other approach to overcoming the size limitation on the electric motor is to supply the heat pump with a supplementary heating system which is used when the heat pump cannot meet the heat demand. In a properly designed system this will only occur on a small number of days each year, and the easiest way of providing such heat in an electrical system is by resistance heating. This does not cause difficulties for the electricity supply authorities as their restriction on electrical motor size is aimed at limiting the surges caused by the starting transients rather than at limiting the actual total current drawn by the installation.

As an example, consider a design heat load of 8 kW at $-1\,^{\circ}$C, which is a suitable choice for the UK (where, over most of the country, the mean temperature falls below zero on only nine days in the year, and below $-3\,^{\circ}$C on only one day in the year). The performance of this house is shown in Fig. 4.8 as the line with negative slope. If the heat pump is assumed to have a capacity of 7 kW at $1\,^{\circ}$C, and to have a COP$_h$ of 2.2 at $0\,^{\circ}$C and 2.5 at $7\,^{\circ}$C, then it can be seen that the provision of 3 kW of boost heating will allow the heat pump system to meet the demand down to $-3\,^{\circ}$C, after which further supplementary heating will be necessary. Since the number of occasions throughout the heating season on which the temperature normally falls below $1\,^{\circ}$C is small, the total energy consumed for supplementary heating is also small.

The relationship between the amount of supplementary heating required throughout the year and the capacity of the unit is shown in Fig. 4.9. This shows that if the heat pump is sized to meet 50% of the design heat load ($-1\,^{\circ}$C) then only 15% of the total heat demand need come from the supplementary heater. If the heat pump capacity is increased to 75% of the design heat load, then the supplementary heat requirement falls to only 5% of the total heat demand. This provides a clear indication that, for a well

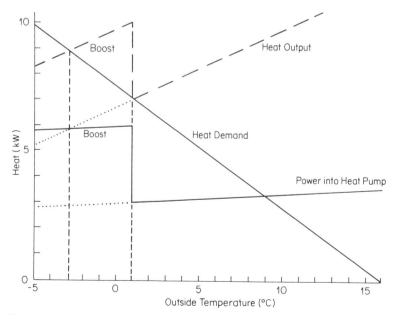

Fig. 4.8. Comparison of heat demand in an 8 kW house with the supply from a 7 kW heat pump and 3 kW supplementary heater

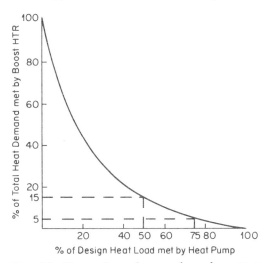

Fig. 4.9. Comparison of seasonal supplementary heating requirement with fraction of design heat load met by main heating system

designed system, supplementary heat requirements are relatively small and need not cause excessive concern over their influence on seasonal performance.

There has been considerable interest in the possibility of introducing thermal storage into electrical heat pump systems. The attraction of this lies in the ability to make use of the cheaper night-rate tariffs and so make the running costs even more competitive. The disincentives include operation during winter nights when performance is worst and the consequent possible requirement for an increase in the size of the heat pump. Since off-peak tariffs are generally offered only for an eight-hour period (or less in the case of Economy 7), the same heat demand from the house has to be provided in these eight hours instead of twenty-four if a completely storage based system is used. Consequently, the heat pump must have at least three times the capacity of a continuously running system, and we run up against the electric motor size restrictions. This means that storage systems should be examined very carefully before receiving too enthusiastic an acceptance. Despite these comments, storage does have a part to play, as the incorporation of a small component would allow unusually heavy heating loads such as the morning warm-up period to be met in a convenient manner. In fact, to continue this line of reasoning, an alternative view could be that the incorporation of a (relatively small) amount of storage would allow the heat pump to run for 24 hours rather than 15, would allow the cheaper night-time tariffs to accumulate heat for use during the day, and might even allow a slightly smaller unit to be installed. This is a complicated problem whose resolution depends almost as much on the electricity tariff structure as on purely technical factors.

The other important aspect of domestic heating systems is the selection of the heat distribution system to strike an acceptable balance between the radiation and convection components, and to decide whether a wet system or a warm-air system would be preferable. Even with the wet system there is a choice between low temperature ceiling or floor heaters, or the more conventional upright panel radiators. With the newer building insulation standards, the heat output of about 70 W/m^2 that can be achieved with a floor system at 30 $^\circ$C is fully adequate. As a consequence, there has been a rapid growth in the use of floor heating systems, notably in Germany, using plastic pipes buried in a light concrete screed.

It may seem that air-distribution systems should be more suitable than conventional panel radiators for use in new heat pump installations, but this is not necessarily so. The desired low air temperatures can be achieved only with the movement of large air volumes, and refrigerant-air heat exchangers are less efficient than their refrigerant-water counterparts. As a result, the difference in perfomance is usually not significant.

Another feature of the panel radiator system concerns retrofitting. In situations where an existing oil-fired system is to be converted to a heat pump system, it is often suggested that, because of the lower design water temperature, the radiators will be too small and will therefore need to be changed as well. However, most domestic boilers are grossly oversized for their application. Typically, a 100 m^2 house with a heat demand of 5–7 kW in cold winter weather might have a boiler of 18 kW capacity, and the radiators are sized to dissipate this heat output at a radiator temperature of about 80 $^\circ$C. If the radiator temperature is reduced to the 50 $^\circ$C more appropriate to heat pump operation, then the heat dissipation falls to about 7 kW. Thus, in general terms, existing radiators can be expected to be adequate for use with a new heat pump unit.

4.6.2 Heat Recovery

As has already been noted, larger scale operations frequently introduce opportunities for heat recovery and the specific example of a department store has been mentioned. Many other possibilities exist, one elegant use of heat pumps being the Versatemp system developed by Temperature Ltd for application to office blocks. The system involves the use of individual room water–air heat pumps connected to a water distribution system kept at 27 $^\circ$C by a central boiler. Each heat pump is reversible, that is, it can act either as a heater or as a cooler. If a room is too cold, then the heat pump removes heat from the 27 $^\circ$C water flow, cooling it slightly. If the room is too warm, however, the heat pump switches to cooling mode, removes heat from the room, and rejects it to the water system, raising its temperature slightly. Thus, it is possible to transfer excess heat from warm areas such as workshops or south-facing rooms to colder, perhaps north-facing, areas.

An excellent example of both sensible and latent heat recovery is given by a municipal swimming pool which combines many of the ingredients required of a successful application. There is a large and fairly stable heat demand, a requirement for mechanical ventilation both for comfort reasons and to keep air—moisture levels down so as to protect the fabric of the building. There is also a centralized plant installation. Over recent years there has been a change in attitudes to the temperature levels deemed acceptable in municipal swimming pools, and this has had serious consequences for their energy demand. Older pools were designed for water and air temperatures of about 21 °C , while the present standard is for a water temperature of 28 °C and an air temperature of 27 °C. In addition to the increased energy demand implied by the new standards, a higher ventilation rate is required to maintain spectator comfort and to protect the building. The presently accepted ventilation rate is about ten air changes per hour, or about $0.015 \, m^3/s$ per m^3 of wetted area. The corresponding evaporation rate is about $2400 \, l/m^2a$, and results in a latent heat loss of about $1650 \, kWh/m^2a$. Typically, a modern municipal pool may suffer a heat loss of about $650\,000 \, kWh/a$ merely to evaporate water off the wetted surface areas (this includes wet tile areas as well as the pool surface). If 75% of this can be recovered together with, say, 50% of the sensible heat input to the pool hall air, then savings of up to $2\,000\,000 \, kWh/a$ are possible. This makes the introduction of a heat pump with its evaporator in the exhaust air stream and its condenser in the incoming air stream an attractive proposition. One such system is shown in Fig. 4.10.

While industrial heat pump applications can involve heat recovery of exhaust air as discussed above, the main attraction lies in the recovery of waste process heat. For an extensive discussion of this topic, the reader is referred to Reay.[2] In summary, the waste heat can be at a low or a high temperature, it can be air or water, or it can even be in the form of manufactured products that must be cooled. It can also arise either continuously or from a batch process. In general, industrial applications must therefore be examined and assessed individually. However, a small number of common factors exist.

The standard industrial process heat medium is steam. Because of this, there is considerable interest in the recovery of low temperature waste heat to produce steam, and the heat pump is the only way of achieving this aim. At present, there is only limited operating experience with heat pumps at the appropriate temperatures but it is an area which will reap high rewards for successful developers.

Lower temperature applications may also be attractive, particularly where there is a requirement for warm water or space heating or other similar demands. A particularly interesting area is where a factory has a legal or other requirement to cool its effluent to an 'acceptable' temperature before its

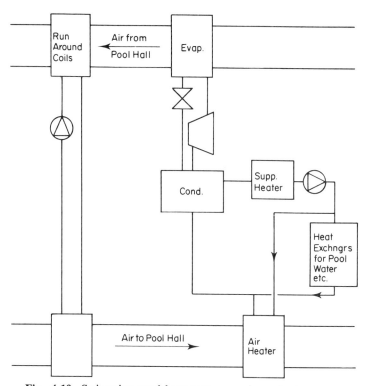

Fig. 4.10. Swimming pool heat recovery using a heat pump

discharge to the environment. In such situations, the heat pump offers the opportunity of recovering, for useful applications, the heat which has to be removed.

A good example of this combined heating/cooling theme is provided by milk chilling operations on farms, the heat rejected by the chillers being used to produce warm water for washing. A recent study by Ubbels *et al.*[4] considered systems in which a milk precooler transferred heat from the warm milk (34 °C) to the supply water, bringing the milk to 21 °C and the water to 20 °C. A heat pump then chilled the milk and supplied hot water at temperatures of up to 80 °C. The study indicated that, in the Netherlands where milk production is about 11 Mt/a, the present electrical requirement of about 500 GWh/a could be reduced by about 200 GWh/a by using this type of heat pump system.

4.6.3 Drying

One example of moisture removal has already been discussed—that of the removal of airborne water vapour as in a swimming pool hall or department store. A related application is that of removing liquid water (or any other liquid for that matter) from some material. The differences between the two cases are obvious. In the first, the heat has already been supplied from somewhere to evaporate the water and increase the air humidity, whereas in the second, the water is still present as liquid and the drying process must supply the latent heat of vaporization. This is usually achieved by passing a heated or dried air stream over the material to be dried and then exhausting the resulting moist air. Thus, the drying process relies largely on the transfer of water from the liquid state in the material to the vapour state in the air stream. The heat pump can save energy in this process by condensing the water vapour out of the air stream, collecting the latent heat of condensation and transferring it back to the incoming air stream as sensible heat. In this way, the heat needed to evaporate the water is recycled. Practical experience has been encouraging. One attractive area is timber drying, where there is not only an energy saving, but there is the further advantage that the degree of control over the drying process is increased, with a resultant improvement in the quality of the final product.

Figure 4.11 shows a schematic diagram of a typical drier. It consists of a kiln containing the material to be dried, and over which passes a warmed, dried stream of air. The air picks up moisture from the sample, and then passes to the evaporator of the heat pump where it is cooled and the water is condensed out again. From there, the cooled, dried airstream passes back to the condenser of the heat pump where it picks up the latent and sensible heat that it lost at the evaporator together with the work supplied to the compressor. It then passes back over the drying material again, and so on.

Another important area is distillation or liquid concentration. This can be tackled by using a heat pump as above, except that the condenser is immersed in the liquid and the evaporator is set in the exhaust vapour stream so that the latent heat of evaporation is passed directly back to the liquid. Alternatively, a mechanical vapour recompression system can be used, as shown in Fig. 4.12.

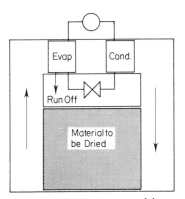

Fig. 4.11. Heat pump drier

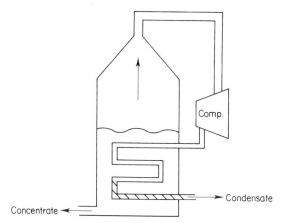

Concentrate

Condensate

Fig. 4.12. Mechanical vapour recompression

This is an attractive technique which is long established and which has a proven high efficiency. The system is extremely simple. The liquid is heated, giving off vapour which is removed and compressed by a compressor, raising the temperature. The hot vapour is then passed through a heat exchanger immersed in the liquid and condenses, releasing its latent heat to maintain the liquid temperature. Only low temperature differences and compression ratios are required which makes the process extremely effective.

4.7 FUTURE PROSPECTS

Over the period since the first practical machine in 1930, heat pumps have passed in and out of fashion several times. The last period of serious interest was in the 1950s when their development was effectively curtailed by the low price of fossil fuels—energy saving was not seen as being worthwhile. It is unlikely that this particular impediment will arise in the current development period, even though occasional short-term oil gluts occur. Because we can foresee the end of the oil era, even if the date is uncertain, its demise means that the market position has changed markedly, that fuel prices are set to rise steadily in real terms and that, as a consequence, energy conservation measures will become more and more attractive.

The other reasons why the popularity of heat pumps is not greater than it is are more complex. Possibly of greatest importance is their high capital cost. It will have to be clearly demonstrated that the heat pump will show an acceptable rate of return against other systems before market penetration can be expected to improve substantially. The present situation is a classic industrial marketing problem; large sales volumes bring down costs, but large sales cannot be generated until the costs come down (unless other steps are taken or fashions dictate). The domestic heat pump will simply have to fight its way through this market limitation.

On the industrial side, the economics are easier to evaluate and heat pumps are beginning to make their way into many aspects of industrial heat recovery. The contrast with the domestic market is marked in that industrialists do their accounting in such a way that some assessment of the total cost of the system can be made, while private individuals do not. Thus, slowly but steadily, heat pumps are beginning to appear in appropriate industrial applications and this is a process that will accelerate as the early installations prove their worth.

There are a number of semi-technical and non-technical obstacles to the introduction of heat pumps on a large scale. These include such factors as the unwillingness of some electricity supply authorities either to allow large single phase electric motors to be connected to their system, or to install domestic three-phase supplies. Another example is given by the strict trade codes of West Germany, rigorous enforcement of which would require up to four different tradesmen, including the chimney sweep, to be involved in installing a heat pump system.

Research and development efforts will undoubtedly reduce the capital cost of heat pump installations to a level that is more acceptable in the domestic market. However, the long term success of heat pumps really depends on improvements in the control of their response to changing environmental conditions and to developments in the absorption cycle and engine driven machines. It is through these developments that the heat pump will be able to extend significantly into the larger dwelling domestic market and into rural areas where the existing electricity supply system does not have the robustness needed to support large electric motors.

Another area where development is needed is in producing heat pumps that can be used in the high density redevelopment areas of the cities. Here, large apartment blocks or closely spaced low rise buildings cannot support the burden of evaporator fans that would be associated with supplying each dwelling with its own air source unit. Instead, a district heating approach seems appropriate and some experiments have already been tried in meeting this need—with a large degree of success. Indeed, one experimental scheme has successfully used the municipal sewage drains as the heat source for a district heating heat pump. It is through projects like this that the heat pump might eventually break into the high density city market. It is interesting to note that city areas offer an advantageous operating regime for heat pumps because of the heat island effect wherein large cities are some $3\,^{\circ}C$ warmer than the rural areas surrounding them. The heat pump would help towards reducing this effect, and would take advantage of its existence to produce a slightly better seasonal performance than it would outside the city.

Commercial pressures will ensure that industrial heat pump applications evolve, particularly as the amount of available expertise and experience builds up. At the present time, there is an acute shortage of trained heat pump engineers. There are many refrigeration engineers, but heat pump practice is sufficiently different that retraining is needed before they can handle heat pump equipment. The problem is that so much of the technique is the same in the two industries that the differences, which are all associated with the fact that the unit is a heater and not a cooler, cause practical and conceptual difficulties at the field level. Thus, extensive training of field staff will be required if there is not to be a maintenance problem over the next few years. This, however, is not a new situation, but one that faces any emergent industry.

All in all, it would seem that the heat pump is beginning to gain commercial respectability and it now looks as though the mid-1980s will see the beginning of a healthy growth of heat pump sales and applications in the domestic, commercial, and industrial markets.

4.8 REFERENCES

1. McMullan, J. T., and Morgan, R. (1981). *Heat Pumps,* Adam Hilger, Bristol.
2. Reay, D. A. (1979). *Heat Recovery Systems,* E. and F. N. Spon, London.
3. Heap, R. D. (1979). *Heat Pumps,* E. and F. N. Spon, London.
4. Macmichael, D. B. A., and Reay, D. A. (1979) *Heat Pumps, Design and Applications,* Pergamon, Oxford.
5. Ubbels, J., Meulman, A. P., and Verheij, C. P. (1980). The saving of energy when cooling milk and heating water on farms. In A. Strub and H. Ehringer (eds.), *New Ways of Saving Energy.* D. Reidel, Dordrecht.

Energy—Present and Future Options, Volume 2
Edited by D. Merrick
© 1984 John Wiley & Sons Ltd

C. A. McAULIFFE
Reader in Chemistry
UMIST

5

Hydrogen and Energy

5.1 INTRODUCTION

Ten years ago the subject of hydrogen as an all-purpose universal fuel was one of (often heated) debate. A hydrogen economy was greeted with almost religious fervour by a few, but also scorned by a number of short-sighted people who misinterpreted what the more sensible of the proponents of the hydrogen economy put forward.

Molecular hydrogen is an attractive *energy vector*. It is not naturally occurring—it has to be manufactured, i.e. it is a synthetic fuel. Modern society is extremely dependent on the energy vector electricity, and misinterpreters of hydrogen economy proponents do not understand that the debate is really about which energy vector will society best use once the naturally occurring liquid fossil fuels are exhausted.

My own views[1] about what the 'hydrogen economy' might turn out to be are that it is quite unlikely that a single all-purpose fuel, even one as attractive as hydrogen, almost totally dominating the energy economy, will ever emerge. This is because the synthetic fuels industry will ensure the continuation of a hydrocarbon economy by producing liquid and gaseous fuels from coal and biomass. The present reduced level of funding of the US synfuels programme by the Reagan administration does not alter the fact that it is still a major effort, that coal liquefaction technology is highly developed in South Africa, and that future US administrations will have little alternative but to increase Federal spending on synfuels because of both strategic and environmental pressures. However, largely because the manufacture of synthetic hydrocarbons almost invariably needs the input of some reducing species such as hydrogen, there will certainly be a growth in the hydrogen manufacturing industry. This is where the main thrust will be, but ancillary developments, for example the potential use of hydrogen as an aviation fuel, will occur and synergistic developments will ensure growth in the overall importance of hydrogen as an ingredient in the production of other fuels, and also as an energy vector itself.

If the characteristics of an ideal fuel are listed:

(1) plentiful, easily extracted;
(2) transportable;
(3) storable;
(4) many uses;
(5) energy dense;
(6) safe.

then it becomes clear why liquid hydrocarbons and natural gas are so convenient and proved so popular during the era of low prices in the 1950s and 1960s (see Fig. 5.1). Some of the characteristics of an ideal fuel become

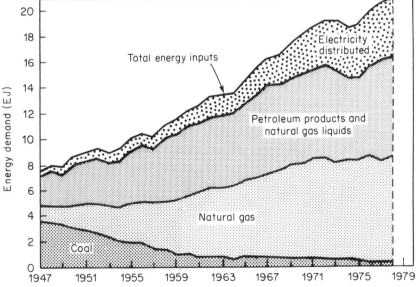

Fig. 5.1. Energy demand by fuel in the household and commercial sectors in the US 1947–78. *Note:* During the 1974–78 period, total energy inputs to the household and commercial sectors increased at an annual rate of 3.3%. Total electrical energy consumption to these sectors increased at an annual rate of 8.1%, natural gas consumption at an annual rate of 6.4%, and petroleum products and natural gas liquids increased at an annual rate of 3.9%. Consumption of coal to these sectors declined at an annual rate of 7.7%. In 1978, natural gas consumption constituted 39% of total energy consumed in the household and commercial sector.
Source: US Department of the Interior (1976), US Department of Energy (1979)

obvious only during a crisis. For example, the closure of the Suez Canal in 1956 and the Arab oil embargo in 1973 emphasized to all in Europe that fuel has to be transportable and storable.

It is ironic that electricity has few of the ideal characteristics; in particular, it is neither storable nor easily transported (Fig. 5.2). The 19 million homes in the UK in 1972 consumed 1.518 exajoules (1 EJ = 10^{18} Joules), an average of 80 GJ per home. Table 5.1 shows that 85% of this was consumed for space heating and water heating (thus offering large opportunities for energy conservation via better insulation). The use of electricity for space heating is questionable on economic grounds, although almost 21% of the total energy consumption of homes is derived from electricity. (In the same period, 67 million homes in the US consumed 12 EJ, or 180 GJ per home; as in the UK, 85% of this was for space heating and water heating.)

Perhaps the greatest problem facing an electric economy, apart from high transmission energy losses, is the varying demand. This is illustrated in Fig. 5.3, which shows that the same utility company has to provide generating plant which can respond to the peak demands at 1800 h in winter and the minimal demands at 0400 h in summer. The provision of the margins for high winter demand is extremely expensive in terms of capital investment, and in terms of energy economics is equally appalling since energy is needed to construct and maintain idle generating equipment. None the less, demand for electricity has

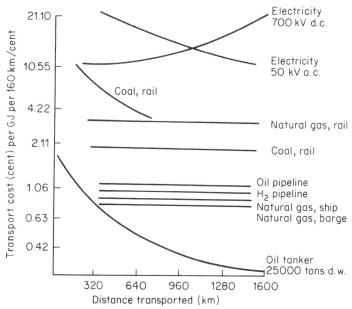

Fig. 5.2. Transport costs of fuels. (Adapted from reference 2 with permission)

Table 5.1 Total household energy consumption in the UK by function and fuel type in 1972 (for 19 million households). (Adapted from ref. 3) (EJ)

	Space heating	Water heating	Cooking	Other	Total
Solid fuel		568		2	0	570
Oil		159		1	0	160
Gas		376		96	4	476
Electricity	109		74	33	96	312
Total		1286		131	100	1517

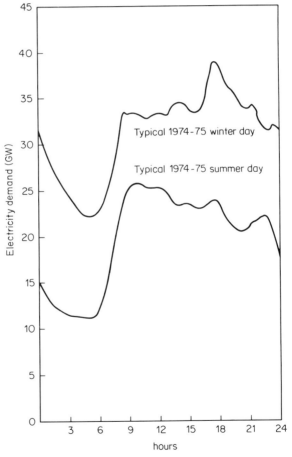

Fig. 5.3. Typical daily variation of electricity demand.
Source: Central Electricity Generating Board (1974)

grown at a faster rate than demands for any primary fuel. Forecasting peak demand is difficult, as such forecasts have to be made several years ahead and have to allow for the fact that demand could be higher than expected, that an unusually severe winter can produce high demands, and that plant maintenance can be made at times outside of peak demand periods. Not surprisingly, there is a relationship between the wealth of nations and their consumption of electricity, Fig. 5.4.

At this point, it is not inappropriate to look at a similar graph of overall per capita energy demand and gross national product, Fig. 5.5. Two facts emerge clearly from Fig. 5.5. The first, most obvious one, is that there is a broad linear relationship between energy consumption and the wealth of nations. However, a somewhat more subtle, but nevertheless profound, relationship is also present, i.e. that some nations (e.g. Canada and Sweden,

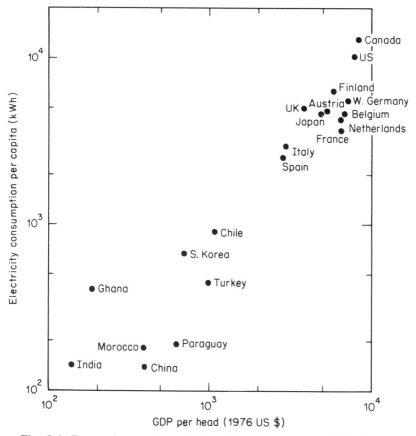

Fig. 5.4. Economic growth and electricity consumption 1976. *Source: Economist* (1978)

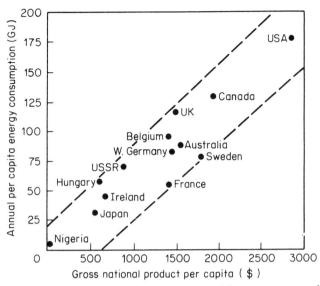

Fig. 5.5. Correlation between commercial energy use and gross national product. (Data taken from reference 4 with permission)

the UK and France) have great disparities in energy demand for the *same* per capita gross national product. This would indicate that energy profligate nations (Canada, UK) could reduce their energy consumption substantially and yet retain a high standard of living. This was pointed out earlier when reference to Table 5.1 showed that 85% of home electricity demand is for space and water heating. Government intervention to encourage a high level of home insulation is highly desirable. Energy conservation is undoubtedly the most important immediate step which Western nations can take to reduce primary fuel demand, extend the life of existing resources see (Table 5.2), and thereby provide time for an adequate synthetic fuels industry to be developed. M. K. Hubbert, a most respected energy forecaster, has published figures of energy available from non-fossil sources, Table 5.3. This shows the immense potential for a solar energy economy. Although solar radiation incident on earth falls mainly in areas where capture is not practical, the fact that 72 minutes of solar energy incident on the earth is equivalent to all the energy the world used last year should be enough to make us realize that the difference between what we need and what is available from solar energy presents a realistic possibility for an energy rich future. Hydrogen has a part to play in almost any scenario past AD 2000, and various aspects of hydrogen and energy will now be examined.

Gregory[6] has made the case for hydrogen well, pointing out that it is the

Table 5.2 Estimated world energy reserves, consumption and 'lifetimes' (Derived from ref. 5)

Source	Reserves (EJ)		Annual consumption (EJ)		Proved reserves Consumption‡ (years)
	Proved	Ultimate	1979 (%)	2000 (%)*	
Natural gas	3000	8000	50 (16)	80 (13)	60
Petroleum	4000	20 000	140 (43)	} 190 (31)	30 }
Oil shale	2000	23 000	—		} 60
Tar sands	3000	10 000	—		
Coal	20 000	300 000	70 (22)	150 (25)	280
Uranium	11 000	3×10^8	10 (3)	100 (16)	1100
†	$> 10^{12}$	$> 10^{17}$			
Renewables	—	—	50 (16)	90 (15)	—
World total	—	—	320 (100)	610 (100)	—

* Assumed demand growth to year 2000 = 2.9% per annum.
‡ Consumption at 1979 rate.
† Breeder reactors.
EJ = Exajoule = $J \times 10^{18}$

Table 5.3 Available non-hydrocarbon energy sources (from M. K. Hubbert, *Scientific American,* Sept. 1971)

Source	Potential (GW)	
Solar	177 000 000	(equivalent to $10^5 \times$ present installed electric power capacity)
Water	2900	
Tidal	64	
Geothermal	10	
Nuclear	very large if breeder reactors used or if fusion becomes commercial	

easiest fuel to produce synthetically, and many other synthetic fuels need hydrogen as a factor of production. Hydrogen is the cleanest fuel, producing only water when burnt correctly, with no carbon monoxide, carbon dioxide, hydrocarbons, or solid residues formed on combustion. Hydrogen is much used in the chemicals industry (see later). Gregory[6] contrasts the present fossil fuel cycle with a potential hydrogen fuel cycle, Fig. 5.6. The vast amount of water at the earth's surface can be affected only insignificantly by the removal of H_2/O_2 from water by electrolysis.

Hydrogen is readily transported and stored, is easily ignited and burns smoothly and evenly in properly designed open-flame burners, both on small

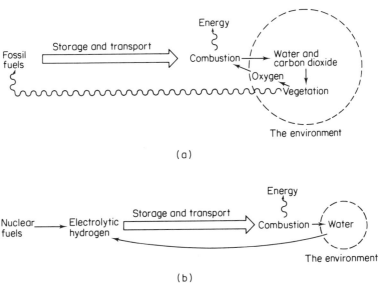

Fig. 5.6. The environmental effects of the fossil fuel and hydrogen fuel cycles. (a) The present energy cycle. (b) The hydrogen energy cycle

domestic appliances and on large industrial units. It is also well suited for catalytic oxidation.

Hydrogen is now made mainly from natural gas, but it costs approximately the same to make hydrogen from coal as to make methane from coal. Thus, it would seem that hydrogen might well act as a bridge between the fossil fuel age and a nuclear or solar era. Almost all types of energy source can be used in a hydrogen-energy system.

5.2 PRODUCTION OF HYDROGEN

Most of the hydrogen produced industrially derives from fossil fuels, although some is produced by electrolysis of water. There is also interest in closed thermochemical cycles for water splitting and some basic research is being done in water photolysis by visible radiation and by biological methods. These approaches are described below.

5.2.1 From Fossil Fuels

Briefly outlined, these include:

(1) Removal of methane and other non-hydrogen constituents from refinery tail gases or coke oven gas at low temperatures.

(2) Reforming of natural gas (or other hydrocarbons)

$$CH_4 + H_2O \longrightarrow CO + 3H_2$$

followed by water-gas shift

$$CO + H_2O \longrightarrow CO_2 + H_2$$

This is followed by carbon dioxide removal using physical or chemical absorption techniques.

(3) Direct production of synthesis gas by reaction of coal with oxygen and steam

$$3C + O_2 + H_2O \longrightarrow CO + H_2$$

followed by water-gas shift and carbon dioxide removal.

(4) Partial oxidation of hydrocarbons

$$CH_4 + \tfrac{1}{2}O_2 \longrightarrow CO + H_2$$

followed by water-gas shift and carbon dioxide removal.

Of these methods, that of (2) is the most widely employed. Desulphurized natural gas is steam reformed in the presence of a nickel oxide catalyst. After cooling to about 375 °C, the product gases undergo the water-gas shift reaction, usually with an iron/chromium catalyst:

$$CH_4 + H_2O \longrightarrow CO + 3H_2$$

$$\downarrow H_2O$$

$$H_2 + CO_2$$

Overall: $CH_4 + 2H_2O \longrightarrow 4H_2 + CO_2$

A second shift reaction may be carried out at about 200 °C over a copper/zinc catalyst. The carbon dioxide is removed by physical or chemical adsorption. Overall, the thermal efficiency of the process approaches 70%.

It is clear that it is water which represents the inexhaustible supply of hydrogen in the future; even in the steam reforming of methane, half of the product hydrogen is derived from added steam. We will thus now examine the second method.

5.2.2 Electrolysis of Water

Physical principles and theory of electrolytic hydrogen production

Water electrolysis is accomplished by passing a direct current between two electrodes immersed in an electrolyte (usually potassium hydroxide solution); hydrogen forms at the cathode and oxygen at the anode. The amount of hydrogen produced is directly proportional to the current passing between the electrodes, and is given by Faraday's law as one kilogramme of hydrogen and eight kilogrammes of oxygen per 26500 ampere-hours of electricity.

The energy which must be supplied to the cell to cause the reaction H_2O (liquid) $\longrightarrow H_2$(gas) $+ \frac{1}{2}O_2$ (gas) to proceed is the enthalpy of formation of water, ΔH, and is equal to 286 kJ/mol at 25 °C and 1.013 bar pressure. However, only the free energy of this reaction, ΔG, equal to 237 kJ/mol, has to be supplied to the electrodes as electrical energy; the remainder is required as heat, and this can theoretically be provided as thermal energy from the surroundings, or from electrical losses within the cell.

According to a fundamental law of thermodynamics, the electrical work, ΔG, done on or by a cell is equal to the free energy change occurring, or

$$\Delta G = -nFEa$$

where n is the number of electrons transferred during electrolysis, E is the reversible voltage of the cell, and F is Faraday's constant. By using this law, the minimum theoretical electrical energy requirement can be measured in terms of the applied voltage, and for the electrolysis of liquid water solution at 25 °C it is 1.229 V, or 118 MJ/(kg hydrogen). A perfect cell would operate at this voltage and energy input, but would require the additional input of thermal energy equivalent to another 24.6 MJ/(kg hydrogen). In order to provide all the necessary energy as electrical energy, the corresponding voltage is 1.484 V (143 MJ/Kg). A practical cell can approximate to this voltage at low output rates, although this still corresponds to a 20% loss of efficiency compared with the 'ideal' situation. Under usual operating conditions, commercial electrolysis plants require much higher power levels, because the power losses in the electrolyte and the electrodes are greater than 20%.

The theoretical reversible voltage (defined by the free energy change) decreases with temperature as shown in Fig. 5.7. Raising the electrolyte temperature lowers the voltage at which water can be decomposed. Moreover, higher temperatures mean that electrode processes occur at faster rates with lower losses; on the other hand, the entropy change for the formation of water is negative, so that, in the case of fuel cells, the useful energy output falls with increasing temperature. The voltage corresponding to the enthalpy change,

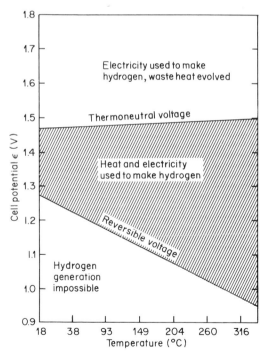

Fig. 5.7. Idealized operating conditions for water electrolysis. (From reference 6 with permission)

also called the 'thermoneutral' change, increases only slightly with increasing temperature, Fig. 5.7. There are three parts to this figure:

(1) no hydrogen is evolved
(2) hydrogen is made at greater than 100% electrical efficiency
(3) hydrogen is made at less than 100% efficiency, with the production of waste heat

In practice, an electrolyser cannot operate at all at the theoretical 'reversible voltage' since, at this condition, the rates of the electrode processes are zero. A higher voltage must therefore be applied for water decomposition. This excess voltage, or 'overvoltage', is related to the current that passes through the cell. Higher currents per unit area of electrode require higher overvoltages and hence lower efficiencies. Overvoltages are reduced by increasing the operating temperature, by proper design of electrodes, and by the incorporation of catalysts into the electrodes.

The efficiency of water electrolysis may be defined as the chemical energy stored in the hydrogen (ΔH) divided by the electrical energy required to produce the hydrogen. There are two values of the chemical energy, i.e. the 'high heating' value (HHV) and the 'low heating' value (LHV). The difference, about 20%, is the heat available as latent heat of condensation. Throughout this work, the LHV is used, since in most end uses the latent heat is not productive. Conventional commercial electrolysis plants operate at electrical efficiencies between 57% and 72%.

Installed electrolysis plant capacity throughout the world was estimated to be 1400 tonne of hydrogen per day in 1980, and this is mainly used in the manufacture of ammonia for fertilizer. The low cost of producing hydrogen from fossil fuels means that hydrogen production via electrolysis accounted for only about 3% of the total amount of hydrogen consumed in the USA. Electrolysis plants are mainly located in areas where there is significant demand for fertilizer, plentiful supply of low-cost electricity, and a limited availability of hydrocarbon fuel. Areas where the above conditions prevail include Egypt, Chile, India, and Norway.

Table 5.4 lists some major electrolysis hydrogen plants, together with their operating parameters and capacities. It is significant that a large number of plants have been operating for a considerable number of years. Figure 5.8 shows the efficiency of such plants in terms of amount of energy required to produce unit mass of hydrogen. An electrical efficiency of 100% is achieved if performance reaches the ΔH(g) line illustrated, assuming the LHV of the hydrocarbon is adopted.

Economics of electrolytic hydrogen production

In the design of an electrolysis plant two factors predominate in the determina-

Table 5.4 Summary of electrolytic plant equipment (prepared by Teledyne Isotopes, Inc.)

Company location	Cell name	Current Type	density (A/m^2)	Emf (V/cell)	Module size (kg/d H$_2$)	Pressure (bar)	No. of plants	Largest size (t/d H$_2$)	Earliest plant year	Best known plant year
A. Norsk Hydro Notodden, Norway	Hydro-Pechkranz	filter press	1500	1.773	820	0.07	3	129	1927	Rjakon, Norway 1965
B. Lurgi Frankfurt, Germany	Zdansky-Lonza	filter press	2150	1.832	1900	30	32	10	1955*	Cuzco, Peru 1958
C. DeNora Italy	DeNora	filter press	3000	2.00†	1860	0.07	2	50	1958	Nangal, India 1958
D. Pintsch-Bamag Germany	Bamag	filter press	2470	1.788	1180	0.9	200	—	1935	UNK
E. Electrolyzer Corp. Canada	Stuart	tank	2150	2.04	18	0.002	1000	0.5	1930	Teledyne Wah Chang. Alabama, USA 1971
F. Cominco Canada	Trail	tank	860	2.142	17	0.007	1	35	1939	Trail Canada 1939
G. Teledyne Isotopes USA	EGGS	filter press	4300	2.1	29	4.8	2	0.023	1968	Teledyne Isotopes USA 1972
H. Demag Elektro-metallurgic GMBH Duisburg Germany	Demag	filter press	990 to 3000	1.75 to 1.95	410	0.07	57	80	1945	Aswan Dam Egypt 1960
I. Electric Heating Equipment Co. USA	Kent	tank	1240	2.2	13	0.007	100	0.7	1920	Hobart, Tasmania 1949
Cells being developed										
J. Teledyne-Isotopes USA		filter press	4300	1.65	6	138	Designed for military aircraft application			
K. Teledyne Isotopes		filter press	2690	1.64	43	207	Designed for nuclear submarine application			
L. General Electric		solid electrolyte	3500	1.2 to 1.8		0.07	2000 $^\circ$C not now under development			
M. Westinghouse		solid electrolyte					Used for CO$_2$ electrolysis in spacecraft atmosphere control system.			

* First Zdansky-Lonza plant
† DeNora has indicated an ability to achieve 1.61 V on new cells.

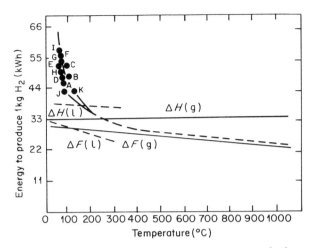

Fig. 5.8. The efficiency of various hydrogen producing
plants: for the significance of A–K see Table 5.4

tion of the cost of the hydrogen produced:

(1)   the capital cost of the electrolysis plant, suitably discounted
(2)   the operating cost—essentially the electric power cost.

The capital cost is dependent on the size and nature of the electrodes,
a low capital cost requiring the highest possible current density. However, a
higher current density generally results in a lower operating efficiency,
although some increase in current density may also be obtained by using
precious metal catalysts or extremely complicated electrode construction. For
optimized cell designs, the operating parameters vary according to the cell's
application and the cost of power. In 1980, capital costs of large-scale electrolysis plants were $320 (kg hydrogen) per day, or approximately $ 320/kW
based on an energy input of 190 MJ/kg hydrogen).

The major operating cost is the purchase of electricity. Even in the case of
off-peak electricity, say at $0.83/GJ (1980 US average price), power costs can
still represent as much as 75% of the overall cost.

The overall cost of hydrogen production can be represented by an empirical
equation of the following type:

$$C_T = aC_0 + b$$

where C_T is the total production cost of hydrogen (in $/GJ), C_0 is the cost of
electricity (in $/GJ,), b represents a fixed cost factor of capital equipment
depreciated at a given rate per year, a is an energy conversion factor.

In the design of an electrolyser, the factors a and b are variable but C_0 is fixed. Thus, desirable features of any plant are low unit costs, long lifetime and high utilization.

Factor *a* has the greatest effect on overall costs. It represents the efficiency with which electrical energy can be transformed into chemical energy stored in the hydrogen. Although a.c.–d.c. rectification efficiency and cell current efficiency modify overall performance, the major loss of energy is that required to overcome the ohmic resistance of the electrolyte as well as anodic and cathodic over-potentials. The value of *a* is unity if the electrical energy is converted to hydrogen at 100% overall efficiency.

Figure 5.9 shows the variation of production costs with the price of electricity. This assumes capital costs of $330/(kg hydrogen) per day, a fixed charge rate of 15%, a 90% plant factor and an electrical efficiency of 60%.

The future for electrolyser systems

Although water electrolysis is already a relatively efficient process, further improvements are inevitable and desirable. Existing large-scale electrolyser plants are all operated at current densities of about 1000 to 2000 A/m^2 and at voltages of about 2.0 ± 0.1 V. Increasing the operating temperature and reducing some of the internal ohmic losses appear possible, and a reduction of 25–35% in power requirement is certainly realistic. The importance of electricity costs exerts a considerable leverage for decreasing the power needs per unit of hydrogen production and in decreasing the cost of power.

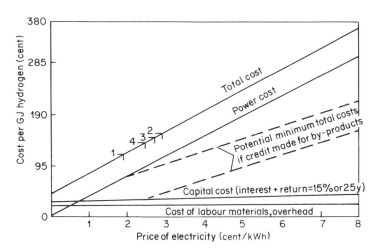

Fig. 5.9. Total cost of hydrogen production: 1, fuel oil at $15.7/m^3; 2, naphtha at $30 per tonne; 3, coal gasification at $7.40 per tonne of coal; 4, natural gas at $0.71 per GJ. (1977)

The solid polymer electrolyte

Although the electrolytic manufacture of hydrogen has been essentially a European technology, in the USA space and military applications have recently caused an increase in interest in electrolyser technology, leading to a number of advanced concepts for electrolyser design and construction which promise to reduce the cost of hydrogen manufacture.

The most important development in electrolyser design is the recent introduction of solid polymer as the electrolyte. In comparison with aqueous electrolyte systems, this allows use of more severe conditions of temperature and pressure with minimum maintenance costs. The solid polymer electrolyte (SPE) is a solid plastic sheet of perfluorinated sulphonic acid polymer similar to Teflon which, when saturated with water, is a good electrical conductor (less than 15 ohm m resistivity). This is the only electrolyte required and there are no free acid or alkali liquids in the system. Ionic conductivity is provided by the mobility of the hydrated hydrogen ions (H^+, H_2O). These ions move by passing from one sulphonic acid group to another. The sulphonic acid groups are fixed, keeping the acid concentration within the electrolyte constant. A thin layer of platinum black (10 to $50 \, g/m^2$) is attached to the polymer surface to form the hydrogen electrode. A similar layer of alloy catalyst forms the oxygen electrode.

In the USA, the General Electric Company is devoting considerable effort to developing an electrolysis process based on a solid polymer electrolyte. The programme is funded jointly by General Electric, the US Department of Energy, and some local electric power producers. The target is to achieve an overall efficiency of more than 85% at a capital cost of less than $100 per kW (1978 money values), with electrolyser cell lives of 40 000 hours or more and a total system life of more than twenty years. They announced that if these criteria could be met, a scale-up to a 5 MW demonstration unit would follow. Hydrogen from such a system is estimated to cost $5.00/GJ, without a credit for by-product oxygen. In 1977, the Illinois Gas Institute estimated that a nuclear power plant could produce electrolytic hydrogen at $5.36/GJ without oxygen credit ($4.73/GJ with oxygen credit). Thus the General Electric scheme looks attractive so far. Recently, they have achieved cost reductions through development of moulded carbon and phenolic separator/current collectors which replace the original transition-metal screens. Improved gasket materials have also been developed permitting operation at higher gas pressures (up to 40 bar).

5.2.3 Thermochemical Water Splitting

Current status

Thermochemical processes producing hydrogen are not in commercial use today but several have reached the pilot plan stage. However, it must be admitted that most schemes are still at the laboratory stage and the main areas of research are presently concerned with the evaluation of thermodynamic and kinetic properties of possible cycle reaction steps. All of the cycles are subject to the Carnot cycle limiting efficiency.

During the last 5 to 10 years, a number of groups around the world have searched for acceptable water-splitting cycles. The largest effort by far is at the Euratom Laboratory at Ispra, Italy. This project began in 1969 and is presently employing fifty people. Other organizations involved include Julich (KFA), the University of Aachen, the Institute of Gas Technology, Argonne National Laboratory, Los Alamos National Laboratory, General Electric Company, and General Atomic Company. About thirty cycles have been published in the literature and undoubtedly many more have been considered by the research groups. A representative few will be discussed here.

Some of the most promising cycles have been based on the chemistry of halide compounds. One of these processes using calcium bromide, and which unfortunately also involves mercury, was named 'Mark 1' by De Beni[7] in 1970;

$$CaBr_2 + 2H_2O \longrightarrow Ca(OH)_2 + 2HBr \qquad 730\,^\circ C$$
$$Hg + 2HBr \longrightarrow HgBr_2 + H_2 \qquad 250\,^\circ C$$
$$HgBr_2 + Ca(OH)_2 \longrightarrow CaBr_2 + HgO + H_2O \qquad 200\,^\circ C$$
$$HgO \longrightarrow Hg + \tfrac{1}{2}O_2 \qquad 600\,^\circ C$$

The advantages of this cycle include: easy product separation and practically 100% recovery of chemicals for recycling. Moreoever, the highest reaction temperature of $730\,^\circ C$ is well within the range of present high temperature gas-cooled reactors. De Beni and Marchetti[8] estimate the efficiency of the Mark 1 process to be 40–60%. The main disadvantages involve difficulties in handling hydrobromic acid and the contamination possibilities inherent in the use of mercury.

In every thermochemical cycle, at least one element is required which is able to change its oxidation state. It is well known that iron tends to a higher valence when it is an oxide than when it is a chloride. This, then, is the reasoning behind the use of the three elements iron, chlorine, and oxygen in many proposed cycles (one by Professor Knoche at Aachen University,[9] four by Hardy,[10,11] one at General Electric[12] and others elsewhere). Table 5.5 summarizes the work.

The five-step sequence illustrated from Aachen University requires excessive recirculation and high temperatures, making this cycle unattractive. On the other hand, thermal efficiencies evaluated for the Mark 7 and Mark 9 processes appear to show more promise. Of the cycles proposed by General Electric, 'Agnes' is the only process published which uses iron as the transition element and, based on a limiting Carnot efficiency of 58%, an overall efficiency of 41% is thought possible. The use of other metals and transition elements with

Table 5.5 Cycles using iron, chlorine, and oxygen

Knoche, Aachen University

$3Fe + 4H_2O \rightarrow Fe_3O_4 + 4H_2$	$500\,^\circ C$
$Fe_3O_4 + \frac{9}{2}Cl_2 \rightarrow 3FeCl_3 + 2O_2$	$1000\,^\circ C$
$3FeCl_3 \rightarrow 3FeCl_2 + \frac{3}{2}Cl_2$	$350\,^\circ C$
$3FeCl_2 + 3H_2 \rightarrow 3Fe + 6HCl$	$1000\,^\circ C$
$6HCl + \frac{3}{2}O_2 \rightarrow 3H_2O + 3Cl_2$	$500\,^\circ C$

Hardy, Euratom, Mark 9

$6FeCl_2 + 8H_2O \rightarrow 2Fe_3O_4 + 12HCl + 2H_2$	$650\,^\circ C$
$2Fe_3O_4 + 3Cl_2 + 12HCl \rightarrow 6FeCl_3 + 6H_2O + O_2$	$175\,^\circ C$
$6FeCl_3 \rightarrow 6FeCl_2 + 3Cl_2$	$420\,^\circ C$

Hardy, Euratom, Mark 7

$3H_2O + 3Cl_2 \rightarrow 6HCl + \frac{3}{2}O_2$	$800\,^\circ C$
$18HCl + 3Fe_2O_3 \rightarrow 6FeCl_3 + 9H_2O$	$100\,^\circ C$
$6FeCl_3 \rightarrow 6FeCl_2 + 3Cl_2$	$400\,^\circ C$
$6FeCl_2 + 8H_2O \rightarrow 2Fe_3O_4 + 12HCl + 2H_2$	$600\,^\circ C$
$2Fe_3O_4 + \frac{1}{2}O_2 \rightarrow 3Fe_2O_3$	$400\,^\circ C$

Wentorf and Hanneman, General Electric, 'Agnes'

$3FeCl_2 + 4H_2O \rightarrow Fe_3O_4 + 6HCl + H_2$	$450–750\,^\circ C$
$Fe_3O_4 + 8HCl \rightarrow FeCl_2 + 2FeCl_3 + 4H_2O$	$100–110\,^\circ C$
$2FeCl_3 \rightarrow 2FeCl_2 + Cl_2$	$300\,^\circ C$
$Cl_2 + Mg(OH)_2 \rightarrow MgCl_2 + \frac{1}{2}O_2 + H_2O$	$50–90\,^\circ C$
$MgCl_2 + 2H_2O \rightarrow Mg(OH)_2 + 2HCl$	$350\,^\circ C$

Table 5.6 Cycles using other metals and chlorine

Funk and Reinstrom, Allison Div. General Motors[13]

$H_2O + Cl_2 \rightarrow 2HCl + \frac{1}{2}O_2$	$700\,^\circ C$
$2VCl_2 + 2HCl \rightarrow 2VCl_3 + H_2$	$25\,^\circ C$
$4VCl_3 \rightarrow 2VCl_4 + 2VCl_2$	$700\,^\circ C$
$2VCl_4 \rightarrow 2VCl_3 + Cl_2$	$25\,^\circ C$

Knoche, Aachen University[14]

$H_2O + Cl_2 \rightarrow 2HCl + \frac{1}{2}O_2$	$900\,^\circ C$
$2HCl + 2CrCl_2 \rightarrow 2CrCl_3 + H_2$	$200\,^\circ C$
$2CrCl_3 \rightarrow 2CrCl_2 + Cl_2$	$1000\,^\circ C$

Wentorf and Hanneman, General Electric 'Beulah'[12]

$2Cu + 2HCl \rightarrow 2CuCl + H_2$	$100\,^\circ C$
$4CuCl \rightarrow 2CuCl_2 + 2Cu$	$30–100\,^\circ C$
$2CuCl_2 \rightarrow 2CuCl + Cl_2$	$500–600\,^\circ C$
$Cl_2 + Mg(OH)_2 \rightarrow MgCl_2 + H_2O + \frac{1}{2}O_2$	$80\,^\circ C$
$MgCl_2 + 2H_2O \rightarrow Mg(OH)_2 + 2HCl$	$350\,^\circ C$

chlorine has also been proposed (Table 5.6).[12-14] Of these, the Beulah cycle
has the highest thermal efficiency reported (53%).

Hybrid thermochemical–electrochemical cycles.

A promising thermochemical cycle is that proposed early in 1978 by
Westinghouse and known as the sulphur-iodine cycle. This is a hybrid cycle
which employs electrolysis and high-temperature chemistry to decompose
water. Westinghouse is presently evaluating the electrolysis of sulphurous acid
at high temperature and pressure, developing materials for handling corrosive
substances, and estimating the overall economics of the process. The
Westinghouse cycle is not being developed solely for incorporation into a
nuclear power plant, and a number of heat sources, including coal, are being
considered. A similar concept has been under development by General
Atomic.

The most critical problems of the process, Fig. 5.10, are involved with
containing sulphuric acid at pressures of 20 bar and temperatures of up to
$450\,^\circ$C during the vaporization steps. The capacity of such a plant is projected
to be 4 Gm^3/d (STP), and the overall efficiency is estimated to be about 50%.

The only two-step process published[15] consists of a reverse Deacon reaction
followed by electrolytic decomposition of hydrogen chloride. The steps are
outlined below:

$$H_2O + Cl_2 \longrightarrow 2HCl + \tfrac{1}{2}O_2 \qquad 700\,^\circ C$$
$$2HCl \longrightarrow H_2 + Cl_2 \qquad 300\,^\circ C$$
$$\text{(electrolysis)}$$

The thermodynamics of the hydrogen chloride–hydrogen–oxygen–water
system have been widely studied.[16] At one bar pressure and $730\,^\circ$C, the
reverse Deacon reaction proceeds with a 60% conversion of water. Hence,
such a step has been included in many other proposed cycles, e.g. the
vanadium chloride process. Moreover, since electrolysis of hydrogen chloride
requires less energy input than electrolysis of water, further development of
this cycle is likely.

Cycles based on the reaction of water with alkali metals to produce
hydrogen and metal–oxygen compounds have also been proposed.[17] One
employing caesium is as follows:

$$2H_2O + 2Cs \longrightarrow 2C_sOH + H_2 \qquad 100\,^\circ C$$
$$2CsOH + \tfrac{3}{2}O_2 \longrightarrow H_2O + 2CsO_2 \qquad 500\,^\circ C$$
$$2CsO_2 \longrightarrow Cs_2O + \tfrac{3}{2}O_2 \qquad 700\,^\circ C$$
$$Cs_2O \longrightarrow 2Cs + \tfrac{1}{2}O_2 \qquad 1200\,^\circ C$$

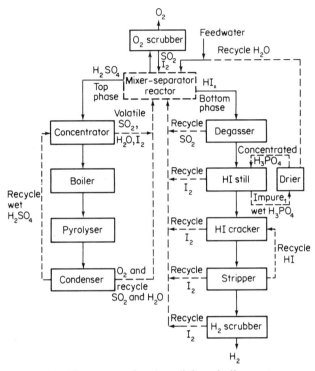

Fig. 5.10. The process for the sulphur–iodine water splitting system involving the reactions

$2H_2O + SO_2 + xI_2 \rightarrow H_2SO_4 + 2HI_x$ (aqueous 298 K)
$2HI_x \qquad \rightarrow xI_2 + H_2$ (573 K)
$H_2SO_4 \qquad \rightarrow H_2O + SO_2 + \frac{1}{2}O_2$ (1144 K)

Adapted from *Chem. and Eng. News*, Nov. 21, 27 (1977)

Table 5.7 presents, in more detail, some of the principal operating parameters of a few selected cycles. Note that the iron–chloride–oxide process refers to the Mark 9 cycle. A heat-to-work efficiency of 30% is assumed. Incidentally, this figure is important when comparison of thermochemical efficiency and electrolytic efficiency is made, as it sets the limit for hydrogen production via the latter route.

Primary energy sources for thermochemical processes

Thermochemical cycles were first conceived with the aim of producing hydrogen from the thermal output of high-temperature gas reactors (HTGR). The successful development of such reactors is of paramount importance as

Table 5.7 Process conditions for some water decomposition processes

	Caesium oxide process	Vanadium chloride process	Calcium bromide process	Fe-chloride-oxide process	HCI electrolytic process
Process heat (kJ/mol H_2)	523.4	649.0	369.7	376.3	57.8
Pumping/separation work (MJ/kg H_2)	10.48	7.15	144.53	not available	28.59
Waste heat (kJ/mol H_2)	52.3	674.1	18.8	265.9	508.7
Electrical work input (MJ/kg H_2)	—	—	—	—	121.4
Total energy input (kJ/mol H_2) Thermal efficiency	596.6	1611.9	477.3	542.2	865.8
HHV	48%	18%	59%	53%	33%
LHV	41%	15%	49%	45%	28%
Highest endothermic reactor temperature (°C)	1050	725	730	650	816
Fraction of process heat at highest temperature	70%	30%	26%	32%	7%
Reactions in closed cycle	4	4	4	5	2
H_2 delivery pressure (bar)	(1)	(1)	15.2	1	19.3

Heat to work efficiency 30% whenever applicable

far as thermochemical splitting is concerned, since the temperatures available from light water reactors are too low for the cycles of interest.

Although originally developed for the production of cheap electricity, the HTGR seems well suited to act as a heat source for the thermochemical production of hydrogen if the moratorium on nuclear power plant construction is lifted.

5.2.4 Water Photolysis

Photolysis is the process by which a compound is decomposed using the energy of incident light. A photon absorbed by a molecule raises an electron into an excited energy state, thereby making it available for pairing with an electron from a neighbouring atom or molecule in an electron-pair bond, and by this process new molecules are formed. The decomposition of a molecule requires the breaking of molecular bonds; in the photolysis of water, light provides the necessary energy for the bond breaking. The net photochemical reaction can be summarized:

$$H_2O(l) + h\nu \longrightarrow \tfrac{1}{2}O_2 + H_2$$

In this process, light energy of 286 kJ is absorbed per mole of water decomposed. This energy may be considered to be stored in the reaction products for subsequent use. Because water is transparent, it can be photolysed with visible light only by using photocatalysts, their function being to absorb the incident light energy. High absorptivity together with broad spectral activity are the most important factors in selecting such 'photo-sensitizers'. Including the photocatalyst (A) in the reaction mechanism results in the following half reactions:

$$H_2O + h\nu + 2A \longrightarrow 2 \text{ (reduced A)} + \tfrac{1}{2}O_2 + 2H^+$$

$$2 \text{ (reduced A)} + 2H^+ \longrightarrow 2A + H_2$$

Note that this is a catalytic process and thus the photocatalyst A is not consumed. Three types of photocatalyst are available:

(1) salt
(2) semiconductors
(3) photosynthetic dyes.

With all three types, the reactions which occur are oxidation-reduction reactions. The work of Grazel at the Swiss Federal Institute is particularly promising.[18]

5.3 TRANSMISSION AND DISTRIBUTION OF HYDROGEN

The movement of gaseous or liquid fuels by pipeline is one of the cheapest methods of energy transmission and, since the system is buried underground, it is also acceptable environmentally. Energy transport from the point of production to the point of consumption will become an important factor influencing any future energy system, and this is especially so in the case of nuclear energy. In order to achieve economies of scale, plants of the future are likely to be larger than those in present-day use and, for safety reasons, will probably be sited considerable distances from population centres.

Since electricity is the chief competitor of hydrogen as a means of carrying energy, transmission and distribution costs for both energy forms will have an important bearing in determining their relative competitiveness.

Natural gas transmission systems in most industrialized countries are well-established and the technology highly developed. But, because of differences in physical properties, the transmission technology for hydrogen in pipelines may differ from that presently used for natural gas. This has an important bearing on whether or not existing natural gas pipelines can be modified to transport hydrogen. Since much experience has been gained during the US space programme in transporting hydrogen as a cryogenic liquid, the possibility of this method of transmission will also be examined.

Present status of hydrogen pipeline technology

The transmission of hydrogen under pressure and in large quantities is established industrial practice, but in most cases only short distances are involved. In the main, hydrogen pipeline systems are confined to refineries where gas is moved from plant to plant and, so far, there has been comparatively little need or incentive to move hydrogen over great distances in this way.

Nevertheless, there are cases where hydrogen gas is transported over considerable distances by pipeline. Perhaps the best known system is the pipeline network in the Ruhr area of Germany which is operated by Chemische Werke Huls AG and has a total length of 200 km. The pipelines, having diameters ranging from 150 to 300 mm, are designed for pressures up to 16 bar. In 1972 the throughput was approximately 0.3 Gm^3, or 3.5×10^{15} J. Figure 5.11 shows a map of the system.[19] Another hydrogen pipeline is located near Johannesburg, South Africa, and is approximately 80 km in length.

None of the existing hydrogen pipeline systems are long enough to require compressor stations along their length. Hence, questions of optimum design parameters relating to compressor staging cannot be answered by reference to the present systems, and theoretical models are being examined instead.

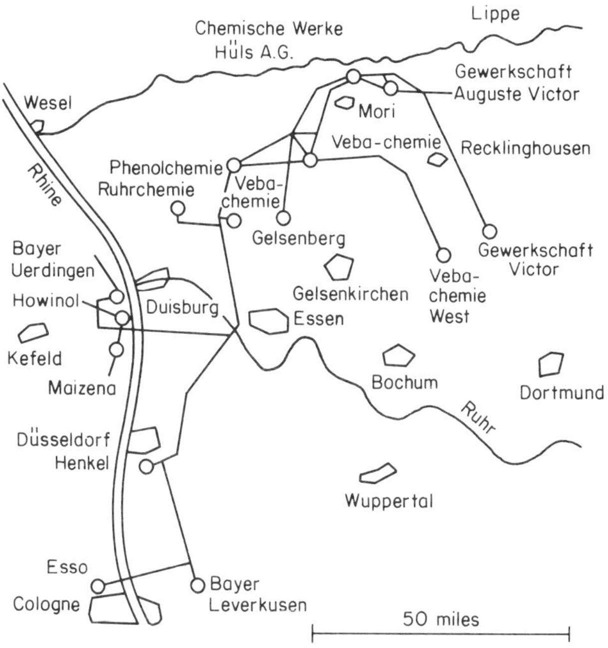

Fig. 5.11. Map of the hydrogen pipeline of the Chemische Werke Huls AG (1970)

However, experience in operating hydrogen pipelines can provide valuable information on other important aspects of transmission, such as safety and material compatibility.

Materials for hydrogen pipelines

Unlike methane, hydrogen can interact with metals under certain conditions and this may limit the materials suitable for construction of hydrogen pipelines. Existing mild steel pipelines have had a good record in this respect,[20] and only under conditions where atomic hydrogen is formed can penetration of the steel lattice occur. This phenomenon has been termed 'intergranular embrittlement'. Molecular hydrogen inside pipelines, at normal temperatures and pressures (less than 150 bar), will in most cases be inert.

However, if the hydrogen is pure then attack at the surface of the metal can take place. For example, the hydrogen evaporated from cryogenic storage vessels used in the USA's space programme has been found to attack welded sections of these vessels. This effect has been termed 'hydrogen environment embrittlement' and presently is the subject of intensive research by NASA.[21] The term 'environment' is used because the hydrogen–metal interaction only occurs at the metal surface.

It has been found that the degree of attack depends upon the nature of the metal and on the prevailing conditions. Aluminium and copper alloys and some stainless steels are not susceptible to attack whereas alloys of nickel or titanium are highly susceptible. Furthermore, attack is most severe at ambient temperatures and under high pressures. On the other hand, an oxygen impurity of as little as 0.6% can completely inhibit hydrogen attack, probably by preferential adsorption on to the metal surface as the oxide.

It thus appears to be extremely important for this phenomenon to be carefully researched, but it is reassuring to note that the extremely good record of existing pipelines towards failure indicates that material compatibility may not be a major problem in implementing a hydrogen economy.

Comparison of electrical and gas transmission costs

The quantity of energy a cable is able to carry is primarily limited by the amount of heat dissipated to the surroundings. Even in the case of overhead lines, where air serves as an effective heat sink, energy capacity is limited to about 2000 MW (direct current). On the other hand, a one metre diameter hydrogen pipeline, under optimal conditions, could carry more than three times this amount of energy.

One of the major drawbacks associated with future overhead transmission lines is the large areas of land needed to satisfy predicted electricity requirements. Thus, there has been much pressure put on the electric utilities to consider underground transmission. However, the costs associated with underground transmission are much greater than those for overhead transmission, mainly because of the costs of the cable which must be insulated and armoured against external damage. Also, the cost disparity is compounded by the reduction in cable capacity arising from the lower rate of heat removal. These factors lead to a cost estimate for underground transmission of $2.60 per GJ per 160 km (Fig. 5.12).

Liquid hydrogen transmission

As an alternative to transmission of hydrogen gas, transport in the form of a cryogenic liquid is also being considered. Considerable experience has already been gained in handling this form of hydrogen in connection with the space programme in the USA. Truck trailers of 37.85 m³ and rail wagons of up to 128.7 m³ have been used to transport liquid hydrogen, but pipelining of liquid hydrogen has been limited to short runs at production plants, and somewhat

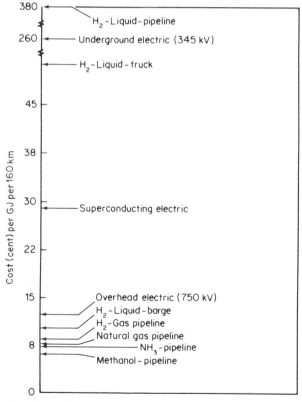

Fig. 5.12. Relative energy transmission costs in the USA (1973), assuming 100% use and 15% fixed charge rate

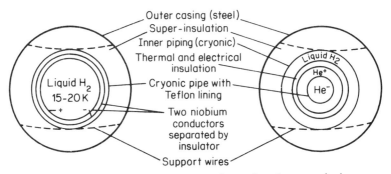

Fig. 5.13. Two possible cross-sections of an 'energy pipe'

longer lines at launching areas. No technology presently exists for moving liquid hydrogen in bulk over large distances, and since transport costs were relatively unimportant during the space programme, existing lines are poorly insulated. The result is relatively poor performance and high costs. In fact, moving liquid hydrogen in a container by truck is considerably cheaper than moving it by pipeline, and transportation by barge is cheaper still.

Hence, it can be concluded that at present transport of hydrogen as a cryogenic liquid is economically unjustified, except in unusual circumstances; for example, where no liquefaction facilities are present for small-scale storage then hydrogen would have to be moved in this form.

Energy-pipe concept

Although too expensive to transport on its own, it is possible that liquid hydrogen could be transmitted with electricity in a common pipe. The low temperature of the hydrogen (less than 29 K) would be ideal for low-resistance electricity transmission. At such temperatures, the electrical resistance of certain metals and alloys diminishes rapidly and they become superconductors.

The electrical utility industry has already considered cryogenic transmission, but in most studies liquid helium has been considered as the coolant. However, hydrogen also possesses suitable characteristics and has the additional advantage that more energy can be transported for a given capital investment.

A proposed system is shown in Fig. 5.13. It consists of a steel pipe lined internally with a superconducting alloy and insulated both thermally and electrically from the outer casing of the pipeline by an annulus of 'super-insulation' (see reference 22 for more information).

5.4 STORAGE OF HYDROGEN

If new sources of energy are to be exploited fully, then associated energy storage systems must be developed. This is because most energy demands are periodic in nature, whereas energy supply systems operate most economically on a constant output basis.

The supply and demand patterns of the electric utility industry illustrate the point well. Since no economic storage method for electricity is currently available on a large-scale basis, generating capacity must be sufficient to meet the maximum demand, which can be more than twice minimum demand (see Fig. 5.3). The result is that fixed charges contribute significantly to electricity costs and consumers pay heavily for a guaranteed supply.

At present, the relative ease with which fossil-fuels are stored is taken for granted. The energy associated with such fuels is in the form of chemical energy which can be released only on combustion. In the case of new energy sources, however, the form of the energy is different. It is usually kinetic (wind, tidal) or heat (nuclear, geothermal, solar) energy. Also, for the renewable energy sources, the energy is not available uniformly, but rather on a cyclic or random basis.

For effective storage of these new energy sources, conversion into a secondary energy form is necessary. Electricity and hydrogen are the two main secondary energy forms available. However, electricity suffers from the disadvantage that it is almost impossible to store effectively. At present, only hydro-electric storage systems have been utilized. Although these can be operated at a relatively high efficiency, the extent of their contribution to utility electricity storage is limited by geographical factors. Compressed air energy storage is a promising concept, but is still under development and requires some combustible fuel during the generating mode of operation. Storage of electricity by means of batteries is not practical on a utility scale of operation.

In contrast to electricity, hydrogen closely resembles our present fuels, especially natural gas. It can be moved and stored in the same way as conventional hydrocarbon fuels. Moreover, its energy is chemical in nature, being released during oxidation of hydrogen to water.

As far as storing hydrogen is concerned, many useful comparisons can be drawn with the storage of natural gas. This is practised extensively in most industrialized countries, where as much as 25% of the demand in winter is met from stored reserves.

Hydrogen can be stored in three forms; as a gas, as liquid, or as a solid combined chemically with a metal. The first two methods are applicable to natural gas storage but the third is special to hydrogen. In fact, the existence of metal hydrides has been known for some considerable time, but it is only recently that their potential for hydrogen storage has been recognized. It is even possible to store hydrogen chemically combined with non-metals; ammonia is perhaps the most promising example, but much would depend upon the relative ease of formation and decomposition of the chemical, and the economics seem unpromising.

5.4.1 Gaseous Hydrogen Storage

In comparison with underground storage, storage of hydrogen above ground on a large scale is expensive. The gas has an extremely low density (about $0.089 \, kg/m^3$) and even at high pressure large volumes will be required, resulting in high material costs. Alternatively, if the storage space already exists, as it does in depleted gas or oil fields, use of such space would lead to low capital costs. Natural gas is stored in this manner, and in the USA storage of natural gas underground has reached the point where it is by far the largest storage capability in use. At the end of 1972, 337 gas storage locations were operational with a total capacity of some $160 \, Gm^3$, representing about 25% of the annual consumption of gas.[23]

The ability to store gas underground depends critically on the nature of the rock strata. Porous, permeable rock is required to hold the gas, whilst sealing of the system is accomplished by the capillary action of water in the caprock. Certain areas are more suited for this type of storage than others and Fig. 5.14 shows which regions in the USA are geologically favourable for gas storage.

The pressure of gas existing within such an underground reservoir is not constant, but varies with depth. In many instances, natural gas is stored at a greater pressure than that originally present in the field. This technique is referred to as 'over-pressuring' and can increase storage capacity several-fold.

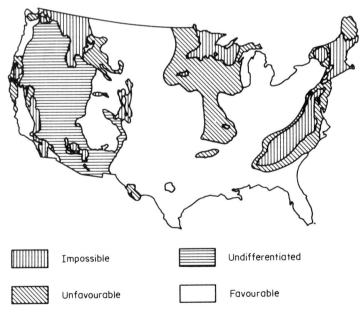

▦ Impossible	▤ Undifferentiated	
▨ Unfavourable	☐ Favourable	

Fig. 5.14. Areas in the USA suitable for underground storage of gas in porous rock

However, certain limits on pressure gradients must be observed and it is generally accepted within the gas industry that pressure gradients of 0.2 bar/m set a limit for storage.

In addition to the use of depleted gas fields, it has been found that natural gas can be stored in porous rock simply by displacing the water present in the pores and creating artificial storage volume. This is the aquifer storage approach and, even though native gas or oil were never present, it permits underground storage in sedimentary structures that have suitable caprock containment.

Since underground storage of natural gas has been operated successfully, the question naturally arises of the extent to which hydrogen is compatible with such means of storage. Perhaps the most important difference between natural gas and hydrogen is the value of their respective diffusion coefficients. Under similar conditions, the hydrogen molecule is sufficiently small to diffuse through an orifice three times more quickly than a natural gas molecule, and this can make sealing of hydrogen storage systems more difficult. Fortunately, where gas is stored in porous rock, then sealing is achieved by capillary action of water, which fills all the voids in the caprock structure. Provided gas pressures are not too high, the caprock will act as an effective barrier to the passage of the gas. Hence, in these cases the sealing efficiency will be independent of the gas diffusivity. Unlike depleted gas field and aquifer storage systems, cavity/cavern storage involves large open spaces to be filled with gas. Even in the case of natural gas, effective sealing is a difficult obstacle to overcome, and the problem would be greater for hydrogen. Nevertheless, the exploitation of porous rock strata presents no sealing problems and this is confirmed by the successful storage of helium—a gas with similar diffusion characteristics to hydrogen—underground in Texas.[24]

The storage potential of underground systems is enormous, with depleted oil or gas reservoirs being economically most attractive for exploitation, since man-made caverns involve excavation costs. In addition to this cost, there is the requirement of effective sealing which can be difficult and costly. Conventional mining costs are dependent on local geological conditions. In easily sealed formations, mining with nuclear explosives could lead to low capital costs but probably would be unacceptable for safety reasons.

5.4.2 Liquid Hydrogen Storage

Storage of hydrogen as a cryogenic liquid is presently the only large-scale method employed. Spherical, vacuum-jacketed containers of up to $3200\,m^3$ capacity have been built and are operated by NASA to ensure a continuous supply of liquid hydrogen fuel for the Apollo space programme.

Liquid hydrogen provides a far more compact method of storage than does the gaseous form. In fact, at NTP, the relative density factor is about 850.

Superficially, storage of hydrogen as a liquid seems attractive, but there are several drawbacks. Firstly, liquefaction requires the expenditure of large amounts of energy, and secondly liquid hydrogen needs to be kept at low temperatures (less than 20 K), which demands sophisticated insulation. The extent to which liquid hydrogen storage is employed will depend largely on the end-use of the fuel.

Much background data on storage of cryogenic liquids can be derived from natural gas systems. Although practised on a much smaller scale than gaseous storage, liquefied natural gas (LNG) storage is a rapidly growing technique. There are about eighty LNG storage facilities in the USA at the moment, and more are being constructed.

LNG is usually stored in flat-bottomed insulated tanks near centres of natural gas demand and is used to even out daily variations in demand, i.e. for peak-shaving purposes. It can also be employed to provide base-load supply, where facilities for pipeline transport do not exist. Some ships designed to carry the cryogenic liquid are currently in operation, and if the proposed scheme for importing natural gas into the USA goes ahead, LNG shipping capacity will increase considerably. The present role of LNG for peak-shaving purposes is likely to correspond to the role of liquid hydrogen in the hydrogen economy. However, base-load use of LNG, involving the transport of the cryogenic liquid by ocean-going tankers, will probably have no counterpart in a hydrogen economy. Tankers will be employed only in areas where an off-shore pipeline is not feasible.

As hydrogen will be produced and transmitted as a gas, then liquefaction facilities may be necessary at the place of storage. In fact, the process of liquefaction has important bearings on the viability of liquid hydrogen storage.

5.4.3 Metal Hydride Storage

Metal hydride storage depends on the reversible reaction between some metals and hydrogen. In general, this can be represented by:

$$\frac{2}{x} M + H_2 \rightleftharpoons \frac{2}{x} MH_x + heat$$

Hydrides have an extremely high packing density, see Table 5.8, but they are costly and heavy.

The systems of interest react with hydrogen reversibly in so far as the gas can be recovered by lowering the pressure of the absorption process. At a given temperature, each hydride is in equilibrium with a fixed pressure of hydrogen, known as its 'decomposition pressure'. If hydrogen is withdrawn and the pressure drops, then decomposition occurs until the evolved hydrogen has built up to the decomposition pressure again. This pressure is a function

Table 5.8 Hydrogen content of various media*

Medium	wt% H	Volumetric density N_H (atoms H/mL, $\times 10^{-22}$)
H_2, liquid	100	4.2
H_2, gas at 100 atm	100	0.5
MgH_2	7.6	6.7
UH_3	1.3	8.3
TiH_2	4.0	9.1
VH_2	2.1	11.4
$Mg_2NiH_{4.2}$	3.8	5.9
$FeTiH_{1.74} \rightarrow FeTiH_{0.14}$	1.5	5.5
$LaNi_5H_{6.7}$	1.5	7.6

* Does not include container weight or void volumes.
[*Source:* CHEMTECH, Dec. 1981.]

not only of the temperature but also of the amount of hydrogen in the solid phase. This quantity is not usually constant but can often vary within rather wide limits. The way in which the dissociation pressure changes with the composition of the solid is shown in Fig. 5.15 for an idealized system $M-H_2$. As hydrogen is taken up by the metal and the ratio H : M increases, the equilibrium pressure increases rather steeply until a point A is reached. Up to this point, the solid consists of a solution of hydrogen in metal, rather than an actual compound. At higher concentrations, however, a second phase

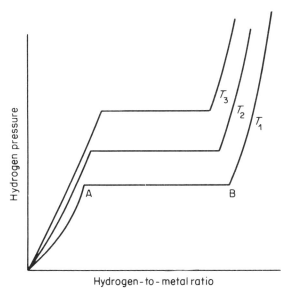

Hydrogen-to-metal ratio

Fig. 5.15. Pressure against composition isotherms for a typical hydrogen–metal system

appears, having the composition B, and the addition of hydrogen does not result in an increase of pressure until all the solid phase has attained this composition. Above this 'plateau' region, further enrichment of the solid in hydrogen requires a steep increase in pressure. The curves T_2 and T_3 illustrate the effect on the pressure against composition relation of raising the temperature. At temperatures above $300\,^\circ C$, hysteresis is usually absent and the equilibrium pressure is the same, whether hydrogen has been added to or removed from the system. It is convenient to characterize the relationship between the pressure and the temperature of a hydride system by reference to the well known thermodynamic formula:

$$\log p = A/T + B$$

where p is the pressure (bar), T is the absolute temerature, and A and B are constants for any given system. Such an equation is valid over the normal range of pressures (Fig. 5.16).

At several centres, research is being conducted into the application of hydrides for fuel storage. No outright winner has yet emerged, although several promising hydride systems have been derived, and some are described briefly here.

Hydrides of magnesium and its alloys

Of the binary hydrides known before 1960, magnesium hydride was perhaps the most suitable for storage. It contains a relatively high percentage of hydrogen (7.65%), decomposes at lower temperatures than most hydrides (one

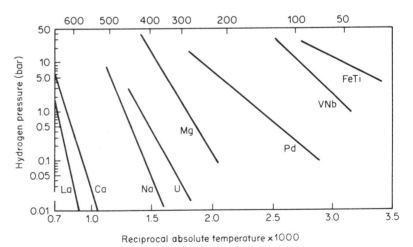

Fig. 5.16. Dissociation pressures of some metal hydrides. The approximate compositions of the solid phases to which these curves refer are: LaH_2, CaH_2, NaH, UH_3, MgH_2, $PdH_{0.6}$, $VNbH_3$, $FeTiH$. The scale at the top of the graph gives the temperature in $^\circ C$.

bar of H_2 at 287 °C) and it is inexpensive. However, it suffers from the disadvantage that it does not form readily by direct combination of the elements; an over-pressure of many times the equilibrium dissociation presure is required. Moreover, the enthalpy of decomposition is high.[25] Of perhaps most interest for the future is the compound derived from alloying magnesium with nickel, which enables the hydride $MgNiH_4$ to be formed.

Iron titanium hydride

Along with the hydrides of magnesium and its alloys, this particular system is the subject of intensive research at Brookhaven National Laboratories in the USA. The intermetallic compound FeTi reacts directly and reversibly with hydrogen to form iron titanium hydride according to the following equations:[26]

$$2.13\ FeTiH_{0.10} + H_2 \longrightarrow 2.13\ FeTiH_{1.04} \qquad \Delta H = -6.7\ kcal$$

$$2.20\ FeTiH_{0.104} + H_2 \longrightarrow 2.20\ FeTiH_{0.185} \qquad \Delta H = -7.5\ kcal$$

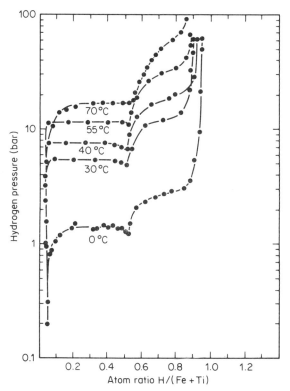

Fig. 5.17. Pressure against composition isotherms for the FeTi–H system

The products of the above reactions are grey metal-like solids which are brittle but non-pyrophoric. The variation of pressure with hydrogen content is shown in Fig. 5.17. Note that high hydrogen pressures can be achieved at moderate temperatures. In addition, there is the advantage that the heat of decomposition is less than half that of magnesium hydride. The main disadvantage is that the elements involved have high atomic weights, so that the energy stored per unit mass is low. For this reason, iron titanium hydride may find use only for storing hydrogen in a stationary facility.

AB$_5$ alloys

An interesting class of hydrides has been discovered by van Vucht and others at Eindhoven.[27] Certain alloys of the formula AB$_5$, where A is a rare earth metal and B is Fe, Co, Ni, or Cu can absorb up to seven hydrogen atoms per AB$_5$ unit. LaNi$_5$ absorbs six hydrogen atoms per formula unit at room temperature and at an equilibrium pressure of 1.5 bar, forming the hydride LaNi$_5$H$_6$. When saturated, the hydride stores 1.3% of its own weight of hydrogen. LaNi$_5$ hydride exhibits several interesting properties. First, the absorption properties are relatively insensitive to impurities in the hydrogen.

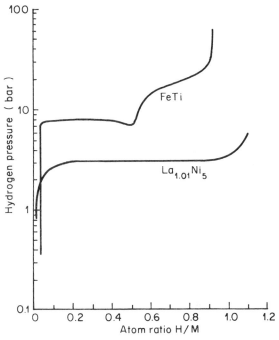

Fig. 5.18. A comparison of the FeTi–H and La$_{1.01}$Ni$_5$–H systems at 25 °C

Secondly, the equilibrium pressure in the plateau region is nearly constant over the whole range and only slightly above one bar at room temperature (Fig. 5.18). Finally, LaNi$_5$ alloy is relatively easy to activate.

Perhaps one of the most important advantages of hydride storage is that the system is comparatively safe; the amounts of hydrogen released can be carefully controlled by varying temperature and pressure. Safety considerations will play a major factor in the use of hydrogen as a mobile fuel.

5.5 USES OF HYDROGEN

The amount of hydrogen produced in the world (currently more than 26 Mt/a) is increasing. For instance, production in the USA was approximately 75×10^9 standard cubic meters in 1980. Most of this hydrogen was produced from natural gas and used in the chemical industry (Table 5.9) for the production of ammonia, plastics, foodstuffs, rubbers and pharmaceuticals, and also as a reducing agent in the metallurgical and scrap-metal recovery industries. Occasionally there arises a special use for hydrogen, as in the recent use of liquid hydrogen in booster rockets for space vehicle launching.

I do not see a rapid revolution occurring, whereby hydrogen will quickly displace other primary and secondary forms of energy. On the contrary, even before there is a 'hydrogen era', there will be a 'coal era', the latter lasting well into the next century. What is clear, however, is that hydrogen production will increase, not only for the increased demands for existing applications (Table 5.9), but because the upgrading of the next, and final, batch of fossil fuels will need massive amounts of hydrogen. Table 5.10 shows some typical industrial hydrogen requirements, and it is obvious that these vital activities will lead to a rapid increase in demand for hydrogen. Whilst this occurs, new methods for hydrogen generation will be developed and, with increased demand, more efficient and thus cheaper production will ensue. Various scenarios can be envisaged but, as well as being incorporated into fuels in the upgrading and conversion of coal to liquid and gaseous hydrocarbons, hydrogen will also

Table 5.9 World consumption of hydrogen (1970)

Application	Consumption (Gm^3)
Ammonia synthesis	100
Methanol synthesis	25
Synthesis of other chemicals	10
Hydrotreating desulphurization	30
Hydrocracking	30
Refinery fuel (low grade H$_2$)	10
Total	205

Table 5.10 Typical industrial hydrogen requirements

Use	Hydrogen requirement per unit of product (m^3)
Ammonia synthesis	1950–2230/tonne NH_3
Methanol synthesis	2.25/kg MeOH
Petroleum refining	109/m^3 crude oil
Hydrotreating:	
naphtha	12/m^3
coking distillates	180/m^3
Hydrocracking	475–595/m^3
Coal conversion to:	
liquid fuel	1070–1250/m^3
gaseous fuel	~ 1560/(10^3 SCM of synthetic gas)
Oil shale conversion to:	
liquid fuel	230/m^3 of synthetic oil
gaseous fuel	1200/(10^3 SCM of synthetic gas)
Iron ore production	560/(tonne of iron)
Process heat	82.4/GJ or
	169/10^3 kg process steam

probably be reintroduced into the general gas supply by merely mixing it with natural gas. As for competition with electricity, Table 5.11 shows how, even when produced by electrolysis, hydrogen may be a cheaper synthetic fuel than electricity, at the point of consumption.

Some of the more obvious uses of hydrogen are outlined in this section, but before embarking on details of these it is important to consider the combustion properties of hydrogen and to establish how they compare with those of existing fuels (e.g. gasoline and Jet-A) and other synthetic fuels (e.g. methanol and ammonia).

It can be seen from Table 5.12 that hydrogen has advantages and disadvantages when compared with natural gas (in Table 5.12 it is assumed that CH_4 represents natural gas). Table 5.13 contains a more comprehensive comparison of fuels and it is clear that hydrogen is capable of carrying more combustible

Table 5.11 Relative prices ($ per GJ) for delivered energy (1970) (US Federal Power Commission)

	Electricity	Natural gas	Electrolytic H_2
Production	2.53	0.16	2.84
Transmission	0.58	0.19	0.49*
Distribution	1.53	0.26	0.32*
Total	4.64	0.61	3.65

* Assumptions made.

Table 5.12 Comparison of some physical and thermochemical properties of methane and hydrogen (data derived essentially from reference 6)

	Methane	Hydrogen
Gas density at 21 °C and 1 bar		
kg/m^3	0.666	0.0833
g/cm^3	6.66×10^{-4}	8.33×10^{-5}
Liquid density at the normal boiling point		
kg/m^3	425	71
g/cm^3	0.425	0.071
Liquid heating value, MJ/kg		
HHV	55.5	142.1
LHV	50.0	120.1
Gas heating value, MJ/m^3		
HHV	37.7	12.1
LHV	33.9	10.2
Compressibility factor at		
1 bar	1.00	1.00
34.5 bar	0.935	1.020
69.0 bar	0.873	1.065

energy per unit mass than any other fuel, but delivers less energy per unit volume. However, the amount of energy released when a fuel is burned is not the only consideration. Another criterion is the air/fuel ratio by weight and the volume per cent in air–fuel mixtures which will support combustion. Some data are given in Table 5.14 from which it is seen that hydrogen has remarkably wide combustion limits. This clearly makes combustion easier, and especially means that the necessity for tuning engines finely is removed; however, it also requires extra care against leakage or spillage because combustion is so much easier. Temperature is also an important factor in flammability limit considerations as these limits become wider at higher temperatures; flammability also depends upon ease of ignition (low for hydrogen in comparison with methane). Not only is hydrogen easier to ignite, it is also much more difficult to quench once combustion is under way. The laminar burning velocity is a measure of flame propagation rates and this too is high for hydrogen; the flame speed for hydrogen at ambient temperature and pressure is about 3 m/s, whereas for methane/air and gasoline/air mixtures the corresponding flame speeds are about 0.4 m/s. The main advantage of this exceptionally high burning velocity for hydrogen is that it permits small combustion chambers.

Schoeppel and his colleagues at the University of Oklahoma have made a

Table 5.13 Comparison data for some selected fuels. J. A. Hoess and R. C. Stahman, *Unconventional thermal, mechanical, and nuclear low-pollution-potential power sources for urban vehicles*, Society of Automotive Engineers Transactions, Paper 690231 (1969)

Fuel	Density kJ/m³	Approximate energy per unit mass MJ/kg	Approximate energy per unit volume GJ/m³	Storage-system requirements for carrying the energy equivalent of 20 gallons of gasoline weight/kg	volume/m³
Gasoline	735	44.4	32.8	83	0.108
No. 2 diesel fuel	838	44.0	37.3	83	0.097
Kerosene	821	44.4	36.3	77	0.097
JP-4 (jet fuel)	777	43.5	34.1	83	0.105
C₂H₅OH (ethanol)	789	27.2	21.5	132	0.165
CH₃OH (methanol)	795	20.5	16.2	176	0.221
CH₄(g)† (methane)	136	50.2	6.9	309	1.066
CH₄(l)‡	425	50.2	21.4	149	0.621
NH₃(l)* (ammonia)	635	18.8	11.9	281	0.516
C₃H₈(l)* (propane)	579	46.5	27.1	121	0.213
C₄H₁₀(l)* (butane)	563	46.1	26.0	116	0.221
H₂(g)† (hydrogen)	17	121.4	2.1	821	3.441
H₂(l)‡	71	121.4	8.6	243	1.515
MgH₂§ (magnesium hydride)	870	8.7	7.5	438	0.449
VH₂¶ (vanadium hydride)	6400	2.6	16.4	1378	0.224

*Liquid at 27°C. †Gas at 27°C and 207 bar. ‡Liquid at cryogenic temperatures and one bar. §Magnesium hydride bed with 40% voids at 260°C. ¶Vanadium hydride bed at 27°C.

Table 5.14 Air-fuel stoichiometric ratios for support of combustion

Fuel	Stoichiometric air fuel ratio by weight*	Approximate volume percent of fuel in combustible mixtures near room temperature†
Gasoline	15.1	1.4 to 7.3
CH$_3$OH (methanol)	6.5	6.7 to 36.5
CH$_4$ (methane)	17.3	5.0 to 15.0
NH$_3$ (ammonia)	6.1	15.5 to 27.0
C$_3$H$_8$ (propane)	15.7	2.1 to 9.4
C$_4$H$_{10}$ (butane)	15.5	1.9 to 8.4
H$_2$ (hydrogen)	34.5	4.0 to 74.2

*Air is here assumed to be 21% oxygen, 78% nitrogen, and 1% argon.
†These data refer to flammability limits in air for upward propagation.

thorough study of the internal combustion engine fuelled by hydrogen and have shown that the combustion of hydrogen/air mixtures produces fewer pollutants than do gasoline/air mixtures. Carbon monoxide, sulphur oxides, lead compounds and unburned hydrocarbons are virtually absent, and nitrogen oxide concentrations can be an order of magnitude lower than with gasoline.

The hydrogen–oxygen fuel cell

In a hydrogen–oxygen fuel cell of the type illustrated in Fig. 5.19, hydrogen and oxygen gases enter at the anode and cathode, respectively, and at the

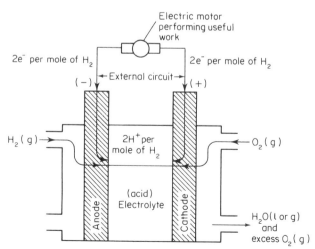

Fig. 5.19. A hydrogen–oxygen fuel cell

anode the following reaction occurs:

$$H_2(g) \longrightarrow 2H^+ + 2e^-$$

The protons migrate through the electrolyte to the cathode and the electrons move in the external circuit. At the cathode the following reaction occurs:

$$2H^+ + 2e^- + \tfrac{1}{2}O_2(g) \longrightarrow H_2O(l)$$

The heat released in the overall reaction

$$H_2(g) + \tfrac{1}{2}O_2(g) \longrightarrow H_2O \qquad \text{(at 300 K)}$$

is either 286 kJ or 242 kJ depending on whether the water formed is in the liquid or gaseous state.

The overall efficiency of a fuel cell can be expressed as follows

$$\eta_{FC} = \Delta\psi/\Delta H = \Delta F/\Delta H$$

where the useful work performed ($\Delta\psi$) is equal to the Gibbs free energy change (ΔF) for the fuel cell reaction, and ΔH is the change in the enthalpy of formation.

Fuels cells have many potential advantages as future power sources. In particular, they do not exhibit the rapid decrease in efficiency with decreasing power output which is characteristic of most other types of power plant. Figure 5.20 illustrates this for hydrocarbon/air fuel cells which have overall efficiencies of 40–50% for power outputs greater than 29 kW. For the

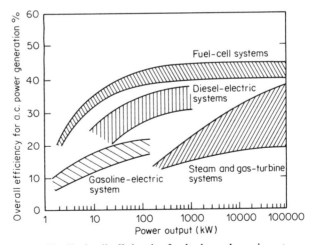

Fig. 5.20. Fuel-cell efficiencies for hydrocarbon air systems producing $H_2O(g)$. (Reproduced from *Hydrogen and other Synthetic Fuels*. US Atomic Energy Commission. TID-26136, US Government Printing Office. Washington DC (1972) with permission)

hydrogen/air and hydrogen/oxygen systems, overall efficiencies are 55% and 60% respectively.

At the beginning of 1978, there were more than eighty 12.5 kW fuel cell stations under test in the USA under the auspices of the gas utility companies' TARGET (Team to Advance Research for Gas Energy Transformation) programme.

Fuel cell stations working on hydrogen would be virtually non-polluting and could be sited even in dense urban areas. Transmission of gaseous hydrogen to these stations could be inexpensive compared with transmitting electrical power to the same urban areas.

The potential for the hydrogen/oxygen fuel cell is considerable. For

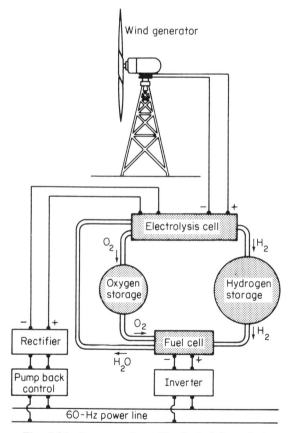

Fig. 5.21. A suggested combined wind-electrolysis arrangement. Wind provides the power for electrolysis of water, and the gases then produce electric current in a fuel cell. Hydrogen and oxygen storage overcomes the problem of variable wind supply

example, hydrogen could be used as an energy storage medium for energy captured from the wind. Wind is an attractive, clean power source, but its capricious nature requires a system which is flexible and which allows for energy storage when the wind blows and the ability to release the energy according to demand. A scheme has been outlined in which wind energy is used to generate electricity to electrolyse water.[28] The oxygen and hydrogen are stored and fed at a constant rate into a fuel cell to produce electricity, Fig. 5.21.

Liquid hydrogen in aviation

It has been suggested[28] that an alternative to jet fuel (Jet A) will have to be found. Of all of the new uses proposed for hydrogen, none is as urgently required as that for fuelling aircraft. During the Arab oil embargo of 1973, the US government assigned commercial airlines a low priority in the allocation of fuel stocks, and the USA's airline industry has had a hard look at the prospects of liquid hydrogen (LH_2) as a fuel and appears to like what it sees.[29,30]

Brewer[28] highlights the problem well when he considers the projected fuel requirements for American airlines in AD 2000. The recoverable oil in the Alaskan North Slope field is presently estimated at $1.53 \times 10^9 \, m^3$. Since the processing procedure of oil refineries currently converts only 5% of a barrel of crude into kerosene (Jet A and Jet A-1), the entire Alaskan field could supply the needs of the airlines for a mere three months. Brewer points out some criteria for a new fuel for aircraft:

(1) What is the preferred fuel from the point of view of cost, noxious emissions, energy efficiency, noise production during combustion, and availability?
(2) What are the technical and financial problems associated with the transition to the new fuel?
(3) How will the new fuel be stored?
(4) Is the new fuel available world-wide?

In 1972, the Lockheed Company began to explore the possibilities of using hydrogen as the alternative to Jet-A in both subsonic and supersonic aircraft. In these studies, the supersonic aircraft was considered to be a Mach 2.7 type carrying 234 passengers and total payload of 220 t; the subsonic passenger aircraft would carry a 400 t payload, including 400 passengers. The aims of the study were:

(1) To assess the feasibility of using LH_2 in commercial aircraft.
(2) To determine the advantages/disadvantages with respect to Jet A fuel.
(3) To identify the technological problems associated with LH_2.
(4) To outline a plan for the development of this technology.

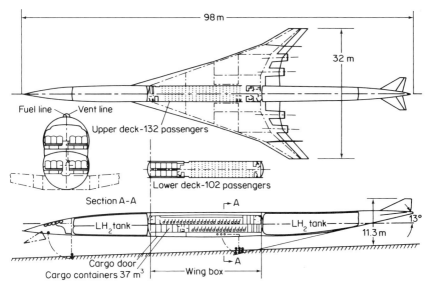

Fig. 5.22. Interior arrangement suggested for a liquid hydrogen fuelled supersonic transport

Operation by 1990 was the goal of the study and it was assumed that LH_2 was to be available at the airport at reasonable projected costs. Turbofan engines based on advanced component technology were 'synthesized' by a computer for both LH_2 and Jet A fuels. Fig. 5.22 shows an outline of the supersonic transport model. It can be seen that a unique feature of this

Table 5.15 A fuel comparison, Jet A versus LH_2, for supersonic transport aircraft (234 passengers, 7800 km, Mach 2.7 speed)

	LH_2	Jet A	Factor (Jet A/LH_2)
Gross weight (kg)	167 000	340 000	2.04
Operating empty weight (kg)	101 200	140 500	1.39
Block fuel weight (kg)	36 940	147 870	4.00
Thrust per engine (kg)	20 900	40 600	1.94
Span (m)	32.2	40.4	1.25
Height (m)	11.4	10.6	0.93
Fuselage length (m)	100.0	90.5	0.91
Wing area (m^2)	639	1005	1.58
L/D (cruise) (kg/h–kg)	6.99	8.5	1.21
SFC (cruise) (kg/h–kg)	0.561	1.51	2.69
Aircraft price (10^6 $)	48.0	67.3	1.40
Energy use (kJ/seat–km)	2434	3474	1.43

Table 5.16 Environmental acceptance parameters for supersonic transport aircraft. The data refer to a General Electric J85 engine in simulated flight at Mach 1.6 and at 16 700 m

	LH$_2$	Jet A
Environmentally perceived noise (dB)		
sideline	105.9	108.0
flyover	104.3	108.0
Sonic boom overpressure (N/m^2)		
start of cruise	63.2	89.5
end of cruise	56.9	67.0
maximum encountered (during climbout)	99.5	119.6
Exhaust emissions		
NO$_x$ (g/kg)	Low	3.7*
CO (g/kg)	None	90*
unburned hydro-carbons (g/kg)	None	0.5*
odours	None	Objectionable
H$_2$O (kg/km)	35.7	18.4

design is that there is no access from the passenger cabin to the flight deck. Access between these was considered neither an advantage nor a disadvantage.

Comparisons of the characteristics of supersonic LH$_2$-fuelled aircraft with those of the equivalent Jet A-fuelled supersonic designs are contained in Tables 5.15, 5.16 and 5.17. The relative economics of the two approaches are summarized in Fig. 5.23.

Table 5.17 Summary of characteristics for supersonic transport aicraft. A cross indicates which fuel is preferable in each case

Characteristic	LH$_2$	Jet A
Weight	X	
Noise	X	
Pollution	X	
Energy use	X	
Production price	X	X*
Operating costs for fuel	X	
Engine size	X	
Physical size	X	X*
Small airport capability	X	

* Subsonic

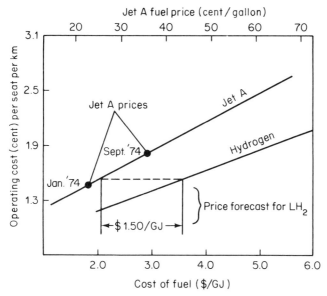

Fig. 5.23. Direct operating costs of supersonic aicraft fuelled by liquid hydrogen and Jet A. Mach 2.7, 7800 km, 234 passengers (1977)

5.6 SUMMARY AND CONCLUSIONS

It is somewhat difficult to put a time-scale on a number of the factors discussed in the chapter. Presently (1983) the world is in a state of non-reality as regards the production and pricing of liquid and gaseous hydrocarbons. In the short-term, and this is what political decisions are usually based on, it appears that there is no energy 'crisis'. OPEC is in disarray and energy costs are, to some extent, falling. This ignores the fact that, for example, the indigenous production of liquid hydrocarbons in the UK will peak about 1990.

Having made these points, it is none the less true that a gradual realization of the need for synthetic fuels will come to all Western governments. Initially this will mean coal liquefaction. This will probably be the breakthrough for hydrogen into the synthetic fuel market, i.e. it will initially be a component in the production of synthetic fuels based on a fossil fuel feedstock. Synergistically, it may well come about that hydrogen's obvious properties as an energy vector will lead to the development of the infrastructure for, at least, a partial hydrogen economy. It is clear that, in many respects, hydrogen is an 'ideal' fuel, and its many other uses (petrochemicals, metallurgical), will certainly give added impetus to its use as a fuel by the early part of the next century.

5.7 REFERENCES

1. McAuliffe, C. A. *Hydrogen and Energy,* Macmillan Press, London, 1980; McAuliffe, C. A. The hydrogen economy, *Chem. and Ind.,* 725 (1980)
2. Smith, M., Rasmussen, G., and Merrick, D. (1981). The economics of hydrogen production and systems, *EEE,* 1, No.1, p. 29.
3. Pullin, D. J. (1977) Approaches to the Structuring of Energy Systems, Ph.D. Thesis, Cambridge Univ.
4. *Resources and Man* (1975). American Association for Advancement of Science, Washington, DC.
5. *'B.P. Statistical Review of the World Oil Industry'* (1978). The British Petroleum Co. Ltd, London.
6. Gregory, D.P. (1973) *A Hydrogen-Energy System,* American Gas Association, Chicago.
7. De Beni, G. *Hydrogen Production Cyclic Process,* French Patent 2 035 558, Feb. 11th, 1970.
8. De. Beni, G., and Marchetti, C. Mark I, A Chemical Process to Decompose Water Using Nuclear Heat, paper presented at 163rd National Meeting of the American Chemical Society, Boston, Mass., 1972.
9. Knoche, K. F., and Schubert, J. (1973). Paper presented in Symposium on Chemical Fuels at 166th National Meeting of the American Chemical Society, Chicago, Illinois.
10. Hardy, C. (1973). Euratom Report 49581.
11. *'Hydrogen Production from Water Using Nuclear Heat'* (1973) Progress Report No.3, Euratom 5059e.
12. Wentorf, R. H., and Hunneman, R. E. (1973). *Thermochemical Hydrogen Generation,* General Electric Company, Schenectady, NY, Report No. 73CR222.
13. Reinstrom, R.M. (1965). *Ammonia Production Feasibility Study,* EDR 4200.
14. Marchetti, C. (1965) Hydrogen and energy, *Chem. Econ. Eng. Rev.,* 5, 5.
15. Hallert, N. C. (1968). Study Cost and Systems Analysis of Liquid Hydrogen Production, NASA CR 73–226.
16. Eagle, W. F., and Waale, M. J. (1962) Recent developments in the oxidation recovery of chlorine from hydrochloric acid, *Chem. and Ind.,* 76.
17. Miller, A. R., and Jaffe, H. *Process for Producing Hydrogen from Water Using an Alkali Metal,* US Patent 3 490 871, 20th January 1970.
18. Gratzel, M. (1981). *Acc. Chem. Res.,* 14, 376.
19. Istings, G. Pipelines now play important role in petrochemical transport, *World. Pet.,* April 1970, p. 40.
20. Nelson, G. A. (1965). Use curves to predict steel life, *Hydrocarbon Processes,* 44, 185.
21. Jewett, R. P. *Hydrogen Environment Embrittlement of Metal,* A NASA Technology Survey, Rocketdyne Division of North American Rockwell, Final Report on Contract NA 58–101(c), Washington DC.
22. Whitelow, R. L. (1974). *Electric Power and Fuel Transmission by Liquid Hydrogen Superconductive Pipeline,* paper presented at THEME Conference, Miami.
23. Gregory, D. P. (1973). The hydrogen economy, *Scientific American,* 228, 1.
24. Trade, M. D. (1966) *Helium Storage in Cliffside Fields,* Paper presented at Society of Petroleum Engineering Meeting, Amarillo.
25. Reilly, J. J., and Wiswall, R. H. (1968) The reaction of hydrogen with alloys of magnesium and nickel and the formation of Mg_2NiH_4, *Inorg. Chem.,* 7, 2254.

26. Reilly, J. J., and Wiswall, R. H. (1974). *Inorg. Chem.*, **13**, 218.
27. van Vucht, J. H. N., Kuijpers, F. A., and Brunning, H.C.A.M. (1970) *Phillips Research Reports,* **25**, 133.
28. Brewer, G. D. (1975). Paper presented at THEME Conference, Miami.
29. Brewer, G. D. (1972). *Advanced Supersonic Technology Concept Study— Hydrogen Fueled Configuration,* NASA Report CR114,718, prepared by the Lockheed California Company, 1972.
30. *Study of Structural Design Concepts for an Arrow-Wing Supersonic Transport Configuration* (1973). Langley Research Centre report to Lockheed California Company.

Energy—Present and Future Options, Volume 2
Edited by D. Merrick
© 1984 John Wiley & Sons Ltd

R. P. SHAH
Corporate Research and Development
General Electric Company
Schenectady, New York, USA

6

Advanced Fossil Fuel Power Generation Systems

6.1 INTRODUCTION

Among the various forms of energy in widespread use, electrical energy is perhaps the most versatile, convenient to use, and economically competitive. Features such as these explain the growth rate of electrical energy all over the world. A major favourable aspect of electrical energy lies in the fact that it can be produced from a wide variety of primary energy sources—fossil (coal, gas, oil, synthetic fuels), solar (wind, hydro, solar thermal, photovoltaic, ocean-thermal-gradient), geothermal, or nuclear. Another distinguishing feature of electrical energy is the great flexibility of generating plant design: individual power generation units may have an output of only a few watts or as many as several hundred MW(e); they can be stationary, floating over the body of the water (e.g. ocean-thermal-gradient plants), or floating in space (power systems for space rockets, Glasser's concept of solar satellite in geosynchronous orbit with microwave power transmission).

Electrical generation systems exist in a wide variety of sizes and types, but here we limit our evaluation to large land-based utilities with individual units in the size range of 25 to 2000 MW(e). We further limit evaluation to fossil-fuelled power generation equipment, which covers essentially all non-nuclear central power station applications. This sector of the energy industry, traditionally known for its steady and tranquil growth, has been subjected to a series of major external forces. The challenges facing the utility industry can be clearly illustrated by examining the situation of United States utilities. Growth rate for electrical power consumption has declined considerably, from an average of 7% per year from 1950 to the early 1970s to about half that rate in the late 1970s. Growing awareness and concern for the environment and for the health and quality of life has resulted in the enactment of laws limiting the amount of pollutants which can be discharged in air or water. The rapid rise in the cost of traditional clean fuels such as petroleum oil and natural gas, coupled with a serious concern about their assured supply, has forced utilities to reduce their use in existing plants and limit new plants designed for these fuels to peaking duty only. Together, these factors have imposed a significant financial burden on the utility industry.

While facing a complex set of difficult issues, the utility industry nevertheless must plan for the future; if it intends to regain its strength and vitality, the utility must look for innovative technologies which would hold a promise of higher energy conversion efficiency and lower cost of electricity while being environmentally acceptable. Utilities must, of course, also design plants to burn the fuel types which are compatible with national policy.

These are formidable challenges, not only for the utility industry, but for the well-being of a nation as well. Recognizing the challenge, the Electric Power Research Institute (EPRI), a non-profit organization supported by funds from private United States electric utilities, and the United States Department of Energy have launched a series of projects to identify, evaluate, test, and commercialize new technological solutions which would benefit the electric utility industry.

The foregoing discussion of issues and actions for the utility industry is not unique to the United States; indeed, challenges and responses similar to those described above are common to most industrial nations of the world.

It was indicated earlier that this chapter would be limited to fossil-fuel based advanced power generation systems using coal or coal-derived fuels. This is because these systems offer a potential for significant improvement in energy conversion efficiency, can be designed to achieve stringent environmental emission limits even when using coal, and because they hold a promise for a lower cost of electricity in comparison with alternative power generation systems of comparable size and flexibility.

The procedure followed in this chapter for the discussion of advanced concepts closely parallels the evolutionary development path of any major new technological system. For each concept, a baseline cycle configuration will be presented, together with its major features, estimated energy conversion efficiency and plant capital cost, a brief discussion of major development items, and a schedule for commercialization. Selection of the major cycle parameters for the baseline configurations was preceded by an extensive evaluation of the impact on plant efficiency and economics due to changes in the values of those cycle parameters within a specified range. The resulting set of parameters represents a baseline configuration which is an optimum combination of efficiency, economics, and permissible degree of developmental risk. However, if any of the principal components or parameters for an advanced cycle concept is changed, an optimum cycle configuration for the resulting concept could be substantially different from those shown here. The example of the gasification/combined cycle system will illustrate this point later. Several advanced concepts are evaluated on a consistent thermodynamic and economic basis in this manner. The results would permit a comparative quantitative ranking of the technologies, which can be used to estimate a potential share in utility market (and hence anticipated revenues) for each technology concept when fully developed. Since the utility industry is the principal market-place for such advanced technologies, attributes of each system characterizing its use by utilities are determined in a quantitative manner; these results also influence the estimate of market share.

In order to realize the full potential of each advanced energy conversion system, considerable investment in research and development is required. Even more significant is the extent of the development time required. Since the utility industry is subjected to several dynamic forces, those advanced systems which could be brought to a demonstration stage early would enjoy a significant marketing advantage. The need for early commercialization would suggest that the demonstration plant should contain fewer developmental components, with a corresponding smaller improvement in energy conversion efficiency, and a higher capital cost than for the fully developed concept. Of course, a parallel effort to develop components for the optimum plant configuration would continue, and it would benefit from the experience gained in operating the first generation demonstration plant. Such a course is being followed for several advanced systems.

6.2 GROUND RULES FOR EVALUATION

It is imperative that a consistent basis be applied in the evaluation of performance and economics of the advanced concepts so that the results can be compared and interpreted with confidence. This can only be achieved by developing clear and detailed sets of ground rules and assumptions. These ground rules should reflect commonly accepted major design guidelines for the components intended for the utility service, and should also incorporate other considerations for plant design, financing, construction, and operation, as applied to the utility industry. The advanced power generation systems discussed in this chapter have been analysed using a consistent methodology and the ground rules discussed below. Estimates of the overall coal pile to busbar efficiency, plant capital cost, and cost of electricity (provided later) are internally consistent and valid for comparative evaluations. For a specific utility, differences in coal type, ambient conditions, design and operational philosophy, project financing, plant construction, etc. could affect the results provided herein, and may even alter the comparative attractiveness of the technologies. The discussion below would permit the effect of financial assumptions different from those adopted for the evaluation in this chapter to be assessed.

Fuel Type

For most advanced systems, run-of-mine coal is delivered to the plant site and used directly. For systems using an on-site coal gasification process, low or medium Btu fuel gas is produced and consumed completely, without any net export. Coal derived liquid fuel is used for one system.

In all cases, Illinois no. 6 coal is used. Composition data for coal and a coal derived liquid fuel are given in Table 6.1.

Ambient Conditions

For air, 288 K (dry bulb), 60% relative humidity is assumed; for cooling water, 288 K.

Table 6.1 Fuel Characteristics

	Coal	Coal derived 'semiclean' liquid fuel
Type	Illinois no.6	Derived from 'H-coal' process
Higher heating value, MJ/kg	25.09	38.84
Chemical composition (wt%)		
Carbon	59.6	88.2
Hydrogen	5.9	7.4
Sulphur	3.9	0.5
Nitrogen	1.0	1.3
Oxygen	20.0	2.4
Ash	9.6	0.2 (<5 μm particle size)
	100.0	100.0

Efficiency

This is expressed on the basis of the higher heating value (HHV) of the fuel. Power plant efficiency is defined as the net electrical output divided by the fuel energy input to the power plant. The overall plant efficiency is the fraction of the energy in the coal pile delivered at the busbar. For systems employing coal directly at the plant site, these two efficiency values are identical. For systems using transportable fuels produced from coal at the off-site, the power plant efficiency is multiplied by the fuel conversion efficiency to obtain the overall efficiency.

The net plant electrical output is calculated by subtracting auxiliary losses and balance of plant power demands from the gross electrical generation. This on-site power consumption is needed to support the plant operation, and could be as high as 7% of the gross output. These power penalties are sometimes omitted by proponents of an advanced concept, and can lead to unfair comparisons.

The energy conversion effectiveness is sometimes expressed by the term 'heat rate' which is equivalent to (3413/efficiency) and has units of Btu/kWh.

Capacity Factor

This is defined as the actual kilowatt-hours output per year divided by the plant rating in kW and the number of hours per year (8760). In practice, the capacity factor of a generating unit is determined by its incremental cost of generation relative to that for other units in a utility system. Units with the lowest generating costs would be used for base-load applications (capacity factor of 0.6 to 0.7). On the other hand, units used for peaking duty (capacity factor of 0.1 to 0.2) can be brought on-line quickly and usually have the highest incremental cost of generation.

In estimating the cost of electricity, a capacity factor of 0.65 is used for all power generation systems in this chapter. A lower capacity factor would imply a higher cost of generation.

Cost Basis

All cost numbers are expressed in terms of 1981 US dollars. The published cost data for various plant concepts are brought to the 1981 level using an average escalation factor of 7.2% per year for both capital and fuel costs.

Capital Cost

This is defined as the total plant cost divided by the rated plant net electrical output, and is expressed as $/kW(e).

Total Plant Cost

This includes all of the cost items associated with the construction of a commercial plant of mature design. In a detailed study of an advanced concept powerplant the cost of advanced or non-commercial components would reflect the 'learning curve' effect of a mature design.

The plant cost estimate would be built up as shown in Table 6.2.

For all cases considered here, a multiplier factor of 1.432 is applied to the 'total installed cost' to arrive at the 'total plant cost'. This cost factor reflects 20% for contingency, 10% for architect/engineer's fee, and 8.5% for interest during construction.

Table 6.2 Total Plant Cost Development Structure

Installed cost
Subsystem I
Cost of major components
Balance of plant material cost
Direct labour cost
Indirect/supervisory labour cost
Subsystem II
Cost of major components
Balance of plant material cost
Direct Labour cost
Indirect/supervisory labour cost
Subsystem III
Cost of major components
Balance of plant material cost
Direct labour cost
Indirect/supervisory labour cost
Total installed cost
Contingency
Architect/engineers fee
Interest during construction
Total plant cost ($)

Fixed Charge Rate (FCR)

This parameter is used to calculate the contribution of the capital cost in estimating the cost of electricity. The parameter reflects the cost of money to the utility (assumed to be 10% per year), amortization period (30 years), income tax (50%), etc. A fixed charge rate of 18% per year is used.

Levelization Factor (LF)

This reflects the effect of escalation in the cost of fuel or labour during the plant operation life. A levelization factor of 1.94 is used, corresponding to an escalation rate (inflation rate) of 6.5%/year for an assumed useful plant life of thirty years, and a present worth discount rate of 10%/year.

Cost of Electricity (COE)

This is expressed as mills/kWh, and consists of three elements:

(1)   Capital element = (1000)*(capital cost $/kW(e))*(fixed charge rate)/
     (8760* capacity factor)

(2) Fuel element = (fuel cost, $/GJ) * (3.6/efficiency)
(3) Operating and maintenance (O and M) cost element.

The levelized cost of electricity would be obtained by multiplying the fuel element and the variable portion of the O and M element of the COE by appropriate levelization factors, and then adding the resulting values to the capital element and to the fixed portion of the O and M element.

Emission Targets

For any coal-based power generation system, the cost of environment emission control measures tends to be substantial. All of the energy conversion systems presented here were designed to meet the emission limit targets shown in Table 6.3.

Table 6.3 Emission Targets

Emission	Fuel	Standard (ng/J heat input
SO_x	Solid	520
	Liquid	347
	Gaseous	87
NO_x	Solid	300
	Liquid	130
	Gaseous	87
Particulates	All fuels	43

In interpreting the emission standards, note that all references to SO_x or NO_x are expressed in terms of equivalent mass of SO_2 or NO_2 per heat input in the primary fuel at the plant site. Thus, for example, the SO_x standard applicable to a combined cycle plant integrated with coal gasifiers would require that the mass of all sulphur compounds emitted in the atmosphere (in the form of SO_2, SO_3, or H_2SO_4), expressed as SO_2, would not exceed 520 ng/J heat input in the coal feed to the gasifier. Similarly, the combined cycle plant using coal-derived liquid fuels would be subject to a SO_x limit of 347 ng/J in the liquid fuel consumed at the plant site.

6.3 ADVANCED CONCEPTS FOR POWER GENERATION

In this section, ten power generation concepts are discussed in terms of their major features, simplified cycle schematics, and estimates of overall plant performance and economics. Under the sponsorship of the US Department of Energy and the Electric Power Research Institute (EPRI), various studies[1-7] have been conducted to evaluate major advanced concepts for power generation from fossil fuels using a consistent methodology to serve as a data base for the support of technology development. In the course of these studies,[1] design parameters and their ranges were identified for each system, and the performance and economics of the resulting system configurations were evaluated. Results of these screening studies provided a basis for the selection of parameter values* for a specific configuration of each system which would

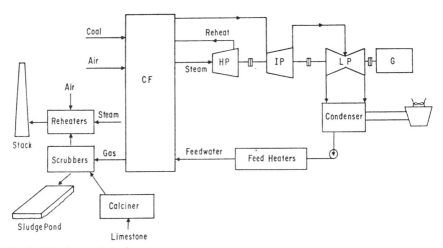

Fuel: Illinois no. 6 coal
Combustor boiler: conventional boiler
Emission control:
   SO_x—wet lime scrubbing of flue gas
   NO_x—combustor design
   Particulates—electrostatic precipitators

Prime cycle: steam
   Throttle: 24.2 MPa, 811 K
   Reheat: 811 K
   FFW temp.: 536 K

Gross output: 820 MW
Auxiliary losses: 73 MW
Net plant output: 747 MW
Power plant efficiency: 31.8%

Fuel consumption: 0.452 kg/kWh
Water consumption: 3.10 1/kWh
Total waste production: 0.122 kg/kWh

Fig. 6.1. Cycle schematic: steam/FGD

---

* For the reader interested in evaluating the impact of parameters different from those presented here, these reports would serve as a useful data source.

provide an 'optimum' combination of efficiency and capital cost to result in a lowest cost of electricity. In the subsequent study of these selected systems, detailed evaluations of plant performance, capital cost, and status of technology development were conducted. The following discussion of the power generation systems is based on the results of the detailed plant configuration studies.

6.3.1 Steam Cycle with Furnace and Flue Gas Scrubber

This case[5] represents the reference steam plant, as shown in a simplified cycle schematic (Fig. 6.1). In contrast to the advanced power generation cycles, this cycle employs a post-combustion method of emission control by removing sulphur dioxide from the stack gas in a wet lime scrubber. All other plant components—furnace/boiler, steam turbine, condenser, and cooling towers—are proven conventional elements.

Limestone is processed in an on-site calciner to produce lime. Lime slurry sprays quench the particulate-free flue gas to 325 K. Sulphur from the gas is removed as calcium sulphite and calcium sulphate, which is precipitated out in the sludge pond. The desulphurized gas is reheated to 394 K by mixing with hot air which is heated in turn by steam extracted from the steam turbine cycle.

6.3.2 Steam Cycles with Fluidized Bed Combustors

Three plant concepts are evaluated in this category:

(1) Supercritical steam cycle with an atmospheric fluidized bed (AFB) heat input system[2] (Fig. 6.2).
(2) Subcritical steam cycle with AFB[6] (Fig. 6.2).
(3) Supercritical steam cycle with a pressurized fluidized bed (PFB) heat input system[2] (Fig. 6.3).

Fluidized bed heat input systems offer important environmental and heat transfer advantages. The turbulent motion of the bed particles provides intimate contact of the sulphur-bearing particles of coal with the sulphur-absorbing particles of limestone or dolomite added to the bed. Fluidized beds also have a uniform temperature, and can be operated at an optimum temperature for sulphur capture and for the control of nitrogen oxides.

The AFB consists of four modules in parallel, each module having a vertical stack of six main fluidized bed cells operating at 1116 K and one carbon burn-up cell operating at 1366 K. A limestone/sulphur ratio of 6.59 kg/kg, which is twice the stoichiometric requirement, is provided to capture 83% of the sulphur in the feed coal. The AFB furnace is basically the same for both steam cycles. The supercritical steam cycle is identical to the one used in the reference

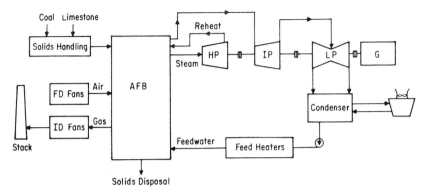

Fuel: Illinois no. 6 coal
Combustor/boiler: atmospheric fluidized bed
Emission control
    SO_x—limestone addition in AFB for in-bed sulphur capture
    NO_x—lower temperature in AFB
Particulates—cyclones, electrostatic precipitators (ESP)

	Case I	Case II
Prime cycle: steam		
throttle	24.2 MPa, 811 K	16.6 MPa, 811 K
Reheat	811 K	922 K
Final feedwater temp.	536 K	583 K
Gross output, MW	878	905
Auxiliary losses, MW	64	62
Net plant output, MW	814	843
Power plant efficiency, %	35.8	37.1
Fuel consumption, kg/kWh	0.401	0.386
Water consumption, l/kWh	2.31	2.27
Total waste production, kg/kWh	0.13	0.09

Fig. 6.2. Cycle schematic: steam/AFB

steam plant with furnace/scrubber. The subcritical steam throttle conditions may offer improved reliability by avoiding the susceptibility of the super-critical steam cycle to transient deficiencies in feedwater quality. A higher reheat temperature of 922 K is specified for the latter case.

The PFB combustion system has four modules each with a vertical stack of six main bed cells operating at 1172 K and one carbon burn-up cell operating at 1366 K. A dolomite to sulphur ratio of 11.79 kg/kg, equivalent to twice the stoichiometric amount, is provided.

A pressure of 1 MPa is maintained in the PFB furnace by an air compressor. The products of combustion leaving the PFB are filtered in a series of cyclone separators and granular bed filters before entering the expansion turbine at 1144 K. The gas turbine exhaust gas is cooled in a gas to feedwater heat exchanger and enters the stack at 394 K.

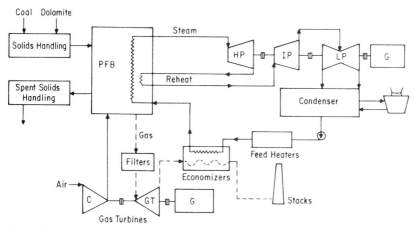

Fuel: Illinois no. 6 coal
Combustor/boiler: pressurized fluidized bed
Emission control:
  SO_x—dolomite addition to PFB for in-bed sulphur capture
  No_x—Lower temperature in PFB
  Particulate—cyclones and granular bed filters

Prime cycle: steam
  Throttle: 24.2 MPa, 811 K
  Reheat: 811 K
  FFW temperature: 396 K
Pressurizing gas turbine:
  Compressor pressure ratio = 10/1
  Turbine inlet temperature: 1144 K

Gross output: 944 MW
Auxiliary losses: 40 MW
Net plant output: 904 MW
Power plant efficiency: 39.2%

Fuel consumption: 0.367 kg/kWh
Water consumption: 2.08 l/kWh
Total waste production: 0.15 kg/kWh

Fig. 6.3. Cycle schematic: steam/PFB/811 K/811 K

6.3.3 Combined Cycle with Coal derived Gaseous or Liquid Fuels

Three advanced power generation concepts in this category are described:

(1) Combined cycle (1589 K) integrated with low-Btu coal gasifier[2] (Fig. 6.4).
(2) Combined cycle (1922 K) integrated with low-Btu coal gasifier[6] (Fig. 6.4).
(3) Combined cycle (1922 K) using coal derived liquid fuel[2] (Fig. 6.5).

The combined cycle power plant includes a number of gas turbine units, the

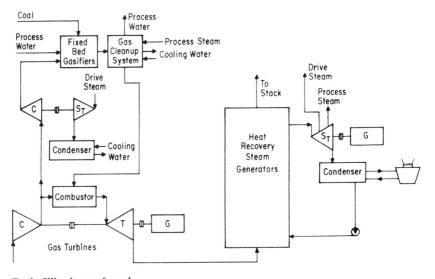

Fuel: Illinois no. 6 coal
Fuel conversion plant:
  Advanced fixed-bed gasification
  Low temperature gas clean-up
Emission control:
  SO_x—H_2S removal from fuel gas
  NO_x—ammonia removal from fuel gas + water addition
       to gas turbine combustor for thermal NO_x control
  Particulates—fuel gas scrubbing

	Case I	Case II
Prime cycle: gas turbine		
Firing temp.	1589 K	1922 K
Turbine hot parts cooling method	advanced air cooling	uncooled ceramic; water cooling
Pressure ratio	12/1	16/1
Bottoming cycle: steam		
throttle	12.76 MPa, 783 K	16.55 MPa, 811 K
reheat:	783 K	811 K
final feedwater temperature	399 K	401 K
Gross output, MW	607	665
Auxiliary losses, MW	22	25
Net plant output, MW	585	640
Power plant efficiency, %	39.6	44.0
Fuel consumption, kg/kWh	0.362	0.326
Water consumption, l/kWh	1.575	1.480
Total waste production, kg/kWh	0.047	0.042

Fig. 6.4. Cycle schematic: gasifier/combined cycle

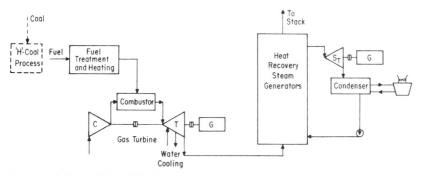

Fuel: coal derived liquid fuel
Off-site fuel production method: 'H-coal' process
Emission control:
SO$_x$—off-site fuel processing
NO$_x$—off-site fuel processing, steam injection in
        gas turbine combustor
Particulates—off-site fuel processing

Prime cycle: gas turbine
  Firing temperature   1922 K
  Turbine hot parts cooling method   uncooled ceramic;
                            water-cooled composite metals
  Pressure ratio   16/1

Bottoming cycle: steam
  Throttle   16.55 MPa, 811 K
  Reheat   811 K
  Final feedwater temperature   401 K

Gross output   873 MW
Auxiliary losses   26 MW
Net plant output   847 MW
Power plant efficiency   51.1%

Fuel consumption   0.18 kg/kWh
Water consumption   1.12 1/kWh
Total waste production   0 kg/kWh

Fig. 6.5. Cycle schematic: liquid fuel/combined cycle/1922 K

same number of heat recovery steam generator (HRSG) units, and a single
steam turbine unit. Hot exhaust gas from a gas turbine enters its associated
HRSG unit where steam is produced for the steam bottoming cycle. The
combined steam flow from all HRSG units is fed to a steam turbine for additional power generation.
    In Case (1), four gas turbine units with a firing temperature (i.e. temperature
at the inlet to the first-stage buckets) of 1589 K are utilized. (There is no
industry wide uniform procedure in use to express the gas turbine conditions;

in some cases, the temperature at the exit of the gas turbine combustor is indicated—this temperature would be higher than the firing temperature defined above by the temperature drop associated with the cooling of the transition pieces and first-stage nozzles ('non-chargeable drop').) Advanced air cooling schemes are used for turbine cooling. The temperature of the gas at the turbine exit is 866 K.

Cases (2) and (3) have gas turbine units with a firing temperature of 1922 K. For these units, the combustion can liners, transition pieces and first stage nozzles are made of uncooled ceramic, and all other nozzles and the rotating buckets are made of water-cooled composite metals. Case (2) employs two gas turbine units, with a turbine exhaust gas temperature of 1066 K. For Case (3), three gas turbine units and an exhaust gas temperature of 1058 K are employed.

The fuel processing schemes convert coal into fuel forms acceptable to the gas turbines and compatible with the emission standards. Cases (1) and (2) have an on-site fuel plant consisting of air-blown advanced fixed bed gasifiers, gas coolers/quenchers, and processes for chemical and physical cleaning of the fuel gas. Integration of the fuel plant with the power cycle component involves the following steps:

(a) Compressed air from the gas turbine compressor exit is extracted, increased in pressure in a steam-driven booster compressor, and delivered to the gasifier.
(b) Steam turbine extractions are used to supply steam for the booster compressor drive turbine and for the fuel gas clean-up steps.

The low steam to air ratio utilized for this advanced gasifier design permits the gasifier to produce sufficient steam in its cooling jacket so that external steam supply from the HRSG or steam turbine is not required.

A 'semiclean' liquid fuel derived from coal in an off-site coal liquefaction/refining plant is delivered to the power plant site for Case (3). On-site fuel treatment includes a water-washing step for removal of the alkali metals followed by the addition of magnesium additives to 'tie-up' the vanadium in the fuel. These treatment steps remove impurities in the liquid fuel detrimental to the gas turbine; however, the concentration level of the fuel-bound nitrogen is not reduced. Compliance with the applicable NO_x emission standard would require that the nitrogen content in the coal-derived liquid fuel be controlled through hydrogenation at the off-site fuel plant.

6.3.4 Closed Turbine (Combined) Cycles with Fluidized Bed Combustors

The two plant concepts evaluated in this category are:

(1) Combined helium gas turbine topping/organic bottoming cycle with an AFB heat input system[2] (Fig. 6.6).

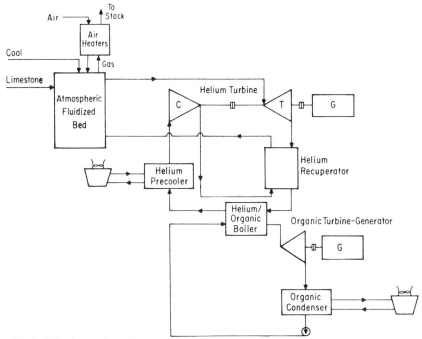

Fuel: Illinois no. 6 coal
Combustor/boiler: atmospheric fluidized bed (two-stage)
Emission control:
  SO_x—limestone addition in AFB
  —sulphur capture in first stage of AFB
  NO_x—lower temperature in AFB
  Particulates—electrostatic precipitator after high
        temperature air preheater

Prime cycle: closed cycle gas turbine
  Working fluid: helium
  Turbine inlet temperature: 1283 K
  Pressure ratio: 3.2/1
  Turbine inlet pressure: 6.89 MPa

Bottoming cycle: organic
  Working fluid: Fluorinol-85
  Turbine inlet: 4.82 MPa, 500 K

Gross output: 515 MW
Auxiliary losses: 39 MW
Net plant output: 476 MW
Power plant efficiency: 39.9%

Fuel consumption: 0.359 kg/kWh
Water consumption: 1.927 1/kWh
Total waste production: 0.119 kg/kWh

Fig. 6.6. Cycle schematic: helium/1283 K

(2) Combined potassium turbine topping/steam bottoming cycle with a PFB heat input system[2] (Fig. 6.7).

In Case (1), a closed-cycle helium gas turbine with a turbine inlet temperature of 1283 K is used. The helium, after being heated in the AFB and

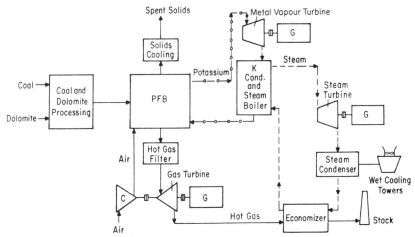

Fuel: Illinois no. 6 coal
Combustor/boiler: pressurized fluidized bed
Emission control:
    SO_x—dolomite addition in PFB
    NO_x—lower temperature in PFB
    Particulates—cyclones and granular bed filters

Prime cycle: closed cycle turbine
    Working fluid: potassium
    Turbine inlet: 105 kPa, 1033 K
    Pressure ratio: 6.3/1

Bottoming cycle: steam
    Throttle: 24.2 MPa, 811 K
    Reheat: 811 K

Pressurizing gas turbine
    Compressor pressure ratio: 10/1
    Turbine inlet temperature: 1144 K

Gross output: 1033 MW
Auxiliary losses: 37 MW
Net plant output: 996 MW
Power plant efficiency: 44.4%

Fuel consumption: 0.323 kg/kWh
Water consumption: 1.575 l/kWh
Total waste production: 0.137 kg/kWh

Fig. 6.7. Cycle schematic: potassium/1033 K

expanded through the turbine, passes to a recuperator to heat the high pressure helium from the compressor. The low pressure helium leaves the recuperator at 589 K. In order to improve the cycle efficiency, energy from the low pressure helium stream is further utilized to vaporize the working fluid, Fluorinal-85 ^{T}, of the bottoming cycle. The organic fluid vapour is expanded through a turbine, condensed, and the liquid condensate is pumped back to the helium/organic boiler in a closed loop system. The organic bottoming cycle provides 22% of the total plant electrical output.

The design of the present AFB furnace is considerably more complex than that discussed previously for the other cases, since it is required to heat helium from 775 to 1300 K. A two-stage AFB configuration is used; the lower stage operates at 1367 K and the upper stage at 1116 K, each stage supplying about the same heat input to the helium. The combustion gas from the high-temperature lower stage is ducted along with the combustion air into the lower temperature upper bed where the desired sulphur capture is achieved. High-temperature air preheaters are provided to cool the exhaust gases leaving the 1116 K bed at 867 K down to 661 K. The plant employs four AFB modules, each with four cell units and each unit containing two stages.

For the second closed cycle case, a potassium topping cycle with a PFB heat input system is used. Saturated potassium vapour from the PFB is expanded in a metal vapour turbine, and the expanded low-pressure vapour at 866 K is condensed in a potassium condenser where the steam is generated. The steam bottoming cycle contributes 63% of the total plant electrical output.

The PFB heat input system is similar in configuration to those discussed previously for the steam cycles; in this case, the main bed cells operate at 1228 K, and the combustion gas temperature entering the expansion turbine is 1200 K.

6.3.5 Open Cycle MHD[2] (Fig. 6.8)

The major components of this plant concept are (1) the combustor, (2) the MHD generator, and (3) the heat recovery section.

The combustor is designed to oxidize crushed coal particles using preheated air to produce a combustion gas at 2811 K. Eighty-five per cent of the coal slag is rejected from the combustor. A fuel-rich condition (107% of stoichiometric fuel/air ratio) is maintained.

A seed material K_2CO_3 is added to the combustion gas, thereby increasing its electrical conductivity. The seeded gas is then expanded in the MHD generator, producing a d.c. electrical output. The expanded gases leave the generator through a diffuser section at a reduced velocity and at 2289 K.

Several components comprise the heat recovery section downstream of the MHD generator.

^{T} Trademark of Halocarbon Products, USA.

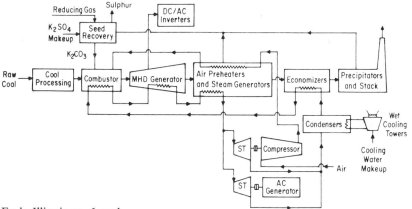

Fuel: Illinois no. 6 coal

Combustor:
  (a) Water cooled
  (b) Inlet air temp: 1644 K

Emission control:
  SO_x—seed (K_2CO_3) reaction
  NO_x—residence time at temperature
  Particulates—electrostatic precipitator

Prime cycle: diagonal MHD generator
  Inlet: 281 K
  Average magnetic flux: 5 Tesla
  Diffuser outlet pressure: 115 kPa
  Potassium seeding: 1%

Bottoming cycle: steam
  Throttle: 24.2 MPa, 811 K
  Reheat: 811 K

Gross output: 1993 MW
Auxiliary losses: 61 MW
Net plant output: 1932 MW
Power plant efficiency: 49.8 %

Fuel consumption: 0.297 kg/kWh
Water consumption: 1.25 1/kWh
Total waste production: 0.037 kg/kWh

Fig. 6.8. Cycle schematic: open cycle MHD

In the first component, the radiant furnace, secondary air is introduced to complete the combustion under conditions that yield acceptable NO_x effluent levels. In this furnace, and in the subsequent components (high and low temperature air heaters, steam superheater and reheater, economizers), the combustion gas is cooled, most of potassium and sulphur are condensed from the gas, combustion air preheat is accomplished, and steam for the steam turbine is generated. Two steam turbine units are used, one driving the main air compressor and one driving a conventional synchronous generator.

In this cycle concept, various means are employed to meet the desired emission standards. The seed material K_2CO_3 absorbs the sulphur in the coal as K_2SO_4. This material is collected from the heat exchange components and from the electrostatic precipitator and returned to a seed recovery system where an intermediate Btu gas reduces the K_2SO_4 to K_2CO_3 seed and H_2S. Elemental sulphur is recovered from the H_2S in an integral Claus plant. The NO_x production rate is controlled by maintaining slightly fuel-rich conditions in the combustor, and then by providing a two-second residence time in the radiant furnace to permit partial decomposition of NO_x.

6.3.6 Molten Carbonate Fuel Cell Integrated with Coal Gasifier[3] (Fig. 6.9)

The power plant cycle consists of three subsystems:

(1) The fuel production process.
(2) The fuel cell plant.
(3) The steam bottoming cycle and turbocompressors.

Coal is reacted with superheated steam and air in an advanced fluidized bed gasifier to produce a low Btu synthesis gas at 1310 K and 1.38 MPa. The hot synthesis gas then passes through cyclone separators, steam generators, and iron-oxide desulphurization beds before it is fed to the fuel cell anode at 922 K.

In the fuel cell plant, synthesis gas is reacted electrochemically with oxygen from the air to produce d.c. power, product water, and waste heat. The d.c. output is converted to a.c. output by self-commutated inverters. The fuel cell consists of an anode, an ionically conducting electrolyte, and a cathode. At the anode, CO is converted to additional hydrogen through a water-gas shift reaction. The combined hydrogen reacts with the carbonate ion $CO_3^=$ to form by-product H_2O and CO_2, with an electric current produced. At the cathode, oxygen from air and by-product CO_2 from the anode reaction combine electrochemically with the electrons to form the carbonate ion $CO_3^=$. Methane in the fuel gas passes through the anode compartments unconverted. The anode exit gas is mixed with air, burned catalytically upstream of the cathode inlet, and then fed to the cathode.

Process air for the gasifier and fuel cells is provided by turbocompressors driven by expanding a portion of the cathode product gas. Waste heat at various places in the cycle is used to generate steam for a steam bottoming cycle which produces 34% of the total plant output.

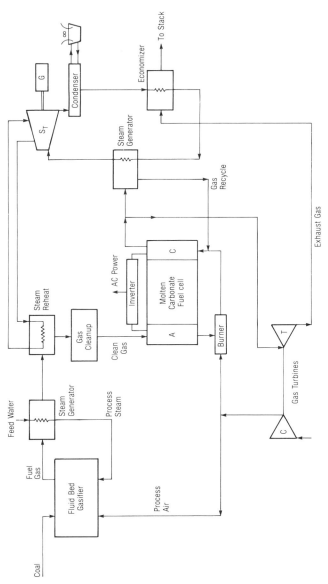

Fuel: Illinois no. 6 coal
Fuel conversion plant:
  Advanced fluidized bed gasification
  High temperature (920 K) sulphur removal process

Emission control:
  SO_x—H_2S removal from fuel gas in hot iron-oxide process
  NO_x—use of catalytic burners for fuel cell
  Particulate—cyclone separators

Prime cycle: molten carbonate fuel cell
Operating conditions: 1.03 MPa, 922 K
Anode: porous nickel
Cathode: porous nickel
Electrolyte: alkali metal carbonate in ceramic matrix

Bottoming cycle: steam
Throttle: 16.55 MPa, 811 K
Reheat: 811 K

Gross output: 654 MW(e)
Auxiliary losses: 19 MW(e)
Net plant output: 635 MW(e)
Power plant efficiency: 49.6%

Fuel consumption: 0.286 kg/kWh
Water consumption: 1.514 1/kWh
Total waste production: 0.032 kg/kWh

Fig. 6.9. Cycle schematic: gasifier/fuel cell

6.4 ADVANCED CYCLES—UTILITY APPLICATION EVALUATION

The ten power generation concepts discussed in the previous section are evaluated in terms of the capital cost, cost of electricity in a base-load application, emissions, and overall development requirements for commercialization. In addition, intangible attributes of the advanced cycles, which would normally enter into utilities' selection processes, are considered, and the results of a cost-benefit analysis are summarized.

6.4.1 Capital Cost and Cost of Electricity

The total plant cost for each cycle concept, calculated by the procedure discussed in Section 6.2, is derived in 1981 dollars and listed in Table 6.4. These costs are also normalized to the reference steam/flue gas scrubber plant. Figure 6.10 plots the normalized capital cost against the power plant efficiency for these cases. The data points in Fig. 6.10 appear to fall in three distinct categories. Within each category, a reduction in plant capital cost correponds to an improvement in the plant efficiency. In the first category are the conventional steam/FGD plant and the steam/AFB plant with conventional steam temperatures. In the second category, two steam/fluid bed cycle concepts and two gasifier/combined cycle concepts are included. These four cycle concepts employ process conditions which would require more technological development than for the first category. The five cycle concepts grouped in the third category have still more advanced technological development requirements than those in the first two categories.

The power plant efficiency values for the advanced cycles shown in Fig. 6.10

Table 6.4 Capital cost estimates

	$/kW(e) (1981 $)	Normalized to steam/FGD
Liquid fuel/combined cycle/1922 K	471*	0.54*
Steam/AFB/811 K/811 K	671	0.77
Gasifier/fuel cell	700	0.80
Open cycle MHD	711	0.81
Gasifier/combined cycle/1922 K	723	0.82
Steam/PFB/811 K/811 K	787	0.90
Steam/AFB/811 K/922 K	826	0.94
Gasifier/combined cycle/1589 K	853	0.97
Steam/FGD (Reference)	877	1.00
Potassium/1033 K	1025	1.17
Helium/1283 K	1403	1.60

* These costs exclude the capital required for the coal liquefaction plant. However, this factor is taken into account in the subsequent calculations by specifying higher liquid fuel cost.

are also the overall ('coal pile to busbar') efficiencies, except for the case of combined cycle burning coal derived liquid fuel. For this case, the power plant efficiency of 51.1% should be multiplied by the coal to liquid fuel efficiency of the off-site fuel plant to obtain the overall efficiency. If the energy efficiency of the fuel plant is 70%, the overall efficiency would be 35.8%. It can be seen that all advanced systems exceeded the 32% overall efficiency projected for the reference steam cycle with the furnace scrubber. However, the plant capital cost estimate for the potassium and helium cycles is greater than that of the reference steam plant.

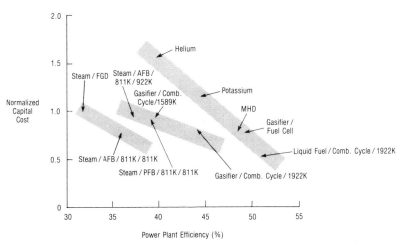

Fig. 6.10. Capital cost versus efficiency of advanced cycles

The estimates of the levelized cost of electricity (mills/kWh) for the advanced cycles are compared to the reference steam/FGD plant in Fig. 6.11. As described in Section 6.2, there are three contributors to the total electricity cost. The capital cost fraction is derived from the cycle capital cost ($/kW(e)) given in Table 6.4, a levelized fixed charge rate of 18%/year, and an assumed base-load capacity factor of 65%. The levelized fuel cost fraction is related to the fuel cost, and the levelization factor of 1.94. A coal cost of $1.44/GJ ($1.52/10^6 Btu) and coal-derived liquid fuel cost of $5.02/GJ ($5.30/10^6 Btu), both in 1981 dollars, are used. The third contributory factor, operating and maintenance cost, is assumed to be 6.6 mills/kWh for all cycles, on a levelized basis. Small errors in this assumption for operating and maintenance cost would not significantly impact the relative ranking of a specific advanced technology plant. However, if the higher operating and maintenance costs are due to a higher plant outage rate, the resulting reduction in the plant capacity factor would have a significant adverse impact on its electricity production cost.

It is seen from Fig. 6.11 that the helium cycle with a capital cost higher than the reference steam plant (Table 6.4) also has a higher levelized electricity cost

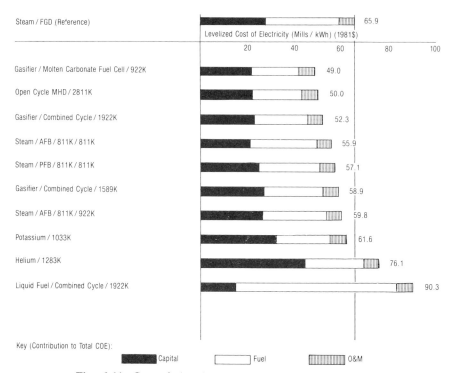

Fig. 6.11. Cost of electricity estimates for advanced cycles

estimate. The other case with higher COE value is the combined cycle using a coal derived liquid fuel. For this case, the penalties of the expensive fuel far exceed the advantages of low capital cost and high efficiency.

A plot of levelized COE versus power plant efficiency is shown in Fig. 6.12. It is seen that most of the advanced cycle concepts can be grouped to suggest a trend of reducing COE as the more efficient power cycle concepts are realized. The three previous noted advanced concepts—the combined cycle with liquid fuel, the helium cycle, and the potassium cycle—again fall substantially outside the band indicating the general trend.

6.4.2 Emissions

All of the advanced cycle plants are designed to meet the emission standards for SO_x, NO_x, and particulates, described in Section 6.2. The combined cycle plant using a coal-derived liquid fuel would have NO_x emissions which exceed the standard. A conventional gas turbine combustor was assumed for the conceptual design of this system. Consequently, a high conversion of the 1.3% nitrogen in the fuel to NO_x resulted in the production of an estimated 430 ng/J NO_x from this source. Furthermore, although steam injection into the combustor was employed to control thermal NO_x (i.e., NO_x arising from

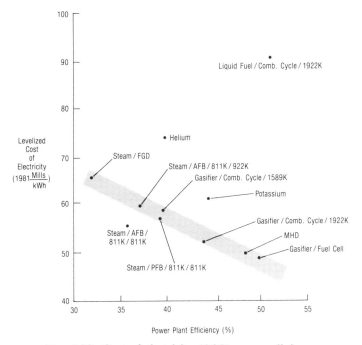

Fig. 6.12. Cost of electricity (COE) versus efficiency

atmospheric nitrogen), the total NO_x was 730 ng/J. Further development of combustion systems will be required to reach the target emission limits for NO_x for liquid fuels. This may include catalytic combustors or further hydrogenation of the coal derived fuel at the off-site plant to reduce the fuel-bound nitrogen content.

Besides airborne emissions, other critical siting factors include power plant solid waste output and water requirements. These streams have been identified for each advanced cycle in Section 6.3. The solid waste output for the cycles using fluidized-bed combustion could exceed the output for the reference steam/FGD plant, depending on the required ratio of calcium (in the sorbent) to sulphur (in the coal) as influenced by the bed design parameters. Both the total water requirement and the waste water production for all advanced cycle cases are smaller than for the reference steam plant.

Table 6.5 Development Requirements

Cycle concept	Key development requirements
Steam/fluidized beds	Full-scale combustion fluid beds with (a) inbed heat exchangers, (b) inbed sulphur capture with continuous feed of coal and limestone/dolomite. In addition, for the pressurized fluidized bed systems, (a) hot gas clean-up suitable for gas turbine operation, (b) integration of PFB with gas turbine.
Helium	Two-stage atmospheric fluid bed with ceramic heat exchanger. Large organic turbine for bottoming cycle.
Potassium	PFB furnace with in-bed potassium boiler tubes. Potassium turbines with large superalloy turbine discs.
Combined cycles	Gas turbines with higher inlet temperatures (a) with advanced air cooling (for 1589 K), (b) with water-cooled and uncooled ceramic parts (for 1922 K), (c) combustors for coal-derived fuels with acceptable NO_x characteristics. Fixed-bed gasifiers with lower steam/air ratio and with total tar recycle.
MHD	Full scale superconducting magnets, coal-burning combustors, and MHD generators. Refractory storage high temperature air heaters. Seed/sulphur removal and seed regeneration. Successful coupling of major subsystems.
Fuel cell	Advanced fluidized bed gasifier with hot gas deep desulphurization. Pressurized multistack fuel cell d.c. module operating with gasifier products. Full-scale catalytic burners.

6.4.3 Development Requirements and Steps to Commercialization

The performance and economic data for the advanced cycles were based on conceptual designs representative of mature commercial components. When these studies were conducted in 1975, the development status of each cycle concept was established and key development requirements, a schedule of development, and the associated development cost estimates were prepared.[2,3] Table 6.5 is a representative list of development requirements identified for these systems.

The steps to commercialization were aimed at the realization of the plant conceptual design. The plan included three major steps: component development; pilot plant design, construction, and operation; and demonstration plant design, construction and start-up. These sequential phases were designed to reflect the specific developmental needs of each individual cycle concept and to bring each concept to a comparable technology level in each phase before proceeding with the next phase; this approach provided a uniform basis for comparison of the various systems.

The estimated time scale for completion of the first commercial plant, as projected in 1975, is shown in Figure 6.13. This projected time scale assumed that development work for all systems would proceed simultaneously at a pace limited only by the technological considerations. The practical considerations

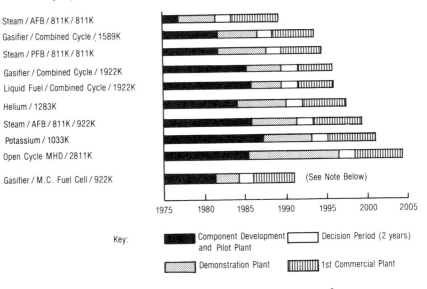

Note: Molten carbonate fuel cell data based on report by United Technologies Corporation[3]. This development schedule is based on the assumption that a suitable gasifier of mature technology will be available when needed; hence no provision is made for gasifier development.

Fig. 6.13. Development steps to commercialization.

Table 6.6 Intangible attributes of advanced cycles

1. Economic viability
   (a) System capital cost
   (b) Cost of electricity
2. Efficiency and fuel conservation potential
3. Natural resource requirements
4. Environmental intrusion
   (a) Atmospheric intrusion
   (b) Requirements for waste handling and disposal
5. Reliability and availability potential
   (a) Forced outage rate
   (b) Planned outage rate
6. Safety
   (a) In-plant safety
   (b) Outside-plant safety
7. Siting flexibility
   (a) Flexibility of siting
   (b) Independence of other systems
8. Life limiting factors
   (a) Life expectancy
9. Flexibility of application
   (a) Load following
   (b) Part load efficiency
   (c) Minimum load
   (d) Unit rating
10. Ease of operation and control
11. Ease of maintenance
12. Potential for factory modular construction
13. Manufacture capability
14. Fuel flexibility
    (a) Adaptability to different fuels
    (b) Adaptability to coal variations
15. Compatibility of fluids and materials
    (a) Working fluid compatibility
    (b) Combustion gas compatibility
16. Working fluid stability
17. Potential for retrofit
18. Opportunity for by-product sale
19. Manpower limitation
    (a) Field labour availability
    (b) Factory labour availability
20. Electrical performance
    (a) Supportive of electric grid
    (b) Start-up power requirements
21. Probability for development success
22. Cost of research and development required
23. Research and development time required

of the limitations on the available development resources and the impact of changing political and institutional factors since the study period would result in a different commercialization schedule if the projections are updated. Nevertheless, Fig. 6.13 points out an important fact: that the cycle concepts with potentially higher efficiency improvements would be available for commercial application towards the end of this century. The impact of this schedule on the market penetration potential of the advanced cycles will be examined later in this section.

A further evaluation of the advanced cycles would include their rating with respect to their intangible attributes,[2,6] where intangibles are defined as those attributes not directly quantifiable in the cost of electricity estimates. These attributes are also termed implementation factors and are listed in Table 6.6. Some of these items also have a tangible component. For instance, high capital cost results in a higher cost of electricity (tangible effect), but it also increases investment requirements which could affect the marketability of a cycle in an economy where capital is not easily available (intangible effect). Criteria can be defined to establish a good, fair, or poor rating or to assign a numerical weighting factor with respect to each item when evaluated for an advanced cycle.[6]

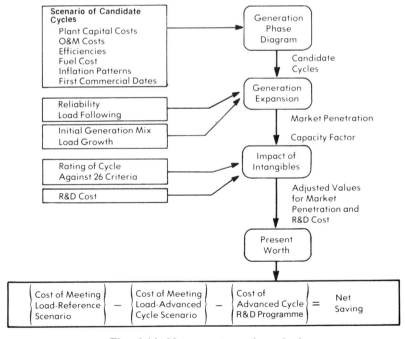

Fig. 6.14. Net present worth method

6.4.4 Cost-benefit Analysis

A rational method of apportioning limited developmental resources among several potentially attractive candidate systems would include a cost-benefit analysis or, equivalently, the assessment of the net current worth of the associated programmes. When applied to the advanced power cycle candidates, this method consists of four distinct computational operations,[6] as indicated in Fig. 6.14.

Generation Phase Diagrams are a screening technique used to separate those advanced cycle candidates which offer a clear economic advantage from those which do not. Construction of the phase diagrams requires a list of cycle candidates which would be allowed to compete in the market-place and their commercial availability schedule as well as their cost and performance data. Figure 6.15 is a generation phase diagram for a 'crowded market scenario,' in which the advanced cycles compete with each other as well as with the conventional technologies.

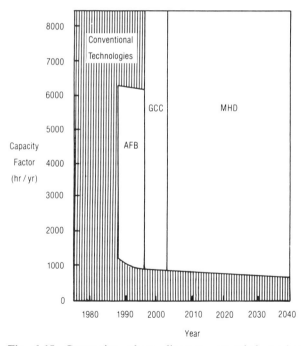

Fig. 6.15. Generation phase diagram—crowded market scenario. The following abbreviations are used in this figure and in Figures 6.16 and 6.17: AFB—Steam/AFB/811 K/811 K; GCC—Gasifier/combined cycle/1922 K; MHD—Open cycle MHD/2811 K

The specific market penetration simulation programme used in the evaluation had provision for up to three advanced cycles, in addition to the conventional power generation technologies. Hence, one example technology for each developmental time frame—near-term, mid-term, long-term—was selected. The three advanced cycle candidates included in this scenario are:

(1)   Steam/AFB/811 K/811 K
(2)   Gasifier/combined cycle/1922 K
(3)   Open cycle MHD/2811 K

In this scenario evaluation, the open cycle MHD is preferred when it is available followed by the gasifier/combined cycle and the steam/AFB. Since the dates of first commercial service are staggered for these cycles, each enjoys a period of time in which it is superior to the previous options and has not yet been superseded by a better candidate. These results are based on the cost of electricity only, whereas market penetration is also determined by the cycle operating characteristics and intangibles.

The generation expansion programme models the utility plant ordering process to obtain estimates of the market penetration for each cycle identified as a winner or a marginal winner in the generation phase diagrams. This second evaluation step requires additional data on plant reliability and load following characteristics, and includes a load dispatch model which simulates

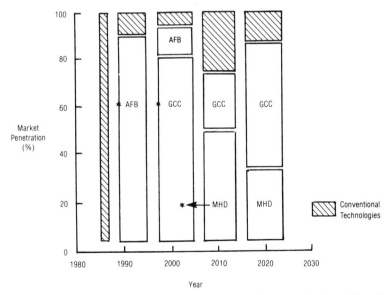

Fig. 6.16. Generation expansion—crowded market scenario. Location of asterisk (*) denotes the first commercial availability of each advanced cycle

the operation of power plants installed in the system. The model projects the annual capacity factor of each plant.

Figure 6.16 is the generation expansion results for the crowded market scenario. Among the three new technologies competing, the steam/AFB dominates the market from 1988 to 1995, and then shares the market with the gasifier/combined cycle system. Open cycle MHD becomes available in the year 2003, but is not purchased until 2006 because of its large size and high forced outage rate (assumed to be 20%, compared to 9% for steam/AFB and 12% for gasifier/combined cycle). MHD also loses part of the potential market because of its poor cycling ability. The gasifier/combined cycle plant is purchased instead to fill the load following needs of the utility system.

Since the generation expansion programme results are sensitive to the unit size and the forced outage rate, the improvement in market penetration for the MHD plant could be estimated if this plant could be built in a 1000 MW(e) size (instead of the 1932 MW size specified earlier) for the same cost and with the same efficiency, together with an improved forced outage rate of 15%. The results, shown in Fig. 6.17, clearly indicate that the smaller plant is purchased immediately as it becomes available in 2003.

The generation expansion programme results are further modified to reflect considerations of the intangible properties of advanced cycles. The last step combines all the results derived so far to project the costs and benefits which accrue to the nation as a result of pursuing each of the possible scenarios. The net present worth of a particular scenario is its net savings as defined in Fig. 6.14. If the crowded market scenario is pursued over the no new technology scenario, a saving (a net present worth) of 197 billion dollars (1975$) is estimated while meeting the future electrical energy demand (1985 to 2054).

6.4.5 First-generation Advanced Cycle Power Plants

The advanced cycle power plants presented in Section 6.3 involve several advanced concepts for components and subsystems. The potential for significant improvements in plant efficiency and/or capital costs for these plant configurations were predicated on the basis that they represent hypothetically mature designs. Mature plant design implies that all performance and cost goals for the developmental components have been met and that the proposed integration of these components can be carried out while meeting the plant operational specifications. In the course of developing these mature plants, it would be instructive to examine the role of the 'first generation' advanced cycle plant systems. Configurations for these first generation plants would be aimed at earlier entry into the market-place and would employ components and subsystems with fewer developmental requirements or with a reduced operational flexibility. Such plant configurations would have a lower potential

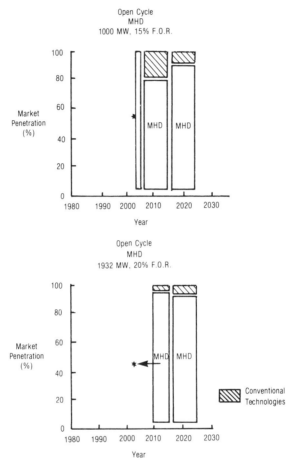

Fig. 6.17. Generation expansion—effect of unit size for open cycle MHD. Location of asterisk (*) denotes the first commercial availability for this plant.

efficiency than their mature counterparts; however, they would also have several advantages:

(1) They have less technical risk than the mature plant and would be ready for demonstration sooner.
(2) They offer an opportunity to gain user acceptance, and provide a 'real world' laboratory to help establish a basis for subsequent improvements.
(3) For those technologies requiring a substantial development time before their commercial introduction, these intermediate demonstration steps are desirable to maintain visibility and to sustain development interest.

A successful demonstration for the first generation version of an advanced cycle power plant could lead to the more attractive mature plant configuration. Some recent activities are reviewed below to illustrate the first generation demonstration routes for three of the advanced cycle power plant concepts.

Open Cycle MHD

In the discussion earlier in this section, it was pointed out that the market penetration potential of this concept was adversely affected by the large unit size and by its late arrival at the commercial stage. The first generation plant configurations are aimed at overcoming some of these shortcomings by incorporating the following features:[8]

(1) Unit sizes 1000 MW(e) and less.
(2) Oxygen enrichment of combustion air in place of high temperature air preheaters.
(3) Use of conventional scrubbers in place of SO_x control by the alkali seed.

Even with these modifications, the first-generation open cycle MHD plant would still contain several advanced components (coal combustor, superconducting magnet, inverter) which have so far been successfully tested only for tens of hours at a typical size of 20 MW(t). However, these modifications would accelerate the commercial availability of the lower efficiency MHD cycle, and could also offer improved plant availability.

Steam/PFB

The approach taken by the American Electrical Power Service Corporation (AEP) for the design of a 170 MW(e) steam/PFB demonstration plant illustrates a first generation plant design.[9] Since the major technical obstacle to the commercialization of this advanced cycle is the PFB combustor–gas turbine interface, the AEP design approach attempts to minimize the erosion/corrosion/deposition effects on the gas turbine blade and vane materials by incorporating the following features:

(1) The gas turbine inlet temperature is set at 1072 K. This results in a combustion bed temperature of 1122 K. The bed operation in this temperature range is intended to inhibit vaporization of the alkali compounds.
(2) A tapered fluid bed, with a gas velocity of 0.9 m/s at the top of the bed, is employed. Two stages of high efficiency refractory lined cyclone dust collectors, followed by a third stage of unlined cyclones, is employed. These measures are hoped to ensure a sufficiently clean gas for reliable gas turbine operation.

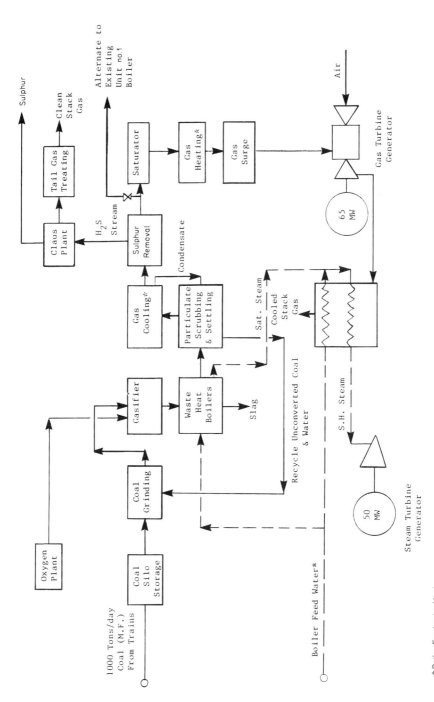

Fig. 6.18. Block flow diagram for gasifier/combined cycle for Cool Water programme

(3) A three-shaft gas turbine is selected for the first generation plant. This turbine unit employs only one impulse stage to minimize blade erosion by not subjecting the blade leading edges to the higher gas velocities associated with an impulse blade design.

AEP has chosen an existing plant site for this steam/PFB plant, and plans to use an existing non-reheat steam turbine (with throttle conditions of 9.0 MPa and 769 K). An overall efficiency of 33% is estimated for this plant.

Gasifier/Combined Cycle System

The many detailed studies of this advanced cycle concept attest to widespread interest in this technology. These cycle concepts would benefit from advances in any of several types of coal gasification systems,[10] could utilize state-of-the-art gas turbines or exploit evolutionary improvements in the turbine inlet temperatures,[11] and would be able to meet stringent emission standards with only a small economic penalty.[11,12]

The first generation gasifier/combined cycle plants could employ state-of-the-art gas turbine units with near commercial second generation coal gasification processes. A detailed study,[11] using a 1358 K gas turbine with a Texaco entrained gasifier or with a British Gas Lurgi slagging gasifier (fixed bed), concludes that such first generation advanced cycle plants could have an 8 to 11% efficiency improvement and an 11 to 15% cost of electricity benefit over conventional coal-fired steam plants with lime/limestone scrubbers. An additional 5% improvement in both the efficiency and cost could be available with the increase in turbine firing temperature to 1422 K.

A first generation gasification/combined cycle plant is being planned for the Cool Water site of the Southern California Edison Company;[13] a block flow diagram is given in Fig. 6.18. This configuration, using an entrained flow gasifier, may be compared with the 'optimum' configuration for a mature plant,[11] and with the configuration using fixed bed gasifiers (Fig. 3.4). The plant would incorporate a Texaco entrained flow gasifier, low temperature fuel gas clean-up, state-of-the-art gas turbine, and a non-reheat steam turbine. The net output of this integrated power plant would be nominally 100 MW(e). The overall efficiency of this plant would be below the level given by Shah *et al.*[11] because of site conditions and design parameters. Other proposals incorporating various types of gasifiers and combined cycle systems are also under active consideration.

6.5 SUMMARY

This chapter provides an overview of the major new energy conversion technologies which are being developed world-wide to utilize coal for electricity generation in a more efficient, less expensive, and environmentally compatible manner.

Section 6.1 introduces the scope of the technologies discussed in this chapter and explains the procedures used in their analysis. Section 6.2 discusses the ground rules and assumptions which have been applied in the evaluation of these technologies so that the results would be consistent and useful for comparative evaluations. Section 6.3 describes ten advanced technologies for power generation and provides cycle schematics as well as details on the performance, emission control method, natural resources requirement, etc. for each case. Section 6.4 provides economic comparisons of these advanced technologies. Major developmental requirements and an estimated schedule for commercialization are discussed. The procedure for the assessment of advanced technologies for utility applications is presented, and results are provided in terms of market penetration potential for three of the advanced technology concepts. Finally, the first generation concepts for some of these technologies, which are being readied for demonstration, are discussed.

6.6 REFERENCES

1. *Energy Conversion Alternatives Study (ECAS)* (1976a). General Electric Phase I Final Report: Vol I, Executive Summary; Vol II (three parts), Advanced Energy Conversion Systems; Vol III (three parts), Energy Conversion Subsystems and Components.
2. *Energy Conversion Alternatives Study (ECAS)* (1976b). General Electric Phase II Final Report: Vol I, Executive Summary; Vol II, Advanced Energy Conversion Systems—Conceptual Designs (Part 1–Analytical Approach, Part 2–Closed Turbine Cycles, Part 3–Open Cycle Gas Turbines and Open Cycle MHD, Part 4–Summary of Results); Vol III, Research and Development Plants and Implementation Assessment.
3. *Energy Conversion Alternatives Study (ECAS)* (1976c). United Technologies Phase II Final Report: Integrated Coal Gasifier/Molten Carbonate Fuel Cell Powerplant Conceptual Design and Implementation Assessment.
4. *Energy Conversion Alternatives Study (ECAS)* (1977). Summary Report, National Aeronautics and Space Administration (NASA), Lewis Research Center.
5. Brown, D. H. (1976). *Conceptual Design and Implementation Assessment of a Utility Steam Plant with Conventional Furnace and Wet Lime Stack Gas Scrubbers*. General Electric Company.
6. Pomeroy, B.D., Fleck, J. J., Marsh, W. D., Brown, D. H., and Shah, R. P. (1978). *Comparative Study and Evaluation of Advanced Cycle Systems,* General Electric Company, EPRI AF–664, Vol I, Vol 2, Parts 1 and 2.
7. Fox, G. R. (1978). *Advanced Power Cycles and their Potential for Electrical Energy Generation,* American Power Conference, 1978.
8. Lippert, T. E. (1981). *Open-cycle MHD Technology Status and Development Perspective,* Westinghouse Electric Corporation, EPRI AP–1864.
9. Markowsky, J. J., Jacob, A. L. and O'Connell, L. P. (1981). *Pressurized Fluidized Bed Combustion of Coal for Electric Power Generation—The AEP Approach,* 1981 Joint Power Generation Conference Paper No. 917–4.

10. Beckman, R., Hsu, W. and Joiner, J. (1981). *Economic Evaluation of Coal Gasification for Electric Power Generation (An Update).* Fluor Engineers and Constructors, Inc., EPRI AP-1725.
11. Shah, R. P., Ahner, D. J., Fox, G. R. and Gluckman, M. J. (1980). *Performance and Cost Characteristics of Combined Cycles Integrated with Second Generation Gasification Systems,* General Electric Company, ASME Gas Turbine Conference, Paper No, 80GT106.
12. Feerrar, S., Joiner, J., Kellard, J. and McElmurry, B. (1978). *Effects of Sulfur Emission Controls on the Cost of Gasification Combined Cycle Power Systems,* Fluor Engineers and Constructors Inc., ERPI AF-916.
13. Walter, F. B., Kaufman, H. C. and Reed, T. L. (1981). *The Cool Water Coal Gasification Program: A Demonstration of Gasification-Combined Cycle Technology,* Conference Proceedings, Vol 2: Synthetic Fuels—Status and Directions. EPRI WS-79-238.

Energy—Present and Future Options, Volume 2
Edited by D. Merrick
© 1984 John Wiley & Sons Ltd

A. F. POSTLETHWAITE
Central Electricity Generating Board,
Generation Development and Construction Division,
Gloucester, UK.

7

Combined Heat and Power

7.1 INTRODUCTION

7.1.1 Scope

This chapter is concerned primarily with the technical principles and practical aspects of combined heat and power (CHP) from the developer's point of view. Steam turbines, gas turbines, and diesel systems are covered, but detailed engineering design of components is not. The chapter concentrates upon concepts which are technically proven and commercially established. Futuristic cycles are not covered in detail.

The chapter is concerned specifically with land-based, central generation of electricity and heat as practised in CHP countries world-wide. Transport and domestic-scale CHP generation are not covered specifically. Also, refuse burning, geothermal heat, industrial waste heat recovery, and heat pump technology are excluded as specialist topics.

CHP generation for district heating and industrial heat supplies is covered in detail in the main text, and low temperature schemes for horticulture, etc. are covered in less detail in an Appendix. Heat bulk transmission and distribution aspects are mentioned and discussed where they affect generation, but detailed engineering and commercial aspects of distribution are not covered. Notation and definitions are summarized in Section 7.5.

7.1.2 Incentives and Problem Areas

There can be several incentives to proceed with a CHP scheme. Their relative importance will vary from scheme to scheme according to local and national circumstances. Possible incentives are: (a) economic, (b) strategic, (c) environmental, and (d) energy management.

Economic Incentives

Economic considerations are an essential part of any CHP proposition. If the proposed scheme is clearly profitable with a relatively low economic risk, there is a major incentive to proceed. Even if the economics are marginal (compared with alternative energy supply systems), there can still be an incentive to proceed if the CHP scheme is relatively risk-free, technically and commercially.

Strategic Incentives

Strategic incentives derive from the fact that most CHP schemes save primary energy. As discussed in Section 7.4, national primary energy consumption can be reduced by some 5 to 10% if the use of CHP in a country is widespread. A country which relies heavily upon imported fuels may therefore have the incentive to develop CHP in order to improve its balance of trade with the rest of the world, and to be less vulnerable to possible interruptions in fuel supplies or to rapid price escalations over which it has little or no control.

A further consideration is that large central plants (both heat-only and CHP) can be designed to burn a wide range of low-grade fuels which cannot be burned economically or for technical reasons in small boilers for individual consumers.

Central plants can also be designed to burn more than one fuel, and large heat supply schemes can be served by several plants which burn different fuels. These factors of fuel diversity and fuel flexibility to meet changing national circumstances can therefore be a contributing strategic incentive.

Environmental Incentives

The most common environmental incentive to introduce CHP is to reduce atmospheric pollution in cities. Coal and oil burned in small boilers serving individual premises can be a major cause of ground-level concentrations of sulphur, soot, and nitrogen oxides, which promote ill-health, corrosive damage to buildings, and generally degrade the environment. Central production plants (both heat-only and CHP) can reduce ground-level pollution in cities by a combination of:

(1) Better control of combustion (continuous monitoring and regular maintenance).
(2) Flue gas clean up (e.g. electrostatic precipitators).
(3) Dispersal of flue gases by means of a tall stack.

CHP also reduces regional and global pollution by burning less fuel. Although this has not been a major reason to employ CHP in the past, it could become more important in the future in response to international concern to reduce pollution levels.

Energy Management Incentives

Central production plants (both heat-only and CHP) provide a service which frees the consumer from the need to operate and maintain his own boiler plant. This applies to all individual heating systems except electricity. In the cases of oil and solid fuels, he is also freed from the need to order and handle fuel, and in the case of solid fuels, to dispose of ash.

The incentive to consumers and to city planners is therefore to raise living standards by eliminating the need to manage individual fuel-burning systems. Building design is also simplified by the elimination of flues and chimneys.

For the city as a whole, it is more efficient (in energy management terms) to buy, handle, and store fuel centrally than at numerous individual premises. Also, with larger and more diverse fuel stocks, and with many boilers supplying the distribution network, a better reliability of heat supply can be expected.

Problem Areas

Disincentives to proceed with CHP are mostly of a commercial rather than technological nature. The technology already exists to build reliable, highly efficient central CHP plants to meet virtually every need of the heat consumer. Any new generation technology can usually be tried out and proven at electricity-only stations which have the back-up of a national electricity transmission grid. The technology also exists for reliable, highly insulated heat distribution pipes, including prefabricated pipe-in-pipe systems for district heating. The commercial snags therefore include the following:

(1) The decision-makers consider that short-term profitability is more important than long-term profitability. Particularly vulnerable are CHP schemes which also require large capital investment in heat distribution. District heating may require thirty years or more to be profitable after repayment of capital and interest.

(2) Capital is in limited supply, and alternative capital projects are considered more important socially, strategically, or economically. Energy conservation is not always a top priority, even if it is profitable.

(3) The energy utility would become overcapitalized. This is particularly true of small utilities who wish to invest in CHP/district heating. Their rate of development may become severely retarded by this factor.

(4) The economic assumptions are too speculative, i.e. there is too much commercial risk. Reasons for this can be:

(a) the market penetration for sale of heat is not established;
(b) the long-term fuel source for the scheme is uncertain in terms of availability or price (or both);
(c) alternative energy supply systems or fuel sources may be just around the corner, e.g. the country expects to discover new fields of oil or gas;
(d) the commercial future of a major industrial consumer is uncertain.

(5) The scheme is unacceptable environmentally. This can relate to the proposed siting of a CHP station in urban areas. A city may also decide that digging up every street to lay district heating pipes would cause unacceptable levels of disturbance to city life.

(6) The will to proceed does not exist, either on the part of the utility, the heat consumer, or local and national government. This is an instinctive reaction to the effect that the nation's or utility's valuable resources can be better deployed in other ways.

In order to identify all the incentives and snags, a proposed CHP scheme should be worked up in stages, starting with 'back-of-an-envelope' calculations, through preliminary design studies and finally, a comprehensive project study. Consultation between the parties concerned should take place at each stage before proceeding with the next, more time consuming stage.

Finding the resources and justifying the cost of more detailed studies can be a problem, especially if the will to proceed has not yet been established. The sharing of the cost of investigations between several parties who would be major beneficiaries is usually an encouraging sign that the goodwill does exist.

7.1.3 Requirements of the Heat Consumer

Field surveys and discussions are usually necessary to establish the requirements of the heat consumer. These may be carried out by the CHP promoter or, if different, the heat distribution authority. In the case of major consumers, such as factories, office blocks, leisure centres, hospitals, etc., heat requirement data are probably already well documented, and can be presented to the CHP developer as a package. All data should relate not just to historical requirements but to present-day and future expectations, covering the economic lifetime over which the CHP scheme is to be appraised.

A problem which both parties must face is that long-term load projections are speculative to some extent. Agreement is therefore necessary regarding the design heat output capacity of the CHP station. The question of whether provision should be made for additional plant to be installed if actual demand turns out to be greater than expected also arises.

What, then, is the nature of heat requirement data which the CHP promoter needs to know? The following items are of major importance:

(1) Heat supply medium (e.g. steam, water).
(2) Heat supply conditions of pressure, temperature, and chemistry, including tolerances.
(3) Heat supply peak demand for design purposes.
(4) Heat demand characteristics, with as much information as possible on hourly, daily, seasonal, and yearly variations.
(5) Return water quantity and quality.
(6) Reliability and availability of heat supply, especially where interruption of supply can result in danger or economic disaster to the consumer.

Some consumers may also wish to take the electrical output of the CHP plant, in which case a similar set of data on electrical requirements will need to be known, including how they correlate (in time) with heat requirements. Such consumers are usually industries which, by choice or necessity, wish to operate independently of the National Electricity Grid System. For the purposes of this chapter, however, only heat demand requirements will be considered.

Conditions for District Heating

The two main applications of district heating are for space-heating and tap-water heating. A supply temperature of at least 70°C is usually required for these purposes in order to be technically suitable for connection to consumers' systems previously supplied from individual boilers. A supply temperature of less than $70\,^\circ$C is possible only for consumers' systems specifically designed for such temperatures, e.g. with larger radiators in rooms. This can be achieved at reasonable extra cost in new premises, but it is significantly more expensive and inconvenient to modify existing consumers' systems. Many major district heating schemes therefore have a design supply temperature in the range 70 to $95\,^\circ$C. This can be maintained nominally constant throughout the year, using a variable mass flow to match heat demand. The temperature difference (Δt) between supply and return pipes would be typically 20 to $40\,^\circ$C and can vary according to season.

An alternative design philosophy is to vary the supply temperature to match heat demand, keeping the mass flow nominally constant. Typically, the supply temperature would be $125\,^\circ$C at the winter peak demand, dropping to about $80\,^\circ$C in midsummer, when demand is at a minimum. The purpose of such systems is to reduce the size of pipes and other distribution equipment. A greater temperature difference (Δt) is achieved at peak demand between the supply and return pipes, sometimes as high as 60 °C. Doubling Δt reduces the pipe bores by a factor of $\sqrt{2}$, with corresponding capital savings. This is offset to some extent by higher operating costs in the winter season because of:

(1) Higher heat losses from distribution pipes.
(2) Higher average annual pumping power.
(3) Less electrical generation from steam turbines.

The optimum supply temperature(s) and flow(s) can be estimated at the design stage by economic appraisal, taking all of the above factors into account. It is necessarily an estimate, since lifetime operating costs (especially fuel prices) are difficult to predict with precision. The design philosophy cannot be changed significantly once the distribution system has been built.

Some heat users in a city require a high temperature supply throughout the year. These include industrial processes, laundries, hospitals (e.g. sterilization), and certain refrigeration plants. The higher the supply temperature, the greater the market penetration of district heating. However, to supply at 120 °C, 150 °C, or 180 °C continuously incurs high operating costs from steam turbines, not just to consumers who need a high temperature, but to all heat consumers in the city. District heating schemes supplying all consumers at high temperature do exist, but it is more usual for separate heat supply methods to be developed for high temperature consumers. These methods include:

(1) Local heat supply schemes, either CHP or heat-only. This includes many industrial schemes.
(2) The use of central CHP plants with two supply temperatures. This can be done with four pipes (two supply and two return), or with three pipes (using a common return). The largest three pipe-system is in West Berlin, supplying at constant 110 °C, and at variable 70 to 160 °C. The control and hydraulic design of three-pipe systems is somewhat complicated.

Most district heating schemes start on a small scale using heat-only boilers. Operating costs from heat-only boilers are virtually independent of supply temperature. Unless the district heating developer has CHP in mind for a later stage of development, he may therefore be tempted to design for a high supply temperature, in order to make capital savings in heat distribution and boiler design. Some commercial and institutional distribution systems have therefore been designed for around 180 °C supply temperature, even though the load is entirely space-heating and tap-water heating.

District heating systems throughout the world are predominantly hot water with twin pipes, supply and return. Some large steam systems have been developed, notably in New York and Paris. However, a relatively high steam pressure is required at the CHP station in order to achieve distribution to all consumers. Large steam distribution systems therefore have an inherent thermodynamic disadvantage compared with hot water systems (whose supply pressure is independent of thermodynamic considerations). Operating pressure is up to 28 bar abs in New York and up to 21 bar abs in Paris (the saturation temperatures of which are 230 °C and 215 °C respectively).

Operating pressure for hot water systems can be up to 10 bar greater than saturation pressure (in order to avoid problems of boiling). Return pipes can operate at lower pressures than supply pipes. A standard design pressure of 16 bar abs is used in many countries supplying water at 90 to 120 °C.

In hilly districts, a CHP station at low elevation may need to operate with a higher pressure margin. High rise apartment blocks present a similar problem, although this is usually solved by using a secondary distribution system for each block, the secondary side of the heat exchanger operating at higher pressure.

Leakages from district heating systems are generally in the order of 5% of the network volume per day. The make-up system is usually located at the CHP station, and will be designed for an agreed maximum make-up rate. This is greater than the average rate, since occasional high leakage rates will be experienced. The required water quality is non-corrosive, non-scaling, non-gaseous, and non-toxic. A mildly depositing chemistry is usually preferred to mildly corroding.

In the USSR, some single-pipe district heating systems exist, whereby the water is put to waste after heating consumers' premises. The system also provides hot tap-water direct. Such systems are used to reduce capital costs where the distance between the CHP station and the consumers is large. One hundred per cent make-up is required, and the raw water source should therefore be of suitable quality to require only minimal pretreatment.

Steam Conditions for Industry

Many industries require steam for direct use in processes. Hot water is sometimes used directly, but this is less common. Other industries require indirect heating for processes, and this is usually provided in the form of steam, although hot water is sometimes used.

Different processes require a wide range of steam pressures, usually in the range 2 to 35 bar abs at the consumer's premises, corresponding to a saturation temperature of 120 to 242 °C. Superheat may also be required, giving a supply temperature sometimes as high as 400°C. Even if superheat is not required for the process, it is usual to provide some 10 °C of superheat at the CHP station, to prevent or minimise condensation in the supply pipeline. The pipeline pressure drop must be added to the consumer's requirements to determine pressure at the CHP station.

Some industries require several steam pressure levels for different processes.

Hot water may also be required for processes or for space-heating or both. The latter may be supplied from the purpose built industrial CHP station, or from a district heating system if it exists nearby. Some industries may also wish to take treated water (unheated) for industrial processes, using the central water treatment plant and storage facilities at the CHP station.

Steam purity may not be important, particularly if all the steam uses are indirect, i.e. via a heat exchanger. Some direct uses of steam, however, may require a particularly high steam purity, which has implications upon the CHP station design.

Condensate return quantities from industry generally lie in the range 0 to 90%, although sometimes are even higher. Direct uses of steam return no condensate, of course, and even indirect uses may require the condensate to be dumped because of the risk of contamination from heat exchanger leaks. It is sometimes possible to make provision for the condensate return to be monitored for quality or to be retreated where there is risk of contamination. The decision to do this rests with the CHP developer, who must assess the economics and the operational risk to his boilers.

Heat Demand Characteristics

The CHP developer needs to know as much as possible about the following aspects of heat demand at the generating station:

(1) Peak heat demand.
(2) Annual demand characteristics.
(3) Daily demand characteristics.

The peak heat demand determines the design output of the generating station. The peak demand may vary with time, because of the connection of new consumers or the expansion of an existing industry. The forecast load growth may therefore be presented as a curve, or as a series of steps. It should be noted that all demand data should include heat transmission losses between the generating station and the consumer arising from conduction, convection, and radiation.

The annual demand characteristics may be presented as average daily, weekly, or monthly heat requirements, for all seasons of the year. The information can also be presented as an annual demand characteristic, plotting MW(th) against hours duration. Five such curves are illustrated in Fig. 7.1, with the ordinate shown as a percentage of peak heat demand. The industrial demand (1) clearly achieves a better annual load factor than the district heating demand (4), being less dependent upon seasonal weather variations. This has important implications for the economics of CHP schemes. The return on

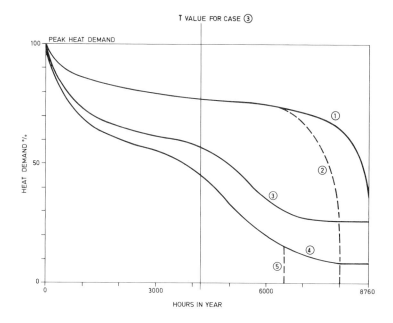

① PROCESS INDUSTRY WITH CONTINUOUS PRODUCTION.
② PROCESS INDUSTRY WITH ANNUAL SHUTDOWN.
③ DISTRICT HEATING WITH HIGH DISTRIBUTION LOSSES.
④ DISTRICT HEATING WITH LOW DISTRIBUTION LOSSES.
⑤ DISTRICT HEATING WITH SUMMER SHUTDOWN.

Fig. 7.1. Annual heat load duration curves

investment is generally better with industrial schemes than with district heating, because of the higher plant load factors which can be achieved.

The curves in Fig. 7.1 can be used to derive an annual load factor, defined as

$$\frac{\text{Area under heat load duration curve}}{\text{Peak heat demand} \times 8760} \times 100\%$$

An alternative representation is the T-value, defined as

$$\frac{\text{Area under heat load duration curve}}{\text{Peak heat demand}}$$

In Fig. 7.1, example 3 has a T-value of 4200 hours per annum, or an annual load factor of 47.9%.

The daily load characteristics can also be shown graphically. Four such curves are illustrated in Fig. 7.2, giving information about the rate of change of load which the generating station must meet. Process industries tend to have more onerous load changes, minute by minute, and this has implications upon the design and cost of the generating plant. A rate of change of ± 1% per

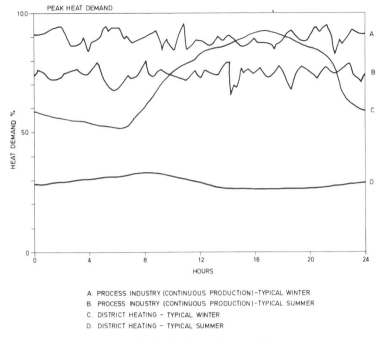

A. PROCESS INDUSTRY (CONTINUOUS PRODUCTION)-TYPICAL WINTER
B. PROCESS INDUSTRY (CONTINUOUS PRODUCTION)-TYPICAL SUMMER
C. DISTRICT HEATING - TYPICAL WINTER
D. DISTRICT HEATING - TYPICAL SUMMER

Fig. 7.2. Daily heat characteristics

minute is easier to meet than ± 5% per minute. Daily demand variations can be smoothed out to some extent by the installation of heat storage facilities at the generating station.

Availability and Reliability Requirements

Some consumers require the heat supply to be continuously available throughout the year. This applies to some process industries, and to most district heating schemes.

Other industries may shut down for planned maintenance, perhaps annually, for a duration of several days or weeks. Some district heating schemes do not include tap-water heating, and these may also shut down during the mid-summer months. Where the heat supply is not required to be continuously available, the generating station can be designed more simply. Planned maintenance of major components can be carried out during the annual shutdown, and isolation equipment and duplication of parts is not required to the same extent.

Reliability of heat generation may be defined as the ability to meet the heat demand from day to day, i.e. at all times other than planned outages of the generating station. Unreliability therefore arises from unplanned outages of plant, such as breakdowns. The higher the required reliability, the more plant and capital investment is required. The same is true of the distribution system, which requires systems of ring mains to achieve high availability and reliability.

The effects of breakdowns in CHP plant depend on the provision of standby generating plant. Where such plant is readily available, it can usually be brought into full service within about 30 minutes following the failure of a CHP unit. For that period, there is therefore a shortfall in heat supply of anything from 10% to 100% depending upon how much of the demand was being met by the CHP unit. Some district heating schemes can accept a shortfall in winter of around 20% for a duration of 30 minutes without catastrophic results to the consumer. Industrial processes, commerce, and hospitals, on the other hand, may be less tolerant regarding the magnitude and duration of the shortfall. This can influence the choice of unit size at the generating station, and may require an element of running reserve (several units running at partial load) at certain times of the year. All this is to limit the effects of a unit breakdown, particularly trips without forewarning.

The consequences of plant breakdown are more severe when there is no standby plant which can be brought quickly into service. This can occur in winter if insufficient generating plant has been installed, or if several units have broken down simultaneously. In these circumstances, shortfalls in supply can be more prolonged, lasting several days or weeks until, for example, a particularly cold spell of weather has passed. It should be noted that one of the causes of insufficient installed plant capacity is that the peak heat demand has been underestimated. District heating is often planned for average winter peak weather conditions (e.g. the average for the previous ten years) whereas, in fact, conditions are significantly worse one winter in five, or one winter in ten. For those particularly cold winters, the authority may take a calculated risk that no generating plant breakdowns occur at times of peak demand.

A rule of thumb which has been used for many district heating schemes is that the installed standby plant capacity shall be equal to (or greater than) the capacity of the largest generating unit. Therefore if the specified peak demand is 1000 MW(th) and the largest unit has an output of 200 MW(th), then the installed capacity of standby plant should be at least 200 MW(th), made up of one or more generating units. In some countries with large district heating schemes, a lower value of standby capacity is now used, for example, 80% of the capacity of the largest unit (160 MW(th) from the previous example). This is a calculated risk which is intended to save the capital cost of plant which may be required to run for only a week or so per annum.

Industrial, commercial, and institutional consumers may require a higher reliability of supply than general district heating. This is because of the dangers or costs of shutting down their processes or business. Such consumers may install additional standby plant of their own, or they may take a separate heat supply from the CHP authority, who will provide preferential or separate generation of heat to the required reliability. Some such consumers may in any case require higher supply conditions than general district heating, and this too helps to justify the installation of a separate supply system.

7.1.4 Fuel Options

A fundamental decision for every CHP development is the choice of fuel type or types. The fuel is required to be readily available at an acceptable price for the economic lifespan over which the scheme is to be appraised. This lifespan can vary from as little as five years for small industrial schemes, to as much as forty years for major district heating developments. The longer lifespans tend to rely upon indigenous fuel sources whose long-term price and availability can be predicted with some certainty. Coal, peat, and nuclear fuels frequently come into this category, also natural gas and oil in a few countries lucky enough to have such long-term resources. Short lifespan schemes can consider imported fuel oil and by-products of other industries whose future may not be planned further than five to ten years ahead.

All central generating schemes (electricity-only, heat-only and CHP) can achieve fuel diversity and flexibility as discussed in Section 7.1.2. It is therefore possible to design a scheme to burn a particular fuel which is available and cheap in the short term, switching to other fuels at a later date. This switch need not be absolute; consider, for example, a city which is supplied from two generating stations, each capable of supplying 60% of the peak heat demand. Station A is oil-fired, Station B is coal-fired, and all fuel (coal and oil) has to be imported. Either station can be used preferentially according to availability and price of fuel, to supply around 90% of the annual heat (GJ) requirements. The other station would have the standby/peaking role, but the roles can be reversed at any time.

A large range of fuels can be burned centrally, including coal smalls, lignites, brown coals, refuse, heavy oil, and by-products of oil, sewage, lumber, and chemical industries. There are clearly advantages in being able to burn cheap low grade fuels or fuels of variable quality. Some of these are available in only limited quantity; refuse, for example, will never supply more than a few per cent of a nation's energy needs, but it can supply a significant proportion of a local demand if it is collected from a wider geographic area.

Nuclear fuels can be used in central generating plants. Only two nuclear CHP stations have operated to date, from a research reactor in Sweden (which is now closed) and from a 48 MW(e) unit at Bilibino in the USSR[3(h)]. A major difficulty in trying to develop nuclear CHP is that most sites acceptable for nuclear development are too far from the cities to justify the high capital cost of bulk heat transmission. A secondary problem is that the high safety standards adopted for nuclear plants may require the whole station to shut down at any time for checking components of a particular generic type. Therefore conventional generating plant may be required to provide 100% standby capacity. This problem could be overcome if the heat were supplied from a nuclear park comprising several quite different types of nuclear station, e.g. PWR, BWR, AGR. Currently, to be economic, a pipe distance of less than 70 km would typically be required for an average annual transmitted heat load of around 1500 MW(th). As fossil fuel prices increase in the future, this distance can be expected to increase or, conversely, the transmitted heat load to decrease.

7.2 TECHNICAL PRINCIPLES

7.2.1 The Heat Engine and Reject Heat

The basic components of a heat engine cycle are illustrated in Fig. 7.3. The four processes are:

A. Compression of a cool fluid (usually air or water) by means of external work.
B. Adding heat to the fluid to increase its pressure and/or volume (i.e. its potential to do work). This involves a phase change in the case of water.
C. Expanding the fluid in an engine to produce as much external work as possible.
D. Cooling the fluid to restore its original condition.

The engine is useful when the work done in C exceeds the work done in A. The cycle efficiency is then expressed as

$$\eta = \frac{\text{net work output (C – A)}}{\text{heat energy input (B)}}$$

The primary purpose of a heat engine is to produce external work, and such devices have proved useful, convenient, and highly desirable to mankind.

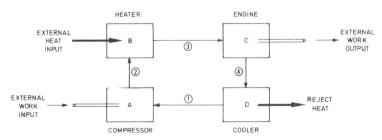

Fig. 7.3. Basic components of the heat engine cycle

Coupling the output shaft of a heat engine to a generator produces electrical power which can be transmitted over long distances.

The heat rejected in D is a necessary by-product of the process, and amounts to anything between about 50% and over 99% of the heat added in B. The heat rejected can be wasted to the environment or it can be recovered for secondary purposes. This chapter is concerned with the ways in which reject heat may be recovered for useful purposes.

7.2.2 Steam Turbine Schemes

General Principles

The basic components of the steam turbine cycle are shown in Fig. 7.4, the letters and numbers of which are consistent with Fig. 7.3. Most modern power stations are based on the steam cycle system, and are known as condensing stations where there is no heat recovery.

The broken line in the state diagrams (P–V and T–S) is the ideal cycle known as the Rankine cycle. This is often used as a yardstick to measure the theoretical efficiency attainable within the temperature limits of the cycle (T_3

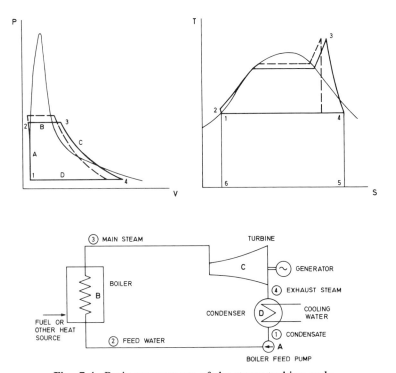

Fig. 7.4. Basic components of the steam turbine cycle

and T_4). The solid line is a real cycle operating between the same temperature limits. In the T–S diagram of Fig. 7.4, and in many ensuing T–S diagrams, the compression line 1–2 is exaggerated, with a slope to the left, whereas in reality, it is much shorter and with a slope to the right.

Cycle efficiency is represented on the T–S diagram by

$$\frac{\text{Area enclosed by } 1\text{--}2\text{--}3\text{--}4}{\text{Area enclosed by } 1\text{--}2\text{--}3\text{--}4\text{--}5\text{--}6}$$

Reject heat is represented by the Area 1–4–5–6.

Since the steam cycle was invented (or discovered), mankind has striven to improve cycle efficiency. These efforts can be put into four main groups:

(1) To reduce friction i.e. to move nearer the Rankine cycle. For example, internal friction in the steam turbine and boiler feed pump has been reduced by improved design features.

(2) To increase the top temperature T_3. This has been achieved in two ways:

    (a) By developing new materials and manufacturing techniques to withstand the higher temperatures and associated higher pressures, affecting the design of turbines, boilers, boiler feed pumps, and associated components. A top temperature of 537 °C is generally used today for

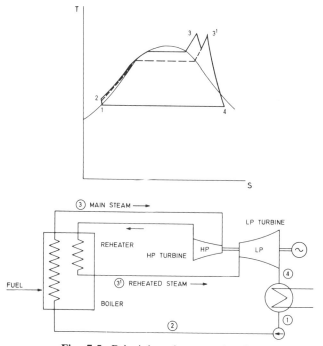

Fig. 7.5. Principles of steam reheating

steam turbine cycles producing 50 MW(e) or greater. Higher temperatures cannot be achieved with oil-fired boilers because of gas-side corrosion problems. With many coals, however, higher temperatures are possible, and some countries, notably the United Kingdom, have gone to 565 °C in unit sizes of about 200 MW(e) and greater.

(b) By using reheat cycles whereby the top temperature is attained in two (or more) stages, some of the work having been done in the HP turbine before reheating. The principle of steam reheating is shown in Fig. 7.5. The broken line shows how the cycle would look without reheat using the same expansion line 3′–4. Although the reheat pipes between turbine and boiler introduce additional friction losses, the net advantage to cycle efficiency is positive.

(3) By reducing the bottom temperature T_4. This has been achieved by exhausting from the LP turbine at sub-atmospheric pressure. Although energy-consuming vacuum pumps or ejectors are required to keep the condenser air free, the net gain to the cycle efficiency is positive at the pressures in general use.

A further development on multiple exhaust turbines has been to arrange condensers in series. This is illustrated in Fig. 7.6 with some typical

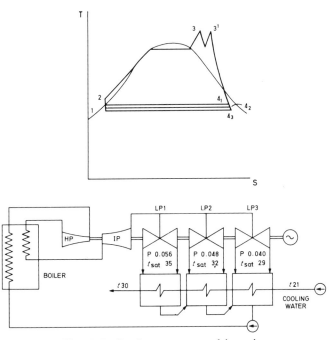

Fig. 7.6. Condensers arranged in series

exhaust conditions. Instead of all three LP turbines exhausting at 35 °C, the series arrangement allows LP2 to exhaust at 32 °C and LP3 at an even lower temperature of 29 °C. By this means additional useful work is done by those proportions of mass flow (about one third each) passing through LP2 and LP3.

(4) By bled steam feed water heating. This is a form of CHP (in the technical sense) which is used in virtually all steam generating stations (regardless of whether they are electricity-only or CHP in the commercial sense). Instead of expanding all the steam to the condenser, proportions of partially expanded steam are bled off from the turbine at convenient stages for the purpose of feed water heating. This conserves fuel in the boiler (see Fig. 7.7). In this example there are five stages of heating and two stages of pumping, separated by the deaerator vessel. The water temperature rise in each heating stage is typically 20 to 40 °C.

All the thermodynamic principles so far described are applied generally in modern condensing stations. In the context of CHP generation, they are equally important, since CHP may otherwise be at a serious disadvantage

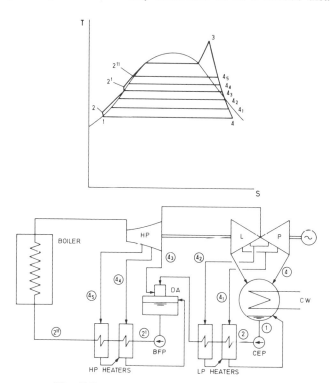

Fig. 7.7. Principles of feed water heating

thermodynamically (and therefore economically). This helps to explain why some countries with advanced designs of condensing station have difficulty introducing small CHP units. For example, a 15 MW(e) CHP unit may be limited by practical design considerations to main steam conditions of around 65 bar abs 480 °C (without reheat), with a final feed-water temperature of around 180 °C. Even allowing for the sale of district heat, such a unit will not normally be able to compete economically with large modern condensing units (burning the same fuel) operating with steam conditions of around 160 bar abs 535 °C/535 °C and a final feed temperature of 250 °C.

Back Pressure Turbines

The term 'back pressure turbine' is used to describe CHP turbines where all the heat in the turbine exhaust steam is recovered for useful purposes. The

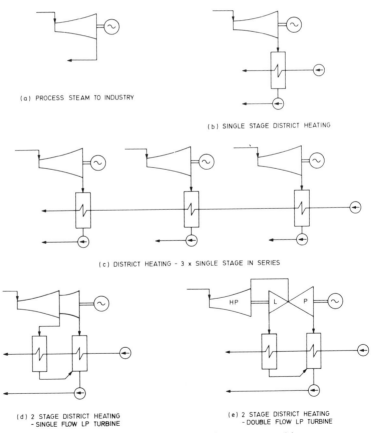

(a) PROCESS STEAM TO INDUSTRY

(b) SINGLE STAGE DISTRICT HEATING

(c) DISTRICT HEATING - 3 x SINGLE STAGE IN SERIES

(d) 2 STAGE DISTRICT HEATING
- SINGLE FLOW LP TURBINE

(e) 2 STAGE DISTRICT HEATING
- DOUBLE FLOW LP TURBINE

Fig. 7.8. Arrangements of back pressure turbines

exhaust steam can be supplied directly (e.g. as process steam to industry) or it can be condensed in a heat exchanger (e.g. to supply hot water for district heating). The three distinguishing features of back pressure (BP) schemes are:

(1) No large cooling water system is required, i.e. no large cooling towers, sea water, or river water source.
(2) There is a loss of electrical output because the steam does not expand to normal condensing conditions.
(3) Electrical output and heat output are directly related to each other, i.e. there is no operational flexibility between the two products.

Figure 7.8 illustrates various BP arrangements. Examples (a) and (b) are the simplest to engineer, comprising a single exhaust to supply process steam and district heat respectively. Single-stage district heating is less common than multistage because it is thermodynamically inferior and therefore more expensive to operate. Some district heating schemes arrange two or three such turbines in series, as shown in Fig. 7.8(c). This achieves multistage heating without complicated designs of LP turbine exhaust. However, such stations are generally less flexible to operate, requiring more than one turbine in operation to achieve the multistage feature.

Examples (d) and (e) are the most usual arrangements, having two stages of heating. The double flow LP cylinder (e) is used in the larger unit sizes (above about 100 MW(e)). With this arrangement, heating is divided equally between the two stages over a wide range of off-design conditions, of which the water temperature rise is the most important. The single flow LP cylinder, on the other hand, is more difficult to design for equal heating load under all conditions, particularly for variations in water temperature rise. One solution to this problem is to use a split LP exhaust, as shown in Fig. 7.9. This is a form of

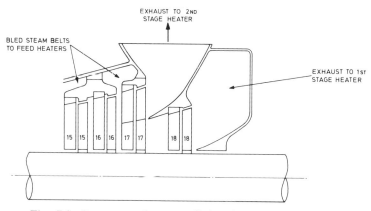

Fig. 7.9. Baumann exhaust applied to back pressure turbine

Baumann exhaust (sometimes used in condensing turbine designs). In this example the steam flow is physically divided between the two exhausts at stage 17.

Back pressure turbines are generally limited to two stages of district heating. More than two stages of exhaust would make the turbine design significantly more complicated, and the incremental thermodynamic gain reduces as each additional stage is added. In other words, the third stage is usually either impracticable or uneconomic. The thermodynamics can be illustrated as follows.

A condensing turbine exhausts at a temperature of t_{cond} with a mass flow of steam m. As shown in Fig. 7.10(a), expansion of the steam is now cut short to

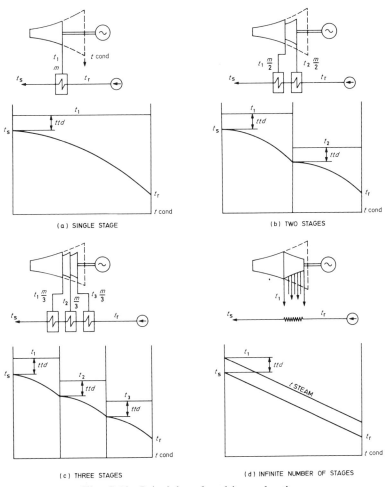

Fig. 7.10. Principles of multistage heating

exhaust at temperature t_1 in order to provide a single stage of district heating (where t_r is the district water return temperature, t_s is the district water supply temperature, and *ttd* is the terminal temperature difference across the heat exchanger, i.e. $ttd = t_1 - t_s$).

As a first approximation, the loss of electrical output ΔE is proportional to the increase in exhaust temperature.

Therefore

$$\Delta E_1 = km \ (t_1 - t_{cond})$$

$$= km(t_s + ttd - t_{cond})$$

where k is a constant. Note that, with single stage heating, the loss of electrical output is unaffected by the return temperature t_r but the supply temperature t_s is of major importance.

Consider, now, two stages of heating as shown in Fig. 7.10(b). The steam flow m is split evenly between the two stages, so that, to a first approximation, the water temperature rise is the same across each heater.

Therefore

$$\Delta E_2 = \frac{km}{2} \ (t_1 - t_{cond}) + \frac{km}{2} \ (t_2 - t_{cond})$$

$$= km \left[\frac{t_1 + t_2}{2} - t_{cond} \right]$$

but

$$t_1 = t_s + ttd$$

and

$$t_2 = \frac{t_s + t_r}{2} + ttd$$

Therefore

$$\Delta E_2 = km \left[\frac{t_s}{2} + \frac{ttd}{2} + \frac{t_s}{4} + \frac{t_r}{4} + \frac{ttd}{2} - t_{cond} \right]$$

$$= km \left[\frac{t_r - t_s}{4} + t_s + ttd - t_{cond} \right]$$

$$= km \left[\frac{t_r - t_s}{4} + \frac{\Delta E_1}{km} \right]$$

$$= \Delta E_1 - \frac{km}{4} \ (t_s - t_r)$$

Similarly, with reference to Fig. 7.10(c) and (d)

$$\Delta E_3 = \Delta E_1 - \frac{km}{3} \ (t_s - t_r)$$

and

$$\Delta E_\infty = \Delta E_1 - \frac{km}{2} \ (t_s - t_r)$$

Two conclusions can now be drawn:

(1) With multistage heating, achieving a low supply temperature t_s is still important, but a low return temperature t_r is also important.
(2) Increasing the number of heating stages from one to two achieves about 50% of the improvement in electrical output which is achievable (theoretically) with an infinite number of stages. The third stage achieves a further increase of only 17%, and further stages even less.

In practice, economic appraisals usually show that the optimum number of heating stages is two. In some schemes, three stages have been installed, often with the third stage used for peaking purposes only.

One of the approximations used above was to ignore the effects of condensate drains. There are two ways to arrange condensate drains from multistage heaters. These are illustrated in Fig. 7.11. Cascading back to the lowest heater is the simpler system. Pumping forward, on the other hand, is thermodynamically superior, but involves additional capital expenditure on pumps and extra maintenance and a change in pumping power. The optimum arrangement should be determined for each scheme by economic appraisal.

A full system arrangement of a back pressure unit for district heating is shown in Fig. 7.12. This would be rated at about 100 MW(e); such units exist in several countries. The useful limit of BP unit size is about 200 MW(e) at present. In larger sizes, too much heat would be produced for most supply systems for most of the year.

Recoolers are sometimes used with back pressure turbines. Such an arrangement is shown in Fig. 7.13. Its purpose is to dump heat not required by the district heating network, and thereby enable the BP turbine to generate the maximum electrical output at any time, e.g. during midsummer when the heat demand is minimal. This overcomes the inflexibility constraint normally associated with BP turbines, and may be a more economic option than to provide a condensing tail to the turbine. The recooler would be rated to give only the required minimum electrical output of the station. The cooling medium can be air (fan assisted) or water from a river (which need not be local

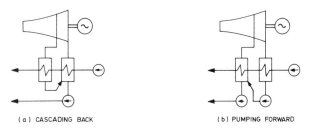

(a) CASCADING BACK   (b) PUMPING FORWARD

Fig. 7.11. Arrangements of condensate drains

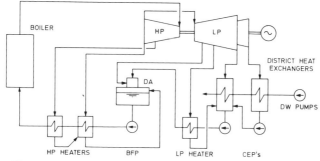

Fig. 7.12. Full system arrangement of back pressure unit

to the CHP station). Recoolers can be important in particular locations which do not have the full back-up of a large integrated electricity grid, either because the grid connection is weak or because a small utility may have tariff constraints on electricity imports.

Existing condensing turbines can be converted to back pressure CHP operation by the removal of rows of blading from the LP end. A modest increase of exhaust temperature to around 70 °C is acceptable for some turbine designs, and others can accept up to 95 °C. The problems of thermal expansion and higher stress levels must be investigated for each individual design. For multi-cylinder designs, the LP cylinder can be abandoned altogether. Typical LP crossover pressures are 1 to 5 bar abs (giving saturation temperatures of 100 to 152 °C) of which some 5 °C must be lost across the heat exchanger.

A converted turbine is usually restricted to one stage of heating, and a typical role for such conversions is for temporary CHP generation during the network construction phase until a large purpose built CHP plant is installed. It is sometimes possible to convert the condenser for district water heating, particularly for temperatures less than 100 °C. This has been done in Poland,

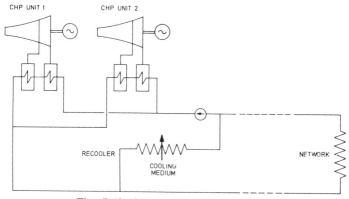

Fig. 7.13. System arrangement with recooler

and required the water boxes to be strengthened to withstand the higher water-side pressure. In other countries, a new district heat exchanger has been preferred, designed not only for the required water pressure but for optimum water velocities in the tubes. (The temperature rise is typically 40 °C for DW compared with about 10 °C for CW).

Extraction Condensing Turbines

An extraction condensing turbine can be defined as a back pressure turbine to which a condensing tail has been added. Alternatively, it can be defined as a condensing turbine from which steam is extracted (or bled) at an intermediate point for heating purposes. The important feature is that steam can be routed either down the extraction pipe for heat production or through the LP cylinder of the turbine for additional electricity production. The machine has in-built flexibility between the two products. Extraction condensing (EC) turbines are also known as intermediate take-off condensing (ITOC) turbines or intermediate pass-out turbines.

The simplest method of extraction is to design a slot (extraction belt) around the periphery of the turbine casing and to connect it to an external extraction pipe. This is uncontrolled extraction and is illustrated in Figure 7.14(a). It is uncontrolled in the sense that pressure at the extraction point varies according to the turbine load (inlet steam flow) and to the quantity of steam extracted. For the extraction of up to about 10% of the mass flow, the loss of pressure can be small enough not to matter. Larger extractions, however, can have two detrimental effects:

(1) The extracted steam is seriously degraded thermodynamically, making it less useful for heating purposes. This is because saturation temperature (which is a function of pressure) decreases as the extraction quantity is increased.
(2) The pressure drop increases across certain rows of turbine blades, causing high stress. The rows immediately upstream of the extraction point are particularly prone to this design limitation.

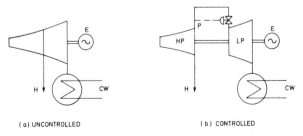

(a) UNCONTROLLED   (b) CONTROLLED

Fig. 7.14. Principles of bled steam extraction

Uncontrolled extraction is used for bled steam feed-water heating, taking typically a 5% extraction per stage. Some industrial heat supplies may also come from uncontrolled EC turbines, usually where the heat load is essentially steady throughout the year. District heating, on the other hand, has large seasonal variations of load, and extraction quantities well in excess of 10% are usually required, particularly in winter.

Controlled extraction is used for extraction quantities in excess of 10%. The principle is illustrated in Figure 7.14(b). The turbine is split into the HP and LP sections, with a pass-through valve between them. Steam is extracted at the outlet of the HP section, but extraction pressure is maintained constant at all extraction quantities by means of the pass-through valve and the control loop shown. Up to about 95% extraction can be achieved by this means, the remaining 5% being required to cool the LP turbine. Some utilities go even further, achieving 100% extraction (i.e. pure BP) during the winter season by putting a bladeless shaft in the LP cylinder and blanking off the pass-through valve. A few days' outage is required, of course, each time the LP rotor is changed, and the flexibility feature is lost when running in this way.

Figure 7.15 illustrates the state diagram for an EC turbine. The expansion line a–b–c represents full condensing operation, i.e. with zero extraction. As extraction is increased, the expansion line moves to the right, either as uncontrolled extraction (a–d–f) or as controlled extraction (a–b–e–f). The important pressures are as follows:

P_1 Extraction pressure (controlled).
P_2 Extraction pressure (uncontrolled).
$P_1 - P_2$ Loss of extraction pressure due to lack of control.
$P_1 - P_3$ Pressure difference across pass-through valve.

Figure 7.15 shows only the full load expansion lines, i.e. maximum steam

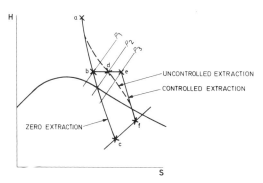

Fig. 7.15. State diagram for extraction condensing

flow at turbine inlet. Under part load conditions, the loss of extraction pressure is even greater for the uncontrolled case.

In practice, the pass-through valve can be an integral part of the turbine internals, arranged as a variable geometry disc which controls the steam flow area. This is shown in Fig. 7.16(b). Only the control actuator is external to the turbine, and only the extracted steam leaves the turbine casing. This provides a simple (i.e. low friction) path for the LP steam, which is important when operating in the pure condensing mode.

Internal valves are suitable for single cylinder turbines with extraction pressures up to about 7 bar abs. At higher pressures, the sliding friction becomes excessive and a conventional external valve must be used. With multi-

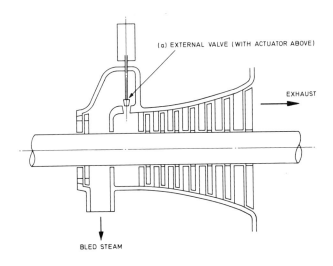

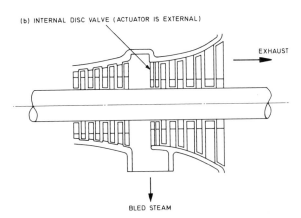

Fig. 7.16. Pass-through valve arrangements

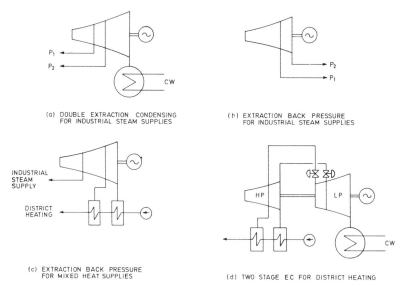

Fig. 7.17. Multiple heat supplies and steam extractions

cylinder turbines (which start at about 80 MW(e) size) all the steam must leave the HP casing anyway, so the pass-through valve can be mounted on the LP inlet flange without making the steam path any more tortuous. Figure 7.16(a) shows this arrangement.

Several extractions can be taken from one turbine, to supply industrial steam at several pressure levels or for multistage district heating, or for a combination of both. Examples are shown in Fig. 7.17. Another example of multiple extraction is feed water heating, and this can be superimposed upon any of the examples shown.

Figure 7.18 shows an EC turbine with two extraction points for industrial steam supplies. The feed water heating system comprises only two stages of heating, utilizing steam from the main extraction lines. Whilst additional

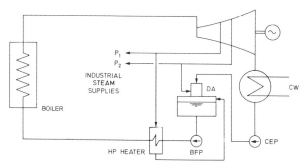

Fig. 7.18. Double extraction condensing with feed heating

extractions for feed water heating may be desirable thermodynamically, many small 'industrial' schemes are designed without additional extraction points in order to keep the turbine casing geometry as simple as possible. This allows standard casing designs to be used, and avoids stress concentration design problems which would otherwise occur with a multiplicity of extraction points in a physically small machine. Such schemes therefore achieve a low cost and good reliability, but at the expense of a poorer thermodynamic performance.

Larger EC turbines for district heating generally have a full set of feed heaters (as would be provided for a condensing unit). Indeed, the HP end of the system can be of standard design, regardless of whether the LP end is EC or condensing. Figure 7.19 shows a full system arrangement for a large EC unit for district heating. The third stage of heating in this example is for stand-by purposes and serves also to recover steam bypass heat during unit start-up.

The flexibility between heat and electricity production provided by EC has already been mentioned. A second important feature is that EC turbines can be made much larger by combining the CHP requirement with that of a straight condensing turbine. This is illustrated in Fig. 7.20.

Case (a) has a condensing tail capable of taking all the steam not required for heat supply H. Electrical output is E_1 with 95% extraction, increasing to $E_1 + E_2$ with zero extraction. The shaded area shows the limits of operating flexibility between H and E.

In Case (b), the values of H, E_1, and E_2 remain the same, but the mass flow at the inlet increases from m_1 to $m_1 + m_3$ to produce an additional electrical output E_3. Because the new (combined) turbine is much larger, its internal

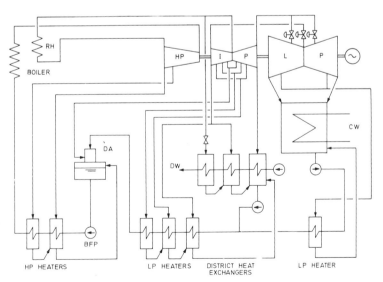

Fig. 7.19. Full system arrangement of extraction condensing unit

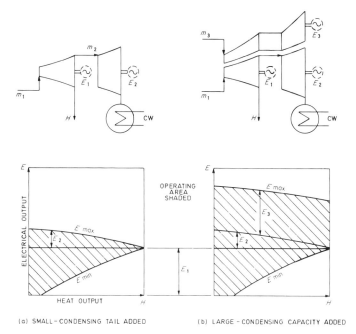

Fig. 7.20. Small and large extraction condensing turbines

efficiency improves, and even more important, it can be designed for higher inlet steam conditions. In other words, the cycle efficiency improves without changing the value of H. If, therefore, the condensing capacity E_3 would have been installed anyway, then it makes good economic sense to combine the two requirements in one large boiler-turbine unit located in the vicinity of the heat load. Such a turbine may therefore be restricted to, say, 50% extraction (although the utility may be tempted to specify it for up to 95% extraction to cater for possible future increases in heat demand).

If, on the other hand, the extra capacity E_3 is not required by the utility, then the extra capital charges of the larger boiler-turbine unit must be debited against the CHP scheme until such time that E_3 is required. In such circumstances, the smaller Case (a) of capacity $E_1 + E_2$ may be the more economic option, or even a simple back pressure unit of capacity E_1. It therefore follows that the optimum timing for CHP developments can be during an era of expansion of the electric power market or during an era of major replanting to replace old generating plant.

To convert an existing condensing turbine to extraction condensing operation, the usual and most convenient method is to extract steam from the LP crossover pipes. Pass-through valves of the butterfly type can be mounted on the steam inlet flanges of the LP cylinder. This produces the system arrangement shown in Fig. 7.14(b). District heating is necessarily limited to one heating stage per turbine, and it is a matter of luck whether the crossover pressure is well-suited thermodynamically for the heating requirements. It is sometimes possible to convert more than one turbine and to arrange the heaters in series for multistage heating. New heat exchangers are always required and a general problem is to find space in the existing turbine house for the additional plant and pipework connections. A new building outside the turbine house is sometimes required.

Most old turbines today would be capable of 50% extraction from the LP crossovers, some considerably more. One design limitation can be the steam exhaust temperature from the LP turbine: as extraction is increased, the LP exhaust temperature increases, causing thermal expansion of the casing, resulting in stress and/or vibration problems above the limiting temperature.

It is more difficult (and less common) to convert an existing turbine to multistage district heating. A complete rebuilding of the LP cylinder would be required. Alternatively, a new double-flow auxilliary turbine-generator could be provided, taking LP crossover steam as illustrated in Fig. 7.21. Such an arrangement would be particularly applicable to plants which have a high LP crossover pressure. For example, this pressure can approach 10 bar abs on PWR nuclear stations. As with any conversion, however, the practical problems of layout, control, and safety need to be investigated thoroughly for each particular turbine design.

7.2.3 Gas Turbine Schemes

Figure 7.22 shows the basic components of the open circuit type of gas turbine cycle. This is in common use throughout the world for land based applications in unit sizes up to 110 MW(e). The operating fluid is air. After compression, the air is heated by the combustion of a fuel to raise its temperature to the design limit of the turbine. Some of the turbine work is used to drive the compressor, the remainder being available for external work such as electricity generation.

The component numbering in Fig. 7.22 is consistent with the general cycle

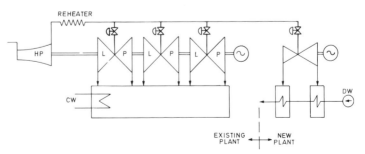

Fig. 7.21. CHP conversion using an auxiliary turbine generator

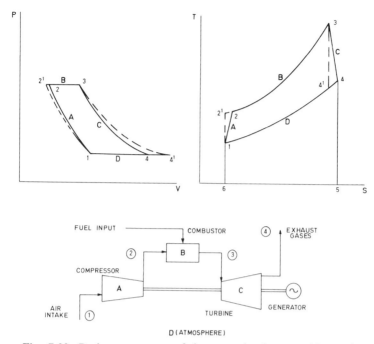

Fig. 7.22. Basic components of the open circuit gas turbine cycle

shown in Fig. 7.3. In the state diagrams (P–V and T–S) the broken line
1–2′–3–4′ is for an ideal fluid and the solid line 1–2–3–4 is for the real fluid.
The net work output is represented by the area 1–2–3–4 and the heat rejected
is represented by the area 1–4–5–6 on the T–S diagram.

In a gas turbine CHP scheme, heat is recovered from the exhaust gases by
means of a waste heat boiler (WHB), known alternatively as a heat recovery
boiler or recuperator. The term 'boiler' is applied generally, regardless of
whether the output is steam or hot water. It is also possible to utilize the
exhaust gases directly (e.g. for drying purposes), but such schemes are
uncommon.

Exhaust arrangements for a GT/WHB system are shown in Fig. 7.23, with
and without gas bypass. The purpose of gas bypass is to allow the GT to produce full electrical output at times when the WHB is shut down or when only
partial heat output is required. Putting steam to a dump condenser is an alternative method of achieving full GT output irrespective of heat requirements.
The gas bypass is also useful during start-up, allowing the GT to achieve full
electrical output quickly, unconstrained by considerations of thermal stress in
the boiler.

With steam raising boilers, it is possible to modulate the gas bypass dampers
to control boiler ouput. With hot water boilers, on the other hand, this is more

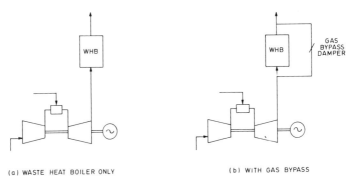

(a) WASTE HEAT BOILER ONLY　　　　(b) WITH GAS BYPASS

Fig. 7.23. Gas turbine exhaust arrangements

difficult to achieve because of the need to control output water temperature within close limits. Bypass dampers on hot water WHBs therefore need to be moved slowly (less than 1 ° angle per minute in some instances).

The gas bypass can rejoin the main flue after the WHB, as shown in Fig. 7.23(b). Alternatively, it can exhaust to atmosphere by means of a second stack immediately after the GT. This is known as a 'blast stack'.

To prevent leakage when shut, double isolation dampers are required for the gas bypass, with sealing air supplied between them. Single dampers tend to leak when shut, the loss being from 2% to 10% depending upon the damper design and the state of maintenance. Such leakage detracts from the potential heat recovery and must be taken into account in performance calculations. Double isolation dampers may also be required at the inlet and outlet to the WHB in order to permit gas-side access for boiler maintenance with the GT on load.

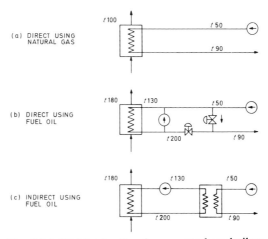

(a) DIRECT USING NATURAL GAS

(b) DIRECT USING FUEL OIL

(c) INDIRECT USING FUEL OIL

Fig. 7.24. District heating from waste heat boilers

Waste heat boilers for district heating are usually of the water-tube type. The simplest arrangement is shown in Fig. 7.24(a), using a clean fuel such as natural gas. The final gas temperature can be reduced to around $100\,°C$ using natural gas. Most fuel oils, on the other hand, contain a significant amount of sulphur and the final gas temperature must be kept above acid dew-point, typically $115\,°C$. This means that the water inlet temperature must be maintained above this value at all times. A typical design water inlet temperature is $130\,°C$. After adding the temperature difference across the boiler tubes, the final gas temperature with oil firing is therefore around $180\,°C$. Figures 7.24(b) and (c) show two methods of achieving the water temperatures required with oil firing. The direct method (b) is lower in capital cost, but the indirect method (c) has operational advantages, allowing pressures and water chemistry to be chosen independently on the two sides of the heat exchanger.

Several types of steam raising waste heat boiler are shown in Fig. 7.25. The shell (or fire-tube) type can be manufactured in sizes up to about 20 MW(th) output. The water-tube type can be made in any size, designed with either natural or assisted circulation. The former frequently has unfinned tubes and a horizontal gas path, whereas the latter usually has finned tubes and can be arranged in any plane of gas path. Once-through coil boilers are also possible.

An economizer can be added to a steam raising WHB (to increase heat recovery), and a superheater can be added (to meet the consumers' require-

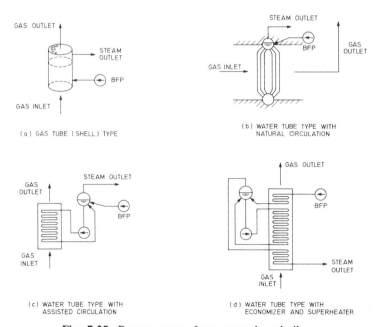

Fig. 7.25. Process steam from waste heat boilers

ments). These additions are shown in Fig. 7.25(d). A positive terminal temperature difference (*ttd*) is required for each section of the boiler, of course. The economic minimum terminal temperature difference is generally around 40 °C for the economizer, 20 °C for the evaporator and 50 °C for the superheater section.

The quantity of heat which can be recovered from the exhaust gases will depend on temperature levels on the water side of the boiler. Figure 7.26 shows some examples, plotting WHB temperature *t* against heat recovered *H*. Maximum recovery H_{dh} can be achieved with district heating, provided that a low return temperature of 50 to 60 °C is achieved. For saturated steam production,

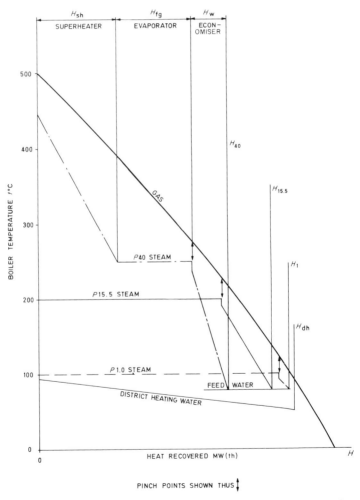

PINCH POINTS SHOWN THUS

Fig. 7.26. Waste heat boiler temperatures versus heat recovered

heat recovery reduces to H_1, $H_{15.5}$, and H_{40} respectively for steam pressures of 1, 15.5, and 40 bar abs. An economizer is included in each case, raising feed-water temperature from 80 °C. Without the economizer, heat recovery would reduce significantly in the 40 and 15.5 bar cases, but not for the 1 bar case.

The 40 bar case has a large superheater, raising the steam temperature from 250 to 450 °C. The effect of adding superheat is to reduce the steam/water mass flow. This reduces the quantity of heat which the economizer can recover, since the proportions H_{sh}:H_{fg}:H_w are fixed by the physical properties of steam and water between the terminal conditions specified.

A steam raising WHB can be designed to produce several independent heat supplies at different temperature levels. Such an arrangement is illustrated in Fig. 7.27, producing superheated steam at 36 bar abs and district heating at 90 °C. By this means, the maximum heat recovery can be approached, although the two independent heat loads will not necessarily coexist in the correct proportions at all times.

Air is the working fluid of open-circuit GT cycles, and only a minor proportion of its oxygen is needed for combustion. Typically the oxygen content reduces from 23% by weight at compressor inlet to 17% at GT exhaust. The turbine exhaust gases are therefore capable of burning further (secondary) fuel, and this is known as afterburning (AB), supplementary firing, or exhaust firing. The important features of afterburning are:

(1) The heat output capacity of the WHB is increased at little extra capital cost. (A higher heat transfer rate is achieved because of the increased temperature of the gases).
(2) The energy in the fuel fired after the turbine can be utilized at a high thermal efficiency. Provided that the stack temperature is not allowed to increase, the only increase in stack loss arises because of the mass of the secondary fuel itself. In practice, the stack temperature may vary, particularly if afterburning is used mainly for peaking purposes.

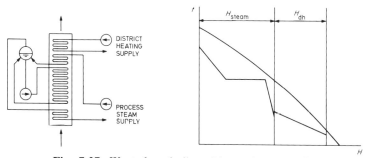

Fig. 7.27. Waste heat boiler with two heat supplies

The disadvantage of afterburning is that the exhaust system and its control becomes more complex and availability can suffer. If the GT has strategic importance for electricity or heat production (or both), the utility may prefer an exhaust system without afterburn. Alternatively, a simple system of in-duct firing may be chosen, as illustrated in Fig. 7.28(a). This provides only moderate boosting of gas temperature (from say 500 to 600 °C), reducing O_2 content typically from 17% to around 14% by weight. The flame has low luminosity and no cooling of the gas duct is required. Such a facility can be used to attain a higher superheat temperature in steam-producing waste heat boilers, and may be particularly useful at partial GT loads.

Some afterburning schemes burn the secondary fuel in a conventional water-cooled furnace. This allows the oxygen level to be reduced to 5% or less (by weight) to achieve furnace gas temperatures in excess of 1500 °C. Some burner designs incorporate a secondary air supply in order to maintain stable combustion under all operating conditions, sized typically between 5 and 10% of the stoichiometric air quantity. This increases the stack heat loss, of course, and the efficiency of use of the secondary fuel therefore reduces.

Figure 7.28(b) shows the 'fired boiler' afterburning arrangement. A gas bypass is usually included in such systems to enable the GT to operate when the boiler is shut down. To enable boiler maintenance to be carried out when the GT is on load, gas isolation facilities may also be incorporated, either double dampers (with sealing air) or blanking plates (which must be fitted with the GT shut down). The complexity of the system in Fig. 7.28(b) with a fired boiler should be compared with the simplicity of Fig. 7.23(a) without afterburn. Nevertheless, systems like Fig. 7.28(b) are used successfully, particularly for industrial heating schemes.

The secondary fuel can be selected independently of GT fuel grade requirements. For example, heavy fuel oil or coal can be burnt in the fired boiler,

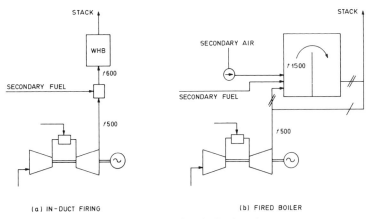

(a) IN-DUCT FIRING   (b) FIRED BOILER

Fig. 7.28. Methods of afterburning

whilst the GT may be restricted to natural gas or distillate. With such arrangements, the utility may require the boiler to operate with the GT shut down. The secondary fuel and secondary air systems must then be designed to a larger size so that full boiler output can be achieved when the GT is shut down. This arrangement is known as 'auxiliary firing'.

Some fired boilers are designed several times larger than required for burning the GT exhaust gases. The fired boiler then becomes the main plant (usually operating at high annual high load factor), and the GT becomes the secondary plant (operating at a low annual load factor), possibly for electrical peaking purposes only. The heat recovery feature can then be regarded as a small operating bonus.

A feature of gas turbine CHP schemes is that the loss of electrical output because of heat recovery is small. Unlike steam turbines, there is no inherent thermodynamic loss of output from heat recovery. The only losses of station electrical output arise because:

(1) Back pressure is imposed by the WHB on the gas turbine exhaust. Careful design of the WHB tube arrangements can restrict this to around 1% of the generated output. The loss can be a little higher with fired WHBs.
(2) Works electricity is required by the WHB and the heat supply system (e.g. pumping power).

The electrical efficiency is determined primarily by the top temperature T_3 which varies according to the turbine design and the fuel burned. Using a clean fuel such as natural gas, a modern gas turbine can be designed for a top temperature approaching 1100 °C. Most fuel oils, on the other hand, contain trace elements such as sodium and vanadium which damage the turbine blades at this temperature. The top temperature must therefore be reduced according to the quantity of undesirable elements present.

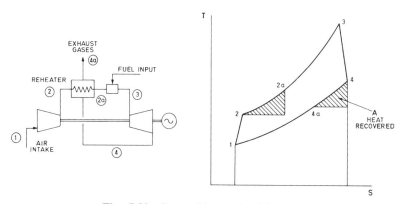

Fig. 7.29. Gas turbine cycle with reheat

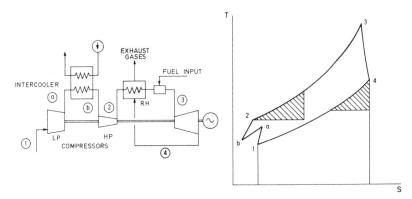

Fig. 7.30. Gas turbine with intercooler and reheat

Attempts have been made to improve the GT cycle efficiency by means of a regenerative heat exchanger, as shown in Fig. 7.29. This is known as reheat. The heat rejected is reduced by the shaded area A. Technically this may be regarded as a form of CHP. Another development has been to add an intercooler between stages of compression, as shown in Fig. 7.30. The purpose is usually to increase the electrical output of a particular GT model.

Neither reheat nor intercooling of GTs has achieved widespread application. Utilities consider that the performance advantages are outweighed by the disadvantages of higher capital and maintenance costs, lower availability, and longer starting times. Simple GT cycles are generally preferred. Only some designs of turbine are in any case suitable for reheat/intercooling. For example, the compact industrial designs derived from aero-engines are generally unsuitable.

The closed circuit GT cycle is less common than the open circuit. The basic components are shown in Fig. 7.31. Very large heat exchangers are required at both the heat sink and the external heat source, and there is a limit to the top cycle temperature T_3, arising from the working temperature of the heat

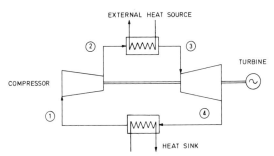

Fig. 7.31. Basic components of the closed circuit gas turbine cycle

exchanger material. The advantages are that the GT is not exposed to the primary fluid which may contain undesirable constituents, and that fluids other than air can be chosen for the secondary circuit, giving improved cycle efficiency. The closed circuit GT cycle may be preferred to steam cycles for some nuclear systems in the future.

Some gas turbines are capable of burning heavy fuel oils (which are significantly cheaper than distillate). However, special fuel pretreatment plant must be provided (principally to remove Na) which adds to the capital costs. The GT must also be down-rated (to prevent high temperature corrosion due to Va), possibly by as much as 10%, and maintenance costs and outage times increase because of ash deposition. Where available, natural gas or distillate are generally preferred for gas turbines.

Coal burning GT schemes are not commercially available as yet, and current and future developments of this nature are beyond the scope of this chapter.

7.2.4 Diesel Schemes

The basic components of the diesel cycle are shown in Fig. 7.32. The inclusion of a turbocharger is common practice in order to increase the output of the engine. Even with this provision, the unit size of diesels is limited to around 25 MW(e) with medium speed four stroke designs, and to around 30 MW(e) with low speed two stroke designs. These types of engine are in general use for land-based electricity generation, and either may be considered for CHP schemes. The four stroke engine is more compact and usually cheaper, whereas the two stroke engine generally requires less maintenance effort and achieves a higher availability because it has fewer cyclinders.

Combustion takes place within the cylinders of the engine. Components which come into contact with the hot gases must therefore be cooled by external means to maintain acceptable metal temperatures. This is achieved by means of jacket water and lubricating oil, which remove significant quantities

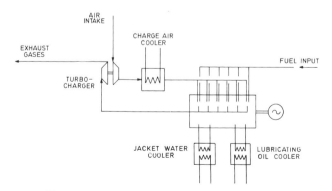

Fig. 7.32. Basic components of the diesel cycle

of heat. This can be rejected to the environment by means of cooling towers, or it can be recovered as low temperature (LT) heat for CHP purposes. Heat from the charge air cooler can also be included in this group. The upper temperature limit for LT heat recovery is usually around 70 to 85 °C, but new developments are now available to give supply temperatures of over 100 °C.

The diesel exhaust gases have a temperature of some 350 to 420 °C, and provide a second source of reject heat (of the same magnitude as the LT source). A waste heat boiler is required, the design and arrangement of which is as described for gas turbines in Section 7.2.3. Most diesel CHP schemes burn heavy fuel oil containing sulphur, so the final gas temperature is normally not lower than 170 °C, and the water inlet temperature must be controlled to prevent acid deposition.

For maximum heat recovery, a diesel CHP scheme therefore comprises LT and HT systems. Figure 7.33 shows such an arrangement, supplying steam for industry and hot water for district heating. The sensible heat for steam raising comes mainly from the LT system, but it is important to understand that this recovers only a proportion of the LT heat available. To recover all of the LT heat, a hot water heat supply system is necessary and, in addition, the return water temperature must be around 40 to 50 °C.

If the required water supply temperature is greater than about 85 °C, then some of the WHB heat must be utilized. This can be achieved, as shown in Fig. 7.33, by a closed circuit economizer loop which can contain either pressurized water or an organic fluid. A diesel heat recovery scheme needs careful design, taking into account the constraints of temperature pinch-points and the quantity of heat available in each component. It is difficult to recover all of the heat available from diesels.

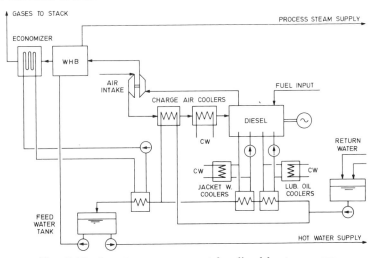

Fig. 7.33. A system arrangement for diesel heat recovery

The recovery system design is further complicated by the fact that the CW system is usually retained so that the engine can operate unconstrained by heat demand considerations. The CW coolers are shown in Fig. 7.33, and there will also be CW pumps and (typically) a fan-assisted cooling tower.

Gas bypass provisions and afterburning can be incorporated in a similar way to gas turbines, as described in Section 7.2.3. The potential for afterburning is somewhat less, however, due to the lower air quantities used by diesels, giving typically 13% oxygen by weight at the diesel exhaust.

A unique and important feature of diesels is that they can make efficient use of heavy fuel oils in small unit sizes. Diesels can burn fuel oils up to 420 cSt viscosity and are being developed to burn even heavier fuel oils which are expected to become commonplace in the future. Whilst it is also feasible to burn gaseous fuels in diesels, thermal efficiency and ratings are lower and they require spark-ignition or a 'pilot' light fuel oil to initiate combustion.

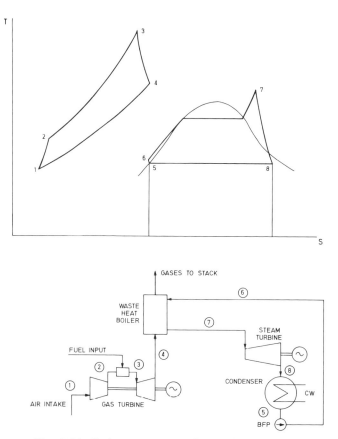

Fig. 7.34. Basic components of a combined cycle

7.2.5 Combined Cycles

The steam produced in a waste heat boiler can be utilized in a steam turbine to produce additional electricity. This is known as a combined cycle. Both GT and diesel combined cycles are possible, but GTs are more common. Figure 7.34 shows the basic components and state diagram for such a cycle.

A feature of combined cycles is the high cycle efficiency which can be achieved, taking advantage of the high top temperature T_3 achievable with gas turbines, and the low bottom temperature T_8 achievable with steam turbines. Many combined cycles burn natural gas in GTs, and this is an excellent way (in terms of capital cost, thermal efficiency, and starting times) to utilize natural gas (if a utility is fortunate and has such a supply). If the steam turbine is CHP instead of condensing, then the fuel utilization and operating economics can be even better.

The system arrangement of a combined cycle with district heating is shown in Fig. 7.35. In this example a number of refinements are incorporated as follows:

(1) In-duct afterburn for use mainly to maintain superheat temperature at partial GT loads.

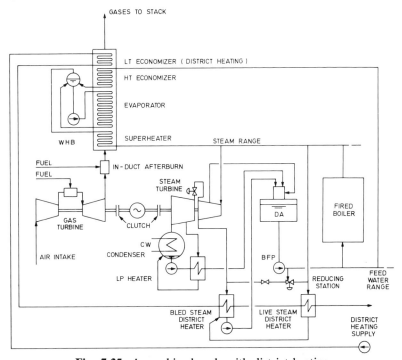

Fig. 7.35. A combined cycle with district heating

(2) A waste heat boiler with four sections, including a low temperature economizer (which is the first stage of district heating).

(3) An extraction condensing steam turbine with controlled extraction for the second stage of district heating.

(4) A common generator capable of taking the combined output of the gas and steam turbines.

(5) A clutch at either end of the generator to enable the GT or ST to operate with the other shut down. The change-overs can be accomplished on-load if the clutches are of the SSS (self-shifting synchronizing) type.

(6) One or more fired boilers for the following duties:
(a) to operate the steam turbine when the GT or WHB is shut down;
(b) to provide a third stage of district heating to meet winter peaks;
(c) to provide all the district heating when both GT and ST are shut down.

(7) Two stages of LP feed heating, including the deaerator. Note that HP feed heating is unnecessary in combined cycles, since it would reduce the heat recovered in the waste heat boiler.

Combined cycle arrangements like Fig. 7.35 can therefore provide excellent flexibility of output between electricity and heat production, which is particularly useful for small utilities who may need to achieve this at a single small generating station. The number of plant items installed varies from scheme to scheme. Some further examples are as follows:

GT	WHB	WHB Drum	ST	Generator	Fired Boilers
1	1	1	1	1 or 2	any number
2	2	2	1	2 or 3	any number
4	4	1	1	3 or 5	any number
4	—	—	1	3 or 5	1

It is also possible to retrofit an existing ST station with GTs replacing some or all of the fired boilers, and requiring no major new buildings. This is known as repowering, and provides additional generating capacity at low capital cost. The STs may already be CHP or they may be condensing sets converted to CHP.

7.2.6 Heat Stores

Heat stores are sometimes incorporated into district heating CHP schemes, using hot water as the storage medium. The value of such stores can be

308   *Energy — Present and Future Options*

twofold:

(1) To improve the economics of plant operation by enabling the CHP plant to generate at maximum electrical output when required irrespective of heat demand.
(2) To provide a standby source of heat which can be brought into service more quickly than standby boiler plant or standby CHP plant.

Heat storage usually follows a daily pattern of operations, sometimes with a different pattern at weekends. The pattern will vary from season to season and from scheme to scheme according to the characteristics of the heat and electrical demands and of the CHP plant installed. The marginal value of electricity produced throughout the day is an essential factor in determining the optimum operating pattern.

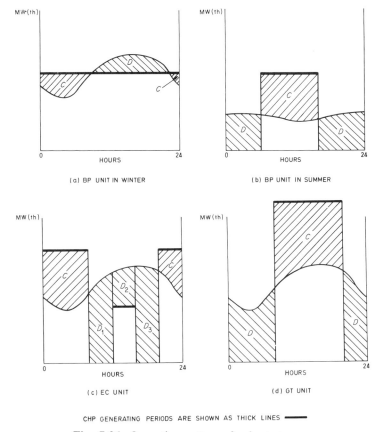

CHP GENERATING PERIODS ARE SHOWN AS THICK LINES ▬▬

Fig. 7.36. Operating patterns for heat storage

Four quite different operating patterns of heat storage are illustrated in Fig. 7.36. The heat demand pattern is plotted as MW(th) against time, and the CHP generating periods are shown as thick lines. In these simplified examples, the charge areas C exactly balance the discharge areas D on a daily cycle.

In the case of the back pressure unit in winter (Fig. 7.36(a)), generation occurs at full load for 24 hours per day, with heat store charging at night. In summer, however, only 10 hours generation at full load is required and Fig. 7.36(b) shows this occurring during the daytime (with heat store charging) with the CHP plant shutdown at night (when electrical demand is low). This is a more economic pattern of summer operation than generation at 40% unit load for 24 hours per day.

Figure 7.36(c) shows an extraction condensing unit which is required to follow the electrical demand pattern. In this example, the unit operates in the full condensing mode for two periods (D_1 and D_3) and with 50% extraction for one period (D_2). Heat store charging occurs at night with full extraction when electrical demand is low.

In Fig. 7.36(d), the gas turbine unit is required to operate during the daytime when electrical demand is near peak. In this example, heat store charging therefore takes place for about 12 hours during the daytime, as shown.

The optimum heat storage capacity to be installed is determined by balancing the expected economic savings against the capital cost of the storage system. Two to three hours' installed capacity (at peak heat demand) is more usual than six hours. Large stores (lakes, reservoirs, or aquifers) have been considered for seasonal heat storage, but none has yet been implemented in major CHP schemes. A small seasonal store exists in Sweden (Ref. 10f) but using solar heat, not CHP.

Heat stores are invariably of the displacement type, whereby hot water enters and leaves at the top and cold water enters and leaves at the bottom. Pressurized stores are suitable for virtually any temperature required by hot water consumers. They are completely full of water and can be connected directly to the main hydraulic system. A system arrangement is shown in Fig. 7.37(a).

Atmospheric pressure stores are lower in capital cost but are suitable only for storage temperatures less than 100 °C. They have a free water surface and it is usual to supply steam to the gas space to prevent oxygen ingress, creating a pressure marginally above atmospheric. To be connected directly to the main hydraulic system a tall atmospheric pressure store must be used as shown in Fig. 7.37(b). The height must be around 10.3 m per bar pressure in the return leg. Such a store therefore becomes a prominent visual feature of the station. It also serves as a head tank to control static pressure.

Where a tall atmospheric store is unacceptable or impracticable, a low-level store can be used. This is arranged as a non-integral part of the hydraulic system, as shown in Fig. 7.37(c). There must be a pressure loss at the entry

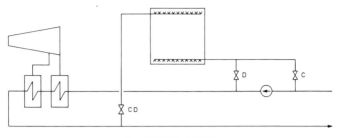

(a) PRESSURIZED STORE

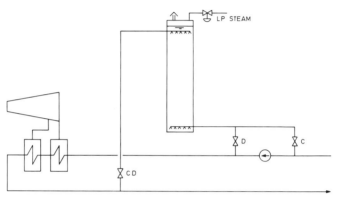

(b) TALL ATMOSPHERIC PRESSURE STORE

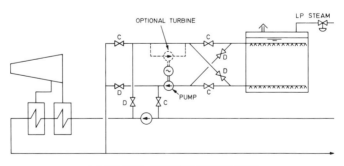

(c) LOW-LEVEL ATMOSPHERIC PRESSURE STORE

VALVES OPEN { C WHEN CHARGING
            { D WHEN DISCHARGING

Fig. 7.37. System arrangements with heat store

to the store and pressure pumps are required at the exit. Some schemes utilize pump-turbines in order to minimize hydraulic losses and to reduce the size of the electric motor. It can be seen that the system arrangement of low level atmospheric pressure stores is somewhat complex to operate and control.

The store can comprise one or several vessels arranged in series or parallel. Construction can be in steel or concrete with appropriate thermal insulation and weather protection. An important consideration is to design for a low velocity at the vessel inlet and outlet. This is necessary to minimize turbulence and mixing between the hot and cold strata, in order to maximize the heat store effectiveness, defined as follows:

$$\text{Heat store effectiveness} = \frac{\text{Practical heat storage capacity (kJ)}}{m(h_{\text{ch}} - h_{\text{d}})} \times 100\%$$

where m is the mass of water in store (kg), h_{ch} is the inlet enthalpy during charge (kJ/kg), and h_{d} is the inlet enthalphy during discharge (kJ/kg).

Figure 7.38 illustrates the temperature profiles in a heat store (with the hot/cold interface about half-way up the length *l*). A perfect store (i.e. with zero mixing) would have the rectangular shape shown, but this cannot be achieved in practice. Real profiles have a mixing band, represented by the inclined profiles. This reduces the effectiveness of the store, since the outlet temperature will start to change well in advance of the mid-point of the mixing band reaching the outlet connection. The effectiveness is also reduced by the physical gap (*1–1'*) between the outlet connections and the ends of the vessel.

The effectiveness of real stores varies in practice between about 60% and

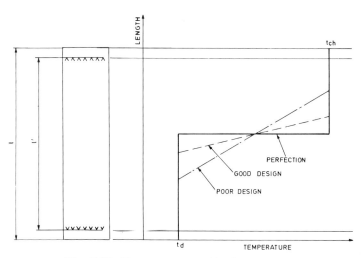

Fig. 7.38. Temperature profiles in a heat store

90%. The best designs have an efflux jet velocity of about 1.2 m/s or less, which is achieved by careful design of the distribution arrangements. A further design requirement is to keep the unusable height (l–l') to a minimum.

7.3 PRACTICAL ASPECTS

7.3.1 Ownership and Organization

Many industrial CHP schemes are privately owned and operated, being 'private' in the sense that they are commercially independent of the public electricity utility. In remote areas, this may arise out of necessity because of the high cost of an electrical transmission connection. A more usual reason, however, is that industrial CHP schemes with a high annual load factor can be highly profitable, particularly if a cheap fuel is available (such as by-products of an industrial process). Industrial schemes may also need to be fully integrated technically with the process plant, making CHP ownership and operation by an outside body impracticable.

The industrial organization may also feel that private generation is more secure in terms of either industrial relations or technical availability. Many private schemes are in any case connected electrically to the national electricity grid, either for additional security or for the import/export of marginal power.

A private industrial scheme may be owned by the firm which owns the process industry or factory. Alternatively, an industrial estate may be served by a central organization which owns and operates the CHP station as well as providing other estate services. In either case, the plant is not restricted to CHP and will often include electricity-only and heat-only plant units.

The private CHP scheme is usually developed and operated with a minimum of professional staff, sometimes a single engineer/manager in the case of small organizations. Outside consultants can be engaged for major projects or extensions, and plant contracts may be of the turnkey type, with the main supplier responsible for all (or most) of the sub-contracts. Client specifications can be short and to the point and much of the detailed engineering decision-making can be left to the supplier. The supplier is then free to include many components of his standard design, (i.e. at low capital cost) but which may not be optimal for the owner in the engineering sense. An element of trust and confidence is necessary between the industrial client and his supplier in order to strike an acceptable balance between low costs and the engineering standards adopted. The use of a firm of outside engineering consultants is intended to secure this balance.

Small CHP schemes for district heating may also be privately owned, in a similar way to the private industrial schemes already described. It is more usual, however, for major district heating CHP schemes to be owned and operated by the public electricity utility. The term 'public' is used in the sense that it serves the community as a whole but ownership can be private, municipal, or nationalized. This depends upon the political set-up of the particular country and there are examples where district heating with CHP has been developed successfully under all three types of ownership, e.g private in the German Federal Republic, municipal in Sweden and Denmark, and nationalized in the USSR and Eastern Europe.

Large public utilities supplying both heat and power are usually organized into several departments or divisions covering:

(1) Generation (of which CHP is a sub-division).
(2) Electrical transmission and distribution.
(3) Heat transmission and distribution.

In some countries, these 'divisions' are quite separate organizations. For example, in Poland the generating authority is nationalized but there is a municipally owned heat distribution authority for each city. Another example is Sweden where municipal heat and power utilities connect to the nationalized electricity transmission grid.

Further sub-divisions of organization cover research, strategic planning, development projects, day to day operation, and maintenance planning. CHP schemes are an integral part of all of these activities, but require extra attention because of the two products which have to meet the requirements of separate sets of consumers.

Large utilities will often carry their own teams of professional staff to develop CHP schemes and to manage construction projects. Alternatively, outside consultants can be engaged to handle some or all of the engineering. In either case, the plant specification is usually more detailed and more rigid than for small industrial schemes. Major CHP plants will be required to operate at high thermal efficiency for a long plant life (around thirty years). Is is therefore worthwhile in many instances to specify the plant to the clients' engineering requirements rather than to accept component designs which the supplier happens to have available. Whilst the initial cost may be higher, there will be a sufficiently long payback period to justify it economically.

Some public electricity utilities supply process steam or hot water across the boundary fence to local industry. Such schemes can have advantages of scale and thermodynamics over small private schemes because the public utility can install much larger plants (in terms of electrical output) without problems of tariff constraints (which may exist when a small private scheme wishes to export surplus electricity to the grid). The tariff for the sale of heat must be fair to both sides. In some countries the electricity industry is required by law only to 'break-even' from the sale of heat.

In centrally planned economies, tariff constraints and the need to demonstrate profit are less of a problem and large industrial consumers are generally supplied from the public generating utility. In the USSR and Eastern Europe, CHP units supplying both district heating and industrial steam are commonplace.

7.3.2 Generating Sites

The choice of site for CHP generation is influenced by the following factors:

(1) Site area available in relation to the CHP plant to be installed.
(2) Site conditions, both at the surface and below ground.
(3) Location with respect to the heat distribution network and major consumers.
(4) Proximity and ease of connection to the electricity grid.
(5) The need for electrical generation in a particular area.
(6) Site access, ease of fuel delivery and dust disposal.
(7) Availability of cooling water, particularly for EC plant.
(8) Environmental considerations, including noise control, air pollution control, and visual aspects.

The easiest sites to develop for CHP are often existing power stations which can be extended or re-developed by demolition of part or all of the old plant. The advantages of such sites are that site conditions and environmental aspects are already well-known and access routes, fuel delivery routes, grid connections, cooling water source, site ownership, etc. are already established. Development plans can therefore be prepared with minimum field-work and with minimum involvement of outside organizations. Time and costs for the investigations are therefore relatively low and planning consents can usually be obtained more readily, probably with little local opposition from a public who are used to having a generating station in the locality.

Existing power station sites may be unsuitable for CHP development if they are too distant from the heat consumer/distribution network. A bulk heat transmission system is then required, which can increase costs significantly. Few bulk transmission systems exist for distances over 10 km, although several have been proposed for nuclear stations over distances as much as 100 km. The size of the transmission system would need to be large (generally in excess of 1000 MW(th)) in order to justify such a scheme economically. Specific capital costs reduce markedly with size, since the pipe diameter is proportional to the square root of the volumetric flow. Larger diameters also realize a higher optimum flow velocity. A high Δt value ($t_s - t_r$) helps to keep size and pipe costs to a minimum. Unfortunately, a high supply temperature (t_s) also reduces the thermodynamic performance of large steam turbine CHP plants, so a compromise is necessary to determine the optimum Δt of heat transmission.

The other problem with existing power station sites (particularly in urban areas) is that they can be too constrained for the CHP plant required, in terms of site area, site access (weight and physical size of plant items), fuel traffic volume, availability of cooling water, electrical fault levels of the local network, permissible building height, and environmental impact generally. Such sites may nevertheless be suitable for CHP units of restricted number, size or type which are less than ideal for the heat supply duty required. Several such sites may therefore be used to supply a whole city. Alternatively, a small urban site can be developed as a mid-merit or as a peak load station (CHP or heat-only), or simply as a network pumping station, with the main CHP generation coming from other more distant site(s) which are suitable for development with highly efficient designs of CHP plant for base load duty.

Existing power stations in urban areas can also be considered for conversion to CHP, usually for use as temporary CHP stations during the early years of heat load growth and network construction. Many non-reheat steam turbines in the size range 20 to 160 MW(e) have been converted to BP or EC operation. Simple conversions can be achieved quickly and at low cost and can be more economic to run than heat-only boilers burning the same fuel. These are usually superseded within a decade by new, highly efficient, purpose built CHP plants upon which the economic justification of the whole district heating development was probably based in the first instance. CHP conversions are usually a temporary expedient.

New generating sites need to be considered when there is no existing power station site suitable for CHP development (or suitably located for the heat supply connection). Greenfield sites (outside the urban area) are generally the most staightforward to develop, but disused industrial sites may be situated nearer the heat load. In either case, considerable field-work is required before development plans can be completed and considerable consultation and negotiation is required with outside bodies concerning land ownership, city planning, environmental standards, access routes, fuel transport, etc. General public understanding and acceptance of the proposed development is a major consideration. Overall, the preparation and planning stage for a new site development takes considerably longer than for an existing site.

7.3.3 Selection of Plant

Heat supply schemes require several sources of heat in order to achieve continuity of supply. It is also usual to install several types of heat source having different characteristics (of cost, fuel type, performance, etc) in order to meet a variable heat load demand in the most economical way. Some of these heat sources may also be required to produce electricity irrespective of heat demand during certain periods of the day or year. The relative characteristics of different types of generating plant are summarized in Table 7.1, covering the main types of CHP plant plus fossil-fired heat-only boilers.

Table 7.1 Relative characteristics of different types of generating plant

Type of plant	Capital cost	Fuel type	Unit size	Electrical efficiency η_e	Flexibility between H & E	Start-up time from cold	Maintenance effort	Construction time
Heat only plant (fossil fired boilers)	Low	No constraint	No constraint	zero	There is no E output	Moderate	Low to moderate	Short, especially for package boilers (≤ 20 MW(th))
Steam turbine conversions	Low unless extensive re-furbishment is necessary for a prolonged life	Depends on existing boilers	Depends on existing plant	Usually poor	Depends on whether BP or EC	Moderate to long	Moderate, increasing with age	Short
Gas turbine plant with simple WHB	Moderate	Gas/distillate or HFO with penalties	Up to about 110 MW(e) 400 MW(th)	Moderate	Good only if gas bypass and afterburn are provided	Short/moderate for E and moderate for H	Moderate	Moderate
Diesel plant with simple WHB	High	HFO ≤ 420 cSt or gas with penalties	Up to about 30 MW(e) 100 MW(th)	High at all loads	Good only if gas bypass and afterburn are provided	Short for E and moderate for H	High	Moderate
Combined cycle plant GT + simple WHB + ST	Moderate to high	Gas/distillate or HFO with penalties	Up to about 250 MW(e) 500 MW(th)	High but deteriorates at part loads	Good only if gas bypass and afterburn are provided	Short/moderate for GT and moderate for ST	Moderate	Moderate
Back pressure steam turbine plant	High	Any except nuclear	200 < 1000 MW(e)—moderate to high η_e but deteriorates at part loads/ < 200 MW(e)–η_e progressively lower and even worse at part loads		None—E is tied to H unless a recooler is provided	Moderate to long	Moderate to high	Moderate to long
Extraction condensing steam turbine plant	High especially nuclear	No constraint	200 < 1300 MW(e)—moderate to high η_e < 200 MW(e)–η_e progressively lower		Excellent	Moderate to long especially nuclear	Moderate to high	Moderate to long

Other possible heat sources which are beyond the scope of this chapter, include the following:

(1) Nuclear heat-only plant (instead of CHP). Such schemes are under construction in USSR, but have not been sanctioned in other countries.
(2) Industrial waste heat, e.g. from steel mill furnaces. This can be utilized for CHP production or as a base load heat-only supply.
(3) Geothermal sources, usually instead of CHP, for district heating.

The high capital cost plant (towards the bottom of the table) would normally be selected for base load duty (i.e. continuous full load operation for most of the year), having relatively low operating costs. Plant towards the top of the list has lower capital costs but higher operating costs and would normally be selected for peaking and standby duty. Fuel availability is often the determining factor in selecting the particular plant type.

The proportion of base-load CHP plant installed is usually in the range 0.45 to 0.65 peak heat demand for district heating, or as much as 0.90 for high load factor industrial heat supplies. This proportion is known as the α value in some countries.

Figure 7.39 illustrates a plant combination with back pressure CHP plant supplying the base load and heat-only plant having the peaking and standby role. In this simple example, the CHP units achieve a running load factor of about 95%, an annual utilization of about 80%, and supply about 70% of the

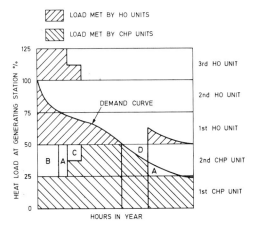

Fig. 7.39. Utilization diagram of CHP and HO units

annual heat demand quantity (after taking into account shutdown periods for planned and unplanned maintenance). The heat-only units, on the other hand, achieve an annual utilization of 15% and supply about 30% of the annual heat demand quantity. For a particular unit,

$$\text{Annual utilisation factor} = \frac{\text{Annual heat supplied} \times 100\%}{\text{Hours in year} \times \text{Heat output rating}}$$

$$\text{Running load factor} \quad = \frac{\text{Annual heat supplied} \times 100\%}{\text{Annual hours run} \times \text{Heat output rating}}$$

The heat only plant usually comprises LP (i.e. cheap) boilers supplying steam or hot water to the heat supply system. Alternatively, HP boilers can be used, supplying via pressure/temperature reducing stations. This can be the cheaper solution where the HP boilers can be used also for electrical generation purposes. The system arrangement of Fig. 7.35 is an example of the use of reducing stations instead of LP boilers.

An alternative type of peaking/standby plant is the gas turbine with waste heat boiler. This can be utilized in combination with steam turbines of the extraction condensing type to meet both heat and electrical peak demands. The large district heating system in West Berlin uses this arrangement, with no HO units at all.

The use of heat stores does not have a marked effect upon the plant ratings installed. The economic optimum storage capacity is usually small compared with the magnitude and duration of the winter peak heat demand. The more important roles of the heat store (as discussed in Section 7.2.6) are to optimize the choice of electrical generation periods and to provide an immediate supplementary heat source in the event of the sudden breakdown of a running unit.

Over a period of several decades, a major utility may install many different types and sizes of CHP and HO plant. It is therefore possible for the utilization diagram of Fig. 7.39 to comprise not five units, but ten, twenty or more, arranged in order of increasing operating costs (the order of merit). This is analogous to the order of merit generally used by electricity utilities to determine the most economic utilization of EO generating units.

7.3.4 System Connections

Having selected the types, sizes, and numbers of generating units to be installed, the method of connection to the heat supply system must be considered, together with the method of connection of the heat consumers.

Steam supply systems are nearly always connected directly to the CHP and HO units at the generating station, arranged in parallel as shown in Fig.

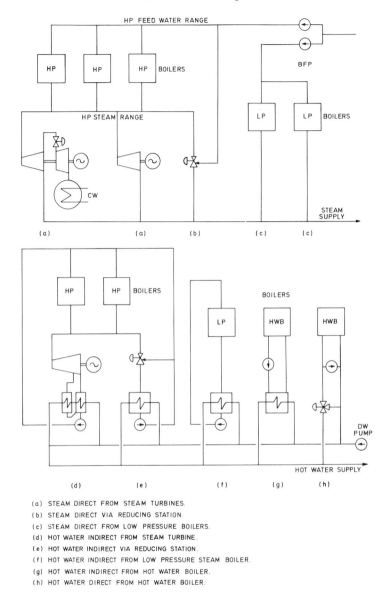

Fig. 7.40. Methods of connection at generating station

7.40(a), (b), (c). At the consumer end, the steam may be used directly (in industrial processes), or indirectly via heat exchangers.

Hot water supply systems are always connected indirectly from steam turbine CHP units, but may be connected directly or indirectly from waste heat boilers and heat-only boilers. Direct connection has cost advantages but indirect connection has design and operational advantages by segregating conditions of pressure, temperature, and water chemistry on the two sides of the heat exchanger. Figure 7.40(d), (e), (f), (g), (h) shows several methods of direct and indirect hot water connection (arranged in parallel). Hot water consumers may also be connected directly or indirectly with similar advantages and disadvantages.

Series connections are possible at the generating station for the supply of hot water. Such an arrangement is illustrated in Fig. 7.41, with the heat-only units in series with the CHP units. For peaking duty, the HO units boost the supply temperature without affecting the thermodynamic performance of the CHP units. This gives an advantage of lower operating costs. The disadvantage, however, is that the HO units are tied geographically to the CHP units. In practice, this usually means that they are located at the CHP station.

With parallel arrangements, on the other hand, operating costs are higher, but HO units can be located strategically around the city to give improved reliability of heat supply with possibly some cost savings because of smaller primary distribution pipes. Whilst the development of several generating sites is more expensive than for the same plant on a single site, there may in fact be no urban site sufficiently large to accommodate all the required plant. Local circumstances therefore influence the choice of the series or parallel method of connection.

Series arrangements are also possible at consumers' premises, having the primary purpose of reducing the return water temperature. Figure 7.42 shows two methods of achieving this at (a) domestic premises and (b) industrial premises. The three-stage domestic consumer station shown in Fig. 7.42(a) has widespread use in Sweden; however, the arrangement and operation is complex, and there are many modes of operation according to the instantaneous demands for space heating and tap water heating. The temperatures shown are typical average annual in Swedish conditions.

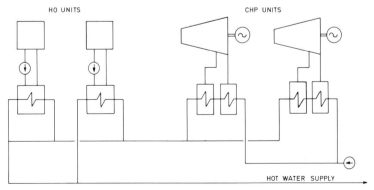

Fig. 7.41. Series connection of generating units

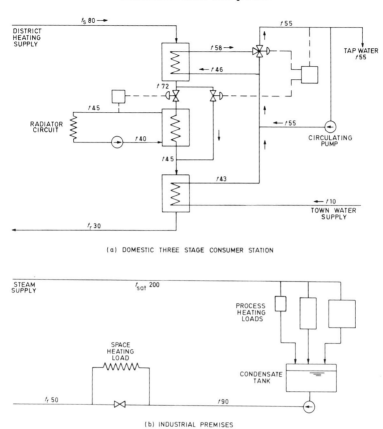

Fig. 7.42. Series connections at consumers premises

The advantages of achieving a low return temperature are that the size and capital cost of the distribution network is reduced and that the electrical efficiency at steam CHP stations is improved, resulting in low operating costs. It is important that CHP development organizations take an interest in consumer equipment design so that the overall system can be optimized. The operation of a CHP system can also be affected by consumers' equipment, e.g. contamination of return water will invariably cause problems of fouling or corrosion at the generating station, causing plant outages for cleaning and repair.

7.3.5 Operation and Control

The five operating states of generating plants are as follows:

(1) Not available, because of planned or unplanned maintenance.

(2) Available unusable, because of temporary or seasonal constraints such as electrical transmission limits or heat transmission limits or possibly low stocks of a particular fuel.
(3) Cold standby, where the plant may require 2 to 12 hours notice to be brought into service (particularly if operating staff are not on duty continuously).
(4) Hot standby, where the plant can be brought into service quickly, taking perhaps 10 to 30 minutes to reach full load.
(5) On load.

The amount of plant in each state will be reviewed periodically. There is usually an annual plan for routine maintenance and perhaps a weekly plan for the remaining categories (which must meet the expected heat demand). This will be adjusted on a daily and hourly basis to meet the actual heat load. The plant will normally be operated according to the 'order of merit' of operating costs. Some 'out of merit' operation may be required for part of the year. For example, a minimum electrical output may be required at a particular generating station because of electrical transmission import constraints (which can be technical or commercial).

The heat supply parameters controlled at the generating station are as follows:

(1) Mass flow.
(2) Pressure.
(3) Network volume and make up quantity.
(4) Chemical conditions.
(5) Temperature.

These parameters may vary independently or in combination. In the descriptions which follow, concerning the control of each parameter, some general knowledge of control principles is assumed. Control does not usually present major difficulties in the design and operation of CHP schemes. Whilst details can be left to specialist control design engineers, it is important for the CHP developer to understand the principles, since they relate also to operational economics.

Mass Flow (Hot Water)

A hot water distribution network comprises pairs of pipes (supply and return) which are installed as ring mains or branches throughout the district to be served. District heating consumers are connected in parallel from this network. Each consumer requires a certain mass flow of water to meet the heating demand. This requires a minimum design differential pressure ΔP to be maintained across the supply and return pipes in all parts of the network.

Because of pipe friction, the actual ΔP is higher near the generating station or pumping station than in the more remote parts of the network. Pipe friction (and therefore actual ΔP) also varies with time according to the mass flow demanded by the consumers. Actual ΔP must therefore be monitored continuously, and it is usual for ΔP to be measured in several critical (i.e. remote) positions in the distribution network. These signals can be telemetered back to the generating station where they can be used to adjust the total mass flow of water by automatic or manual control of the pumps, thus restoring the remote ΔP to its desired value. Figure 7.43 shows the ΔP control loop from remote measurements.

Some district heating schemes have a constant mass flow (variable temperature) operating regime, in which case the mass flow varies only under special circumstances, e.g. if some consumers shut off seasonally. With variable flow regimes, however, the consumer mass flow is controlled automatically by each consumer according to the heat requirements. This can result in significant and frequent changes in the overall mass flow (several per hour) during certain periods of the day.

District water pumps are usually electrically driven and may be of the constant or variable speed type. Constant speed pumps are lower in capital cost but incur throttling losses if fine control of the mass flow is required. Constant flow regimes may therefore control flow only coarsely, with perhaps three constant speed pumps normally in service in winter, reducing to two in summer.

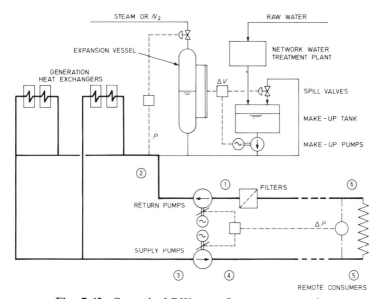

Fig. 7.43. Control of DW mass flow, pressure, volume

Variable speed pumps may be driven by a variable speed electric motor (using a solid coupling) or by a fixed speed motor via a hydraulic (variable ratio) coupling. In Figure 7.43, a hydraulic coupling arrangement is shown for automatic control of the mass flow.

An important consideration in the control of district water flow is the avoidance of pressure surges. This applies particularly to heat transmission/distribution systems having a large mass of moving water, where the change of velocity must be smooth and sufficiently slow to avoid dangerous surges. For large isolating valves in the network itself, the total time (fully open–fully shut) can be as much as 20 minutes in large systems. Safe operating times are determined by calculating the surge pressure and relating this to safe working pressures in various sections of the network.

Pressure (Hot Water)

Although the static pressure varies throughout the hydraulic system according to the mass flow, elevation of the pipework, and location of circulating pumps, it is usual for a constant base pressure to be controlled at the generating station. This can be achieved by means of an expansion vessel, as shown in Fig. 7.43, located between the return pumps and the generation heat exchangers. The operating medium, which can be steam or nitrogen, is admitted to the top of the vessel. This is a reliable method of pressure control and is generally preferred to pressure pumps.

Figure 7.44 shows the hydraulic gradients for a simple district heating system for summer and winter conditions of mass flow. There are two sets of pumps, one in the return leg and one in the supply leg, both located at the generating station. Large district heating schemes may require additional sets

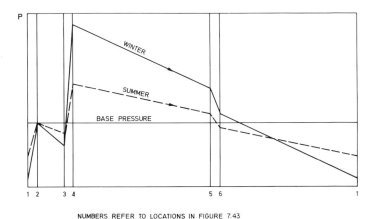

NUMBERS REFER TO LOCATIONS IN FIGURE 7.43

Fig. 7.44. Hydraulic gradients for district heating

of pumps in the supply and return legs, in order to maintain the static pressure within the upper and lower safe working limits in all parts of the system. The hydraulic gradient diagram becomes more complex with additional sets of pumps serving particular sections of the network. Differences in elevation also complicate the diagram, an increase in elevation corresponding to a loss of static pressure. Large changes in elevation arising from the topography or building height can be accommodated by means of local heat exchangers and a secondary distribution system designed for a higher base pressure.

Volume, Make-up and Chemistry (Hot Water)

Small changes in network volume (because of temperature changes) can be compensated for by changes in the expansion vessel water level (see Fig. 7.43). For larger changes, the make-up tank is used, utilizing the spill valves (out flow) or make-up pumps (in flow). Over a period of time, the make-up is always positive because of leaks, outages and, occasionally, extensions of the hydraulic system.

The water treatment plant for make-up is usually located at the generating station. The design output capacity relates to the 'worst case' make-up and may be up to double the average annual leakage rate. Catastrophic leaks, such as a major pipe fracture, are rare and water treatment make-up plants are not designed to meet such situations. Instead, the section which is leaking must be identified and isolated as quickly as possible to avoid a shutdown of the entire system. Recharging sections which have been emptied for repair and maintenance may take hours or days to complete because of limitations on the water treatment make-up rate. The initial charging of major extensions to the network takes a similar time.

The source of raw water can be town water, but a river or lake can be a much cheaper source if the water is of suitable quality and available in sufficient quantity. Sea water or estuary water sources are less common as evaporators are required, and these involve significantly more maintenance. Sometimes, no local source of raw water is available in the quantities required and a new source such as a bore-hole or a new pipeline connection has to be provided, at additional expense.

Many CHP stations have two water treatment plants, one for the HP steam cycle (which requires demineralized make-up water) and one for the network (for which softened water is acceptable). The latter comprises filtration, softening, and degassing units, and has lower capital and operating costs than the former.

Some CHP stations have a single water treatment plant for the two make-up duties, comprising filtration, demineralization, and degassing units. Combining the two plants reduces the total installed capacity required and operation is easier with one plant, but the network make-up costs are greater.

Network water chemistry is monitored and controlled at the generating station on a routine (possibly daily) basis. Control of pH and oxygen content are particularly important for the prevention of corrosion in the network. The boiler water chemistry is no less important and this is controlled to the normal standards for steam power stations.

Temperature (Hot Water)

Constant flow district heating schemes require the supply temperature t_s to be varied to meet consumer demand. One of the problems of this arrangement is that, for large systems it can take an hour or more for temperature changes at the generating station to reach the furthermost consumers. It is therefore desirable to anticipate changes in consumer demand and this can be done by monitoring the ambient air temperature t_a. Figure 7.45 shows a typical empirical relationship between t_a and t_s which has been established for a particular city. Adjustments in t_s can be made manually with a periodic review perhaps every 20 to 60 minutes according to the steadiness of the weather. In some countries, t_s is fixed for a whole day according to weather expectations, but this gives a poorer service to the consumer.

Some variable flow district heating schemes may also operate with an element of temperature variation in order to reduce operating costs. Since load response is achieved primarily by flow control, the time delay problem does not exist. Regular adjustment of the supply temperature is therefore not essential and can be reviewed daily or perhaps weekly.

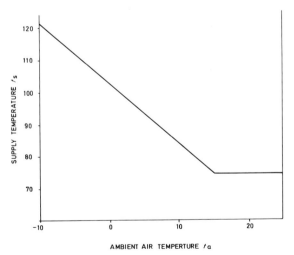

Fig. 7.45. Empirical relationship between ambient and supply temperature with constant water flow

The usual provisions for control of t_s are:

(1) Control of the bypass water flow.
(2) Control of steam flow to the heat exchangers.

These provisions are shown in Fig. 7.46(a) and (b) for back pressure and extraction condensing turbines respectively. With multistage heating, a water bypass on each heating stage is sometimes provided in order to optimize the thermodynamic perfomance. In this figure, the temperature definitions are as follows:

t_r is the return water temperature at generating station.
t_c is the outlet water temperature from CHP unit.
t_s is the supply water temperature from generating station.

The control loops shown can be automatic or manual. The set point of t_s is

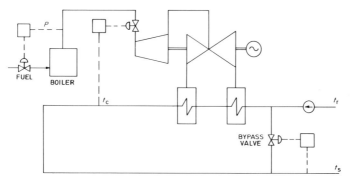

(a) BACK PRESSURE UNIT

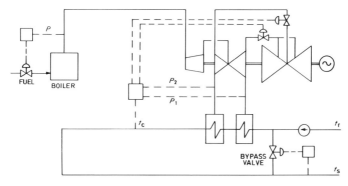

(b) EXTRACTION CONDENSING UNIT

Fig. 7.46. Control of district water temperatures

chosen according to consumer and network requirements (as discussed in previous paragraphs). The set point of t_c can be same as t_s, in which case the bypass valve will be fully shut. During periods of low heat demand, however, the return temperature t_r can rise significantly, i.e. the temperature rise Δt will be low. There is a minimum Δt that the heat exchangers can accept at full load without incurring unacceptably high tube velocities. Under these conditions, t_c will therefore be set higher than t_s in order to maintain an acceptable Δt across the heat exchangers. The water bypass control loop will therefore come into operation to control t_s. Figure 7.47 shows an extreme case of how t_r, t_s, and t_c can vary during the year, and how water bypass provision can be used.

Water bypass provisions can also be used to control t_s from waste heat boilers and from fired heat-only boilers. A second control loop senses boiler water temperature (or in the case of steam boilers, boiler pressure) and adjusts the firing rate.

Control of firing rate on steam turbine units follows load changes by means of a control loop which senses steam pressure at the boiler outlet. This loop is shown in Fig. 7.46. In the case of BP turbines, this happens every time the t_c loop operates. In the case of EC turbines, the t_c loop controls only the extraction steam flow; the operator must therefore intervene to adjust turbine inlet flow if a constant electrical output is required.

Supply Conditions (Industrial Steam)

Steam supply schemes require an immediate response to load demand changes in order to maintain acceptable conditions at the consumers' premises. Load changes are sensed as a pressure change. The pressure control loop then acts

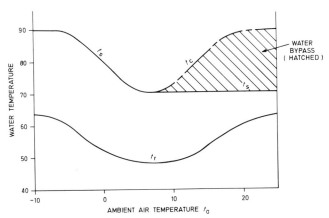

Fig. 7.47. Annual variations of district water temperatures with variable water flow

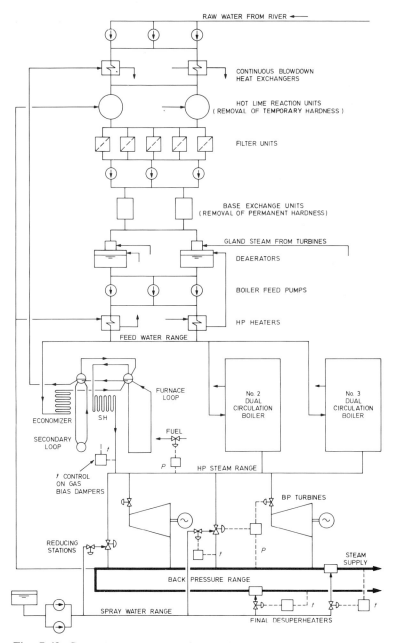

Fig. 7.48. System arrangement and control of an industrial CHP station

on the turbine inlet valve (BP turbines) or the LP turbine inlet valve (EC turbines) or the firing rate (heat-only boilers, gas turbines, and diesels). The pressure control loop therefore controls mass flow to the consumer.

Steam temperature is controlled by separate loops on individual boilers. Final trimming of steam temperature can be achieved by means of desuperheater sprays in the supply pipeline to the consumer. The water for such sprays would normally be demineralized or distillate in order to achieve acceptable steam purity and to avoid chemical deposits after the sprays. Feed heater drain water is sometimes used for this purpose, via a storage tank and spray pump.

Some industrial processes utilize the steam directly, or via heat exchangers which can contaminate the condensate in the event of leaks (e.g. fuel oil heaters). Where the condensate return quantity is low, a large make-up water treatment plant is required and this is a feature of many industrial CHP schemes. Treatment may comprise softening or demineralization units, but in either case, solids will accumulate rapidly in the boiler water.

A further feature of high make-up schemes is therefore a relatively high solids content of the boiler water, together with a high boiler blowdown rate. This may be continuous in quantities as high as 20% of steam output using softened make-up, or around 1% using demineralized make-up. Dual circulation boilers have been used in some schemes so that the boiler solids are allowed to concentrate in a low temperature boiler circuit in order to reduce the quantity of water which needs to be blown down. Heat recovery from blowdown lines is another feature of such schemes and can be used for make-up water heating.

Figure 7.48 shows the basic system arrangement and control features of an industrial steam CHP which has zero condensate return from the consumer. In this example, river water is softened, filtered, deaerated, and preheated before feeding to the dual circulation boilers which have continuous blowdown. Two BP turbine-generators are used for base load duty, with pressure reducing stations for standby/peaking duty.

7.3.6 Performance Calculations and Data

Thermal performance data are required at the planning and design stages of a CHP scheme for the following purposes:

(1) To size components for engineering design purposes, e.g. fuel and ash systems, electrical systems, make-up and cooling water systems, chimneys.

(2) To identify the logistic requirements of the station in terms of peak and average flows of electricity, heat, fuel, ash, raw water, chemicals, etc.

(3)   To provide data for an economic appraisal, particularly the fuel input, electricity output, and heat output under various operating and supply conditions.

Performance calculations are first carried out in working units of MW of energy (electricity, heat, and fuel), and a full set of input requirements to the calculations is given in Appendix A1.

These calculations must be repeated for several unit loads and possibly for several sets of supply conditions, fuel quality, and any other variable which can affect performance significantly. The results can be presented as a graph so that performance at any unit load and with any set of conditions can be interpolated.

Existing stations converted to CHP are usually treated differently in performance calculations. The utility will already know the fuel consumption for various unit loads and operating conditions. All that is necessary, therefore, is to calculate loss of station electrical output ΔE arising from operation in the CHP mode, and the heat output H. The term 'Z factor' is sometimes used for the expression $\Delta E/H$ with respect to stations converted to CHP.

The next step (for either new and converted units) is to determine the annual duration of unit operation at particular loads and with particular operating conditions. Take, for example, the two CHP units shown in Fig. 7.39 (which are back pressure steam turbines for the purposes of this exercise). The annual operating regime can be summarized as follows (per CHP unit):

	Nominal unit load	Hours run p.a. per unit	Supply conditions
Summer base load	100%	1300	t_s70
Winter base load	100%	3500	t_s90
Restriction C	75%	400	t_s90
Restriction D	85%	720	t_s80

Note that these are nominal unit loads. Precise outputs will be determined from the performance calculations. For example, if the design supply temperature is $90\,^\circ\text{C}$, then the summer base load (at $70\,^\circ\text{C}$) may provide around 97% (heat) and 105% (electrical) relative to the output at design conditions. The electrical equipment must be designed to take this high output, of course, otherwise all energy flows on the unit must be reduced to restore the electrical load to 100%.

The need to overdesign components for off-design operating conditions also applies to other parts of the unit, e.g. the fuel system. A well-designed station is consistent in the extent to which components are overdesigned for off-design operation.

It is important to use realistic values of non-availability due to planned and unplanned maintenance (Fig. 7.39 refers). These should relate to operating records of similar plant items and to the arrangement of associated systems and connections. Non-availability values should not be reduced unless there are justifications in terms of engineering design or in terms of improved maintenance effort or improved maintenance planning.

Any restriction in station operation due to external factors (such as electrical transmission constraints or seasonal shortage of a particular fuel) must also be taken into account when working out the annual durations of operation.

The annual unit inputs and outputs of fuel, heat, and electricity can now be calculated. There are two components, on-load energy and off-load energy. On-load energy is obtained by summing the product of hours run and MW of energy. That is

$$N_{on} \text{ (GJ)} = \Sigma \frac{[MW \times \text{Duration (hours)}]}{3.6}$$

Annual off-load energy (N_{off}) is the summation of allowances for energy used when shut down and during unit starting. It is therefore necessary to predict the number of hot and cold starts per year and annual durations of shutdown periods, together with realistic values of the energy consumption appropriate to each.

The annual energy flows of the station are therefore

$$N_{station} \text{ (GJ)} = N_{on} + N_{off}$$

The calculation can cover CHP operation only, but it is more usual to include heat-only operation, in other words to relate all performance data to the heat supply service. The summation can cover one generating station or a group of stations serving a common heat distribution network. The calculations may need to be repeated for expected demand growth and for commissioning of new CHP units.

In the case of CHP units which are also capable of electricity-only operation, additional performance calculations are necessary. The durations of EO operation will relate to the electricity demand characteristic and the order of merit of electricity generating units. It is usual to show the EO annual energy inputs and outputs as separate items. This gives a clearer picture of performance than to aggregate them with the pure CHP and HO data. The condensing component of extraction condensing units can be treated in this way.

The full performance calculations so far described are essential for a full assessment of energy flow values for engineering design, station logistics, and economic appraisal purposes. At the beginning of an investigation, however, it is useful to be able to perform quick performance calculations—indeed, much of the detailed information is simply not available at the beginning.

Quick calculations can be done by using the expected full load performance data of the main plant (e.g. turbine-generator and CHP boiler) and then multiplying by an appropriate 'performance factor' based upon past experience with plant of a similar size, type, and duty. This overall 'performance factor' can therefore take account of many small factors such as d_p, d_b, L_h, L_m, L_g, C_s, C_f, C_t, and EW listed in Appendix A1. Off-load energy losses and part-loading effects can also be incorporated into the overall factor. This method is quick, but not as accurate as the full calculation.

The purpose of quick performance calculations is to provide an early indication of the likely operating costs. Different sizes and types of CHP plant can therefore be assessed at an early stage in order to narrow the field to a few plant options to be evaluated in detail.

A useful yardstick for comparing the performance of different CHP plants is thermal efficiency, which is the ratio of input and output energy values defined as follows:

$$\text{Electrical efficiency } \eta_e = \frac{\text{Electrical output}}{\text{Heat input}} \times 100\%$$

$$\text{Combined efficiency } \eta_c = \frac{\text{Electrical + Heat output}}{\text{Heat input}} \times 100\%$$

These can relate to instantaneous (MW) performance or to annual (GJ) energy values. Both electrical and combined efficiency are commonly quoted. Combined efficiency values of 65 to 85% are possible. These can be impressive to an outsider, but they are of little real use to the utility. Any heat-only boiler can achieve a 'combined efficiency' of 80 to 90%.

To the CHP developer, electrical efficiency is the more useful and meaningful value. A principal objective of CHP is to supply as much electricity as possible for a given heat supply. Some utilities prefer to work in terms of H/E or E/H ratio (instead of η_e) as the CHP performance indicator.

A word of warning is necessary when comparing thermal efficiency values calculated by different utilities, plant suppliers, etc. Definitions can vary widely according to the needs of particular organizations. Apparently large differences in thermal efficiency may turn out to be false when the definitions have been examined. Some of the options are listed in Appendix A2.

In Appendix A3, tables of some typical performance data are given for different types of CHP plant and supply conditions. The data are for guidance only, and variations are possible within each example. A consistent set of boundaries and operating conditions is used for these examples, as listed in the appendix.

7.3.7 Overall Programmes

A realistic programme is required in order to identify key dates and time periods for the initial planning, plant ordering and supply, site construction and station commissioning. The programme affects the incidence of capital expenditure (which can incur considerable interest charges prior to commissioning) and the timing and magnitude of revenue earning capability (i.e. the commissioning dates of successive generating units).

The programmers should avoid bottlenecks of activity, identify potential causes of delay, and organize corrective action to overcome the problems which can arise. The utility may manage the whole programme itself, or may delegate the management to a consultancy organization or a turnkey contractor.

The overall programme for a CHP scheme must relate to the construction of the heat distribution system and to the growth of the connected heat load (even if the CHP authority is not responsible for distribution). Small industrial schemes can be planned and fully commissioned within a decade, whereas district heating in large cities can take fifty years or more to reach full development, with distinct phases of development for HO and CHP plant. Two examples will now be considered.

Example 1—Small Industrial Scheme

Two diesel units with waste heat boilers are to be installed to supply process steam to an existing industrial complex, with no change in heat requirements. Existing HO boilers are to be retained for standby/peaking purposes. Figure 7.49 shows the main elements of the programme which has an overall time span of 44 months (from the start of initial planning to full commissioning).

Example 2—Major District Heating Scheme

A large district heating network is to be constructed. Starting with a few existing group heating schemes in year 1, the connected heat load is expected to grow from 20 MW(th) to around 500 MW(th) in year 16, with further, unspecified, growth thereafter. Figure 7.50 shows the overall programme.

The phases of development can be identified as follows:

(1) Initial planning, consents and design concepts.
(2) Network construction in several zones of the city, supplied from temporary HO stations.
(3) Interconnection of several sub-networks and commissioning of Station 'A', comprising four new 50 MW(th) heat-only units.

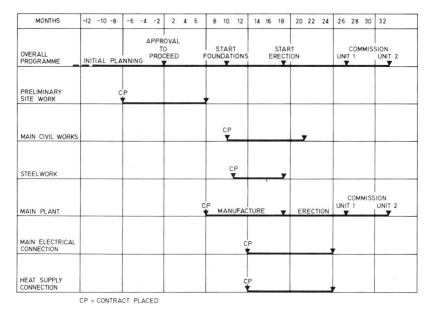

Fig. 7.49. Overall programme—small industrial scheme

(4) Commissioning of temporary CHP Station 'B', comprising three existing 60 MW(e) turbines converted to CHP operation (heat output 3 × 75 MW(th)).

(5) Commissioning of a new (permanent) Station 'C', comprising, in the first instance, one CHP unit rated at 300 MW(th) and two HO units rated at 150 MW(th). Space is left for future extensions as the heat load continues to grow.

The principal objective of phased development in district heating is to utilize newly-constructed sections of the network as soon as possible (using the cheapest available heat source) in order to obtain an early return on the capital investment in the network. The heat sources chosen will vary from city to city, and some of the phases described above may be missing.

The rate of district heating load growth per city varies from 10 to 100 MW(th) per annum in Western Europe. Higher growth rates have been achieved in East European cities, Moscow achieving over 1000 MW(th) per annum at peak. The rate of development can be limited by:

(1) The rate at which a utility is able to invest capital.
(2) The extent of environmental disturbance which is acceptable to a community (in some European cities devastated in the Second World War, no additional disturbance was caused by the start of network construction).

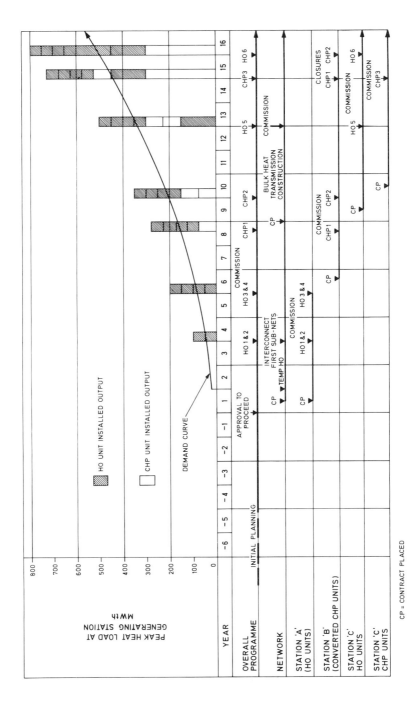

Fig. 7.50. Overall programme—major district heating scheme

(3) The willingness of potential consumers to connect (this is influenced by the availability and cost of other methods of space heating).

Because of these constraints on the rate of network construction and heat load growth, it can take a decade or more before the first major CHP unit is commissioned. Construction will then continue (with further large CHP units being commissioned) until saturation is approached. This can take two or three decades in a small city (less than 500 000 population), but over fifty years in large European cities.

7.3.8 Economic Aspects

For a CHP scheme to be economic, the savings in fuel and other variable costs must balance or exceed the higher capital and other fixed costs.

The fuel cost savings which can be achieved may vary in time because of the following factors:

(1) CHP fuel prices and CHP plant mix.
(2) EO fuel prices and marginal EO plant efficiency.
(3) HO fuel prices for individual heating systems.

To achieve the greatest fuel cost savings, the CHP plant should be designed to burn a cheap fuel (relative to HO and marginal EO plants), and to achieve a high thermodynamic performance. This means that the electrical efficiency and, to a lesser extent, the combined efficiency (heat plus electrical output) must be as high as possible. Typically, the economic value per MW of electrical output is at least three times that of the heat.

The means of achieving high electrical efficiency in CHP stations are discussed in Section 7.2, and performance data are summarized in Appendix A3. Steam turbines would normally be selected for burning coal or nuclear fuels, a gas turbine combined cycle is usually best for natural gas and diesels are ideal in small unit sizes for burning heavy fuel oils. Most CHP schemes already in service today are based upon steam turbine plant, burning coal, heavy oil, and other fossil fuels.

The CHP plant mix can vary with time. As discussed in Section 7.3.7, the early phases of development can be heat-only and/or CHP conversions, neither of which has sufficiently low operating costs to justify the high capital cost of heat distribution. The main (high performance) CHP plant may be vital to the overall economics, but its commissioning may be deferred until a later phase of development.

Marginal EO fuel costs can be computed knowing the characteristics of the national electrical demand and of the EO system, which is the electrical output/availability/fuel cost relationship for each generating unit connected to the electrical grid system. This is relatively easy for the present-day and in the short-term, but it becomes increasingly speculative further into the future (ten to thirty years ahead). Electrical demand levels are affected significantly by the economic and industrial activity of the nation, which in turn is affected by world-wide recessions, restrictions, and booms.

The EO plant installed may also be difficult to predict in the longer terms, being dependent upon the plant ordering and general energy policies of government as well as of the utility concerned. An all or nothing situation can arise on certain issues, for example whether a nation should proceed with major programmes of nuclear or hydro development.

To overcome these problems of prediction, marginal EO fuel costs may therefore be determined for different scenarios of electrical demand level and EO plant installed. The proposed CHP scheme may turn out to be economic with any of the EO scenarios but, where this is not so, the proposed scheme can be subject to delay until a clearer picture of national energy policy and future electricity demand emerges.

For district heating and associated CHP plants, it is appropriate for the appraisal to be carried out over a period of twenty to thirty years or more. The total capital investment is high, but the market for heat is relatively stable and assured, and therefore a long appraisal period is acceptable. Industrial CHP schemes, however, are usually appraised over a much shorter period, often less than ten years. The problems of predicting marginal EO fuel costs are therefore less pertinent to industrial CHP schemes.

Existing HO fuel prices and systems can be established relatively easily and, for comparison, a proposed CHP supply service usually needs to undercut existing services by some 20% (in total annual charges) in order to persuade consumers to change.

Consumer heat loads and the heat load densities of cities can be determined by survey. In the case of district heating, successive economic appraisals are required to determine the extent to which it is worthwhile developing the network. High density load areas are the most economic. The minimum economic load density usually lies in the range 10 to $20 \, MW/km^2$ at present, but can be expected to reduce as fuel prices escalate in real terms. Cities with a high proportion of apartment blocks generally achieve a large high-density load area.

In order to predict the total connected heat load, surveys must establish the proportion of consumers willing to change to the CHP supply, particularly in the high density load areas. Corrections may be introduced to take into account future conservation measures in consumers premises (e.g. building insulation).

Heat losses from the transmission/distribution system must be added to the total connected load in order to determine the heat load at the CHP station. As a percentage of peak demand at the CHP station, these losses can amount

Table 7.2 Items to be included in capital cost calculations

Main generating plant	Turbine-generators
	CHP boilers
	HO units
Auxiliary mechanical plant	Fuel handling
	Water treatment and
	make-up
	Auxiliary cooling
	Pipework connections
	Lifts and cranes
	Workshop equipment
Auxiliary electrical plant	Step-down transformers
	Switches
	Cables
	Control and instrumentation
Civil works	Site preparations
	Foundations
	Buildings
	Culverts
	Chimneys
	Jetties or railways
Main electrical connections	Step-up transformers
	Switches
	Transmission lines
	Control and protection
Heat transmission connection	DW pumps
	Pipework

to a few per cent with industrial schemes, or some 5 to 20% with district heating. In summer, district heating system losses can exceed the heat sold, and some systems shut down in midsummer (with other provisions made for tap water heating).

Capital costs can be determined by analysing all the elements of professional services, station construction, and supply of plant and equipment which are necessary for completion of the station. Some of the important headings are shown in Table 7.2.

For an overall appraisal of the heat supply service, the capital cost of heat transmission, distribution and consumer equipment would be added to the list in Table 7.2.

The precision with which these costs can be estimated will depend on the extent to which suppliers have been involved, also the extent to which systems have been finalized and the knowledge of site conditions. Usually, a contingency sum is added, either globally or to the individual categories above, to cover for unknown factors.

The accounting method for the economic appraisal varies from organization to organization. It is usual to prepare a cash flow tabulation for individual years, covering all capital and revenue items of expenditure and income (including any taxes). The 'discounted cash flow' technique can then be used to bring the individual yearly cash flows to a common date and to express the profitability in terms of the 'pay-back period' or the 'return on capital'.

7.4 WORLD SITUATION AND CONCLUSIONS

7.4.1 Historical Background, Present Status and Potential for CHP

Commercial CHP schemes have existed on a small scale since the late nineteenth century. The New York steam supply system started in 1882, and Hamburg in 1894.

The first phase of major district heating development took place in the period 1922 to 1939. Today's major hot water systems of Moscow, Leningrad, West Berlin, and Hamburg were developed initially during this period. Schemes were also started in Hungary, Czechoslovakia, Netherlands, Canada, Denmark, and France.

Most district heating development, however, has occurred since 1950, including all the systems in Sweden, Finland, and Poland, together with major extensions and new schemes in many other countries. These are predominantly in the central, eastern, and northern parts of Continental Europe. District heating with CHP continues to grow in these countries whilst other countries (with little DH at present) are now investigating carefully the merits of large scale developments.

Global statistics on CHP and heat supply schemes are difficult to acquire. The World Energy Conference has attempted surveys, by sending questionnaires to individual countries. In many cases, the replies are limited by a lack of national statistics or a suitable organization for data collection. Information on industrial schemes is particularly difficult to acquire. Nevertheless, some data have been obtained, and Table 7.3 shows key data for twelve CHP countries. This table is not, however, complete as there are several important CHP countries missing (owing to national data being not readily available at the time of the survey). These include the United States, Canada, East Germany, Hungary, Romania, and Czechoslovakia.

The kW(e) per capita column in Table 7.3 is chosen here as a criterion of the extent of national CHP development. Other, equally valid criteria would be the annual GJ of CHP energy produced (either heat or electricity or combined) and the percentage of the national heating load served by CHP (with or without the HO component).

Table 7.3 Key data for twelve CHP countries (1975)

Country	Population M	Number of CHP schemes*	CHP installed capacity E* MW(e)	H* MW(th)	kW(e) per capita	Climate (Predominant)	Type of scheme Per cent of E Industrial*	DH*	DH + Industrial*
Finland	5	83	2 249	7 897	450	Continental	53	35	12
Denmark	5	26	2 093	4 810	419	Sub-Continental	4	75	21
USSR	246	900	59 200	206 000	241	Continental	—	—	100
W. Germany	62	—	14 520	—	234	Sub-Continental	59	41	—
Sweden	8	246	1 818	—	227	Sub-Continental	—	100	—
Bulgaria	8	25	1 549	5 220	194	Continental	45	—	55
Poland	36	260	3 940	—	109	Continental	46	44	10
Hungary	10	62	834	3 058	83	Sub-Continental	10	—	90
Japan	118	349	7 720	—	65	Continental/Humid Sub-Tropical	100	—	—
UK	57	179	2 721	—	48	Oceanic	99	1	—
France	55	153	1 847	6 435	34	Oceanic	94	6	—
Spain	37	100	454	—	12	Arid/Mediterranean	100	—	—

* These data are from Reference 1. Reproduced by permission of the World Energy Conference

In terms of kW(e) per capita, it can be seen that Finland and Denmark have the greatest concentrations of CHP. The greatest capacity, however, is achieved by USSR. In fact, the USSR possesses about half of the total world CHP capacity. Moscow is the world's largest CHP city, with over 5 000 MW(e) of installed capacity.

Climate is a major influence on the extent of national CHP development, and Table 7.3 shows a broad relationship between climate and per capita CHP capacity. The top eight countries in Table 7.3 all have winters where a continental high pressure system prevails. The long, cold winters give rise to a high space heating demand which favours the economics of district heating, even with the higher standards of building insulation which are also associated with the colder winter climates.

The bottom four countries (except nothern Japan) have much milder winters because of oceanic depressions or sub-tropical influences. The lower space heating demand makes district heating more difficult to justify economically, and nearly all of the CHP in these countries is industrial.

It should be noted that significant industrial CHP capacity exists in all twelve of the countries, being a function of national industrial development rather than the climate. Some countries, notably Hungary, USSR, Bulgaria, and Denmark, supply industrial heat and district heating from the same generating units.

The Regional and World potential for CHP is shown in Table 7.4, as published in the World Energy Conference Report.[1] The results are broad estimates based upon the assumptions listed. The principal limiting factor to the rate of CHP development in all the Table 7.4 regions is the availability of capital.

7.4.2 Energy Conservation

The World Energy Conference Report[1] estimated in 1978 that the potential fuel saving which could be achieved by maximum CHP development (as defined in the assumptions quoted in Table 7.4) is around 6% of the 1975 world consumption of primary energy. Many industrialized countries have made independent estimates in the range 5 to 10% for national fuel savings by the maximum development of CHP.

The overall figures quoted above do not differentiate between grades of fuel. With centralized heat production, it is possible to burn low grade fuels, replacing high grade fuels burned at individual premises. Low grade fuels offer not only price advantage, but often a strategic advantage, for example by replacing imported oil with indigenous coal. In economic and strategic terms, a nation may therefore regard the importance of CHP energy conservation to be greater than the 5 to 10% primary energy saving, particularly if a reduction in fuel oil imports is obtained.

Table 7.4 Potential for CHP electricity production, 1975

Regions	GNP (1) 10^9 US $	Primary energy consumption (2) EJ	Electricity consumption (3) TWh	Space heating energy (4) EJ	Industrial process steam (5) EJ	Potential for CHP electricity production TWh				Present electricity by CHP	
						(6) DH	(7) Ind.	(8) Total	(9) % of 3	(10) TWh	(11) % of 8
North America	1680	75	2270	13.7	11.3	680	390	1070	47	60	6
Europe (excl. USSR)	1970	63	1910	12.4	13.3	620	460	1080	57	135	13
USSR	650	42	1040	2.9	6.2	145	215	360	35	185	51
Japan	500	12	480	1.9	3.3	55	115	210	44	35	17
Other regions	1290	43	740	Not appl.	8.7	0	300	300	41	15	5
Total	6090	235	6440	30.9	42.8	1540	1480	3020	47	430	14

District heating assumptions

Total space heating demand (N_1) (PJ per annum) is as given by each country in the questionnaire replies (or 7 GJ/1000 US $ GNP where no reply was given).

Degree of urbanization = 67% (world average)
Urban DH penetration = 85%
Proportion of annual heat supplied by CHP = 0.7
CHP ratio E/H = 0.45
Potential CHP electricity production
= $N_1 \times 0.67 \times 0.85 \times 0.7 \times 0.45$
= 0.18 N_1 PJ per annum
= 0.05 N_1 TWh per annum

Industrial heat assumptions

Total industrial steam demand (N_2) (PJ per annum) 6.7 GJ/1000 US $ GNP for the world as a whole except USSR for which the figure is 9.4 GJ/1000 US $ GNP.

Proportion of annual heat supplied by CHP = 0.5
CHP Ratio E/H = 0.25

Potential CHP electricity production
= $N_2 \times 0.5 \times 0.25$
= 0.125 N_2 PJ per annum
= 0.035 N_2 TWh per annum

Centralized heat-only generation also conserves fuel and has the same capability as CHP to burn low grade fuels. Appendix B gives two worked examples of energy conservation calculations, one for district heating and one for an industrial heat supply. In both examples, the important strategic saving of fuel oil can be achieved by virtue of installing central coal-burning boilers. The contribution of CHP is to conserve coal rather than oil.

7.4.3 Conclusions

(1) The technology of combined heat and power is already established for a wide range of fuels, and has been proven commercially in many countries.

(2) Forms of ownership and organization vary from country to country. Large distribution systems and concentrations of CHP exist under both public and private ownership. Climate is a major influence affecting district heating CHP development, whereas industrial CHP is a function of industrial development rather than climate.

(3) Central heat generation can save significant quantities of fuel, often replacing high grade with low grade fuels and can contribute to reductions in air pollution. Combined heat and power has a greater conservation effect than heat only generation, but either can be used to reduce high grade fuel consumption and urban air pollution.

(4) Incentives to proceed with CHP development are economic, strategic, environmental, and energy management. The main constraint is the availability of capital, even where schemes are clearly economic.

(5) The world has fulfilled some 17% of the estimated potential for economic CHP generation. The world potential for energy conservation by this means is about 6% of 1975 primary consumption.

(6) It is technically feasible to utilize reject heat from steam condensing stations for low temperature uses such as horticulture and fish farming, but only a few pilot commercial schemes exist as yet. A large potential for low temperature schemes remains to be exploited.

7.5 NOTATION AND DEFINITIONS

Nomenclature

C	heat store charge	J(or multiple)
d	deterioration factor	
D	heat store discharge	J(or multiple)
E	electrical power	MW(e)
EW	works electricity	MW(e)

F	rate of fuel consumption	MW_f
h	specific enthalpy	kJ/kg
H	thermal power	MW(th)
HW	works heat	MW(th)
l	length	m
L	loss (of heat or mass)	J or kg/s
m	mass flow	kg/s
N	energy	J(or multiple)
P	absolute pressure	bar abs
S	specific entropy	kJ/kg K
t	temperature	°C
ttd	terminal temperature difference	K
T	absolute temperature	K
V	volume	m³
W	rate of work done	MW
η (eta)	thermal efficiency	%

Multiples

k	kilo	10^3
M	mega	10^6
G	giga	10^9
T	tera	10^{12}
P	peta	10^{15}
E	exa	10^{18}

Suffices

a	ambient air	m	mass of steam/water
b	boiler	max	maximum
c	combined	off	off load
ch	charge	on	on load
cond	condensing	p	prime mover
d	discharge	r	return
dh	district heating	s	supply
e	electricity	sat	saturation
f	fuel	sh	superheat
fg	evaporation	so	sent out
g	gas	t	transient
gen	generated	th	thermal
h	heat	w	water
in	inlet to prime mover		

Abbreviations

AB	afterburn	GT	gas turbine
BMEP	brake mean effective pressure	H	heat
BFP	boiler feed pump	HFO	heavy fuel oil
BP	back pressure	HO	heat only
CEP	condensate extraction pump	HP	high pressure
CHP	combined heat and power	HT	high temperature
CW	cooling water	HWB	hot water boiler
DA	deaerator	IP	intermediate pressure
DH	district heating	LP	low pressure
DL	diesel	LT	low temperature
DW	district water	NCV	net calorific value
E	electricity	NG	natural gas
EC	extraction condensing	PF	pulverized fuel
EO	electricity only	SH	superheater
F	fuel	S.O.	sent out
FL	full load	ST	steam turbine
GCV	gross calorific value	WHB	waste heat boiler
Gen	generated		

Definitions

Combined Heat and Power (CHP): the simultaneous production of electric power and heat from a single generating plant. The term 'cogeneration' is more usual in North America. In other languages the terms 'combined generation' and 'combined production' are also used and 'heat-power-coupling' is often used in Germany.

District Heating: a public supply of heat, usually as hot water, but sometimes as low pressure steam. Known as 'distance heating' in German and Scandinavian languages, and as 'urban heating' in French.

Group Heating: the supply of heat to a large building or group of buildings, but not developed as a public supply to a whole district.

Industrial Heat Supply: the supply of heat for industrial purposes, usually as steam, but sometimes as hot water.

Low Temperature Heat Supply: the supply of heat at or near the normal reject temperature of an electricity generating plant. Also known as Low Grade Heat supply.

Total Energy: this is an ambiguous term. It was originally used by the American gas industry to market schemes which would meet the total energy requirements of an industry using a single fuel. In 1971 J. P. Harmsworth[8] reported that the term had come to mean 'on-site power generation with heat recovery', i.e. a small CHP scheme which is capable of operating independently of the national electricity supply system. Today, the term is sometimes used to refer to any CHP scheme, but the Harmsworth definition is more usual and should be regarded as strictly correct.

7.6 APPENDICES

Appendix A—Thermal performance

A1. List of Input Requirements for Full Thermal Performance Calculations

In the following general approach, the working units are MW of energy (electricity, heat, and fuel) and relate to pure CHP operation.

Design performance of prime mover
Heat input $\quad\quad H_{in}$
Electrical output E_{gen}
Heat output $\quad\quad H_{gen}$
using full load data which would be guaranteed commercially by the manufacturer, or other data at part loads.

Deterioration factor of prime mover (d_p) due to erosion, corrosion, fouling, etc., which can be estimated for an average lifetime condition.

Losses between prime mover and boiler which may have been excluded from the main plant performance guarantees.

Loss of pressure (ΔP) due to friction. High values can cause GTs and diesels to be downrated.
Loss of heat (L_h) due to convection/radiation from steam pipes, feed-water pipes, or gas ducts.
Loss of mass (steam/water) (L_m) due to leaks, boiler blowdown, and auxiliary uses not accounted elsewhere.
Note that an appropriate value of enthalpy rise Δh must be used with respect to make-up water.
Loss of mass (gas) (l_g) due to gas bypass (used in some GT and diesel schemes)

Design boiler efficiency (η_b) using guarantee data at full load, or other data at part load.

Deterioration factor of boiler (d_b) due to poor combustion, fouling, etc., which can be estimated for an average lifetime condition.

Correction factors for off-design operating conditions, for example C_s for off-design supply temperatures, C_f for off-design fuel quality, and C_t for transient combustion conditions during load changes.

Fuel input to boiler

for ST schemes
$$F = \frac{H_{in} \times L_h \times L_m}{\eta_b \times d_b \times C_f \times C_t}$$

for GT and diesel schemes $\quad F = \left[\dfrac{H_{\text{in}}}{C_{\text{f}}} \right]$ plus any afterburn

Works electricity (EW) to include the following:

(a) Unit auxiliaries required in service at the particular unit load, corrected for average lifetime deterioration (e.g. pump performance).
(b) An allowance for station works electricity, for example fuel handling, lighting, water treatment, workshops, amenity buildings.
(c) On-site DW pumps (off-site pumps, DW or condensate return, may also be included if an overall appraisal is required).
(d) All transformer losses.

Electricity sent out (E_{so})

$$E_{\text{so}} = (E_{\text{gen}} \times d_{\text{p}} \times C_{\text{s}}) - EW$$

Works heat (HW). Taken from the final heat supply line for on-site uses such as fuel oil heating and space heating of ancillary buildings.

Heat supplied (H_{s})

for ST schemes $\quad\quad\quad\quad H_{\text{s}} = H_{\text{gen}} + E_{\text{gen}} (1 - d_{\text{p}}C_{\text{s}}) - HW$
for GT and diesel schemes $\quad H_{\text{s}} = [H_{\text{gen}} + E_{\text{gen}} (1 - d_{\text{p}}C_{\text{s}}) - L_{\text{h}} - L_{\text{g}}] \, \eta_{\text{b}}d_{\text{b}} - HW$

The above calculations must be repeated for several unit loads and possibly for several sets of supply conditions, fuel quality, and any other variable which can affect performance significantly. The results can be presented as a graph so that performance at any unit load and with any set of conditions can be interpolated.

A2. Options of Boundary and Operating Conditions Used in Performance Calculations

Electrical output boundary generator terminals or station terminals or sometimes consumer terminals? It is particularly important to know whether works electricity has been deducted, and whether this includes all works electricity or perhaps just that which is consumed on-load by the CHP plant. Is DW pumping power included, on site and remote?

Heat output boundary CHP plant connections, or station boundary or consumer connections? Is the output of heat-only plant included? Have off-load heat losses and station heat uses been deducted?

Fuel GCV or NCV basis? Does the figure include all fuel used by the station or have deductions been made for such items as off-load consumption, condensing operation, heat-only operation? Has it been apportioned in some way between electricity and heat production? Has it been measured or calculated? If calculated, have station radiation and make-up losses been included?

Conditions Is combustion efficiency inherently poor due to low grade fuel quality? Is the plant condition design, as new, or average lifetime? Do the data relate to full-load operation or to a realistic operating regime, for example what running load factor and how many starts per annum? What are the supply conditions (e.g. water temperatures) and are they design or average annual values?

A3. Tables of Typical Performance Data

Some typical performance data are given in Tables 7.5 to 7.8 for different types of CHP plant and supply conditions of district heating and process steam. The data are for guidance only, and variations are possible within each example. Boundaries and operating conditions are applied consistently as follows:

Plant
  Purpose-built CHP units.
  Combined cycles use back pressure steam turbines.
  No gas bypass on GT and diesel units.
Electrical output boundary
  Station terminals. Works electricity deducted includes on-load CHP consumption, station uses such as fuel handling and on-site DW pumping.
Heat output boundary
  CHP generating station boundary.
Fuel
  Total on-load consumption of CHP plant.
  No EO or HO operation.
  GCV basis of accounting.
  Fuel grades not particularly difficult to burn.
Conditions
  Units running at full load in pure CHP mode.
  Plant in average lifetime condition.
  Losses L_h and L_m taken into account.
  Supply conditions as stated in examples.

Appendix B Worked examples to Calculate Energy Conservation

B1. District Heating Example

Stage 1—a central coal burning HO station replaces domestic boilers of which 50% were coal burning and 50% were oil burning.

Consumer heat requirement = 250 PJ p.a.
If

η_{coal} (domestic)   = 55% on an annual basis

η_{oil} (domestic)   = 65% on an annual basis
and

η_{coal} (central HO)   = 75% (including distribution losses)

Table 7.5 Performance of large steam turbine units

Cycle top conditions:	$P175$, $t535$ with reheat				
Fuel: coal, PF fired.	GCV	22 500 kJ/kg			
	Ash	23%			
	Moisture	11%			
	Volatiles	25%			

Case	Supply conditions	Electricity		Heat S.O. MW(th)	η_e S.O. %	η_c S.O. %
		Gen MW(e)	S.O. MW(e)			
1	$P\,0.050$ Condensing	250	227	—	35.5	—
2	$t_s\,90 + t_r\,50$ 2 stages DH	215	192	352	30.0	85
3	$t_s\,90 + t_r\,50$ 1 stage DH	209	186	358	29.1	85
4	$t_s\,130 + t_r\,90$ 2 stages DH	207	184	360	28.8	85
5	$t_s\,130 + t_r\,90$ 2 stages DH	204	181	363	28.3	85
6	$t_s\,170 + t_r\,90$ 2 stages DH	200	177	367	27.7	85
7	$P15\ t \leqslant 400$ steam	132	112	435	17.5	86

Note: The electrical efficiency deteriorates as supply conditions are raised, but the combined efficiency varies little.

Table 7.6 Performance of industrial steam turbine units

| Cycle top conditions: | $P85$, $t510$ | | | | |
| Fuel: | Coal, mechanical stoking, analysis as for large ST units. | | | | |

Case	Supply conditions	Electricity		Heat S.O. MW(th)	η_e S.O. %	η_c S.O. %
		Gen MW(e)	S.O. MW(e)			
8	$P\,0.050$ Condensing	80	74	—	29.5	—
9	$t_s\,90 + t_r\,50$ 2 stages DH	66	60	141	23.8	80
10	$P1\ t100$ steam	57	52	150	20.9	81
11	$P5\ t \leqslant 200$ steam	43	38	164	15.0	81
12	$P15\ t \leqslant 300$ steam	27	22	180	8.7	81
13	$P40\ t \leqslant 430$ steam	12	7	195	2.8	81

Note: Case 13 is almost a heat-only scheme as far as performance is concerned, i.e. the electrical efficiency is almost zero.

Table 7.7 Performance of heavy duty gas turbine units

Air intake temperature:	15 °C
Cycle top temperature:	1000 °C (base load duty)
Turbine exhaust temperature:	515 °C (base load duty)
Fuel:	natural gas GCV 37.6 MJ/m^3

Case	Supply conditions	Electricity Gen MW(e)	S.O. MW(e)	Heat S.O. MW(th)	η_e S.O. %	η_c S.O. %
14	EO	85	85	—	27.7	—
15	$t_s90 + t_r50$ DH	82	80	152	26.1	76
16	$P15$ sat steam	82	81	139	26.5	72
17	$P40$ sat steam	82	81	133	26.5	70
18	$P40\ t450$ steam	82	81	113	26.5	63

Turbine exhaust temperature raised 100 °C by afterburn:

19	$t_s90 + t_r50$ DH	82	80	190	23.3	79
20	$P15$ sat steam	82	81	176	23.6	75
21	$P40$ sat steam	82	81	170	23.6	73
22	$P40\ t450$ steam	82	81	151	23.6	67

Combined cycle without afterburn:

23	$t_s90 + t_r50$ DH	112	110	115	35.9	73

Combined cycle with 100 °C afterburn:

24	$t_s90 + t_r50$ DH	122	120	141	34.9	76

Table 7.8 Performance of diesel units

Type: Medium speed, 4 stroke, turbocharged.
BMEP: 19 bar abs (base load duty)
Diesel exhaust temperature: 400 °C
Fuel: HFO   GCV   42 740 kJ/kg
               Viscosity $\leqslant$ 420 cSt (3500 Sec Redwood No. 1)
               Sulphur $\leqslant$ 3.5%
               Vanadium $\leqslant$ 150 p.p.m.

Case	Supply conditions	Electricity Gen MW(e)	Electricity S.O. MW(e)	Heat S.O. MW(th)	η_e S.O. %	η_c S.O. %
25	EO	11.3	11.2	—	36.9	—
26	$t_s90 + t_r50$ DH	11.3	11.0	11.5	36.3	74
27	P15 sat steam	11.3	11.0	6.7*	36.3	58*
28	P40 sat steam	11.3	11.0	5.6*	36.3	54*

Turbine exhaust temperature raised by 100 °C by afterburn:

Case	Supply conditions	Electricity Gen MW(e)	Electricity S.O. MW(e)	Heat S.O. MW(th)	η_e S.O. %	η_c S.O. %
29	$t_s90 + t_r50$ DH	11.3	11.0	14.5	33.0	77

Combined cycle without afterburn:

Case	Supply conditions	Electricity Gen MW(e)	Electricity S.O. MW(e)	Heat S.O. MW(th)	η_e S.O. %	η_c S.O. %
30	$t_s90 + t_r50$ DH	12.1	11.7	9.1	38.6	69

Combined cycle with 100 °C afterburn:

Case	Supply conditions	Electricity Gen MW(e)	Electricity S.O. MW(e)	Heat S.O. MW(th)	η_e S.O. %	η_c S.O. %
31	$t_s90 + t_r50$ DH	12.7	12.3	11.4	36.9	71

* With Cases 27 and 28, there is scope for the addition of a hot water supply system ($t_s \leqslant 80$ °C) which would improve the heat recovery and combined efficiency values significantly.

Note: the effect of afterburn is to raise the combined efficiency but to reduce the electrical efficiency.

Then fuel consumption (domestic) $= \dfrac{250}{2}\left[\dfrac{1}{0.55}+\dfrac{1}{0.65}\right] = 420$ PJ p.a.

Fuel consumption (central HO) $\quad = \dfrac{250}{0.75} = 333$ PJ p.a.

Therefore Stage 1 fuel saving $= 420-333 = 87$ PJ p.a.

Stage 2—a coal-fired CHP station is introduced, supplying 70% of the annual heat requirements.

Therefore CHP heat production $\quad = 250 \times 0.7 = 175$ PJ p.a.

If $\quad$ CHP electrical efficiency $\quad = 30\%$ on an annual basis

and $\quad$ CHP combined efficiency $\quad = 75\%$ on an annual basis

then $\quad$ CHP fuel consumption $= \dfrac{175}{0.75 - 0.30} = 389$ PJ p.a.

$\quad$ and central HO fuel consumption $= \dfrac{250-175}{0.75} = 100$ PJ p.a.

If new coal-fired EO plant would otherwise be built, operating with an annual efficiency of 36%, then

$\quad$ EO fuel consumption $= \dfrac{390 \times 0.30}{0.36} = 325$ PJ p.a

Therefore Stage 2 fuel saving $= 325 + 333 - 389 - 100$

$$= 169 \text{ PJ p.a.}$$

In this example, the additional fuel saving due to CHP is therefore about twice the fuel saving due to central HO production.

In terms of different grades of fuel,

$\quad$ distillate oil saving $\quad = \dfrac{250}{0.65} \times \dfrac{1}{2} = 192$ PJ p.a.

$\quad$ high grade coal saving $= \dfrac{250}{0.55} \times \dfrac{1}{2} = 228$ PJ p.a.

$\quad$ Extra burn of low grade coal $= 333$ PJ p.a. due to Stage 1

$\qquad\qquad\qquad$ less 169 PJ p.a due to Stage 2

$\qquad\qquad\qquad = 164$ PJ p.a. net

Note that in this example, the important savings of distillate fuel oil and high grade domestic coal were due to centralized heat production. The contribution of CHP was to save low grade coal.

B2. Industrial Steam Example

A factory takes heat from a central HO station (HFO fired) and electricity from the national grid. A coal-fired CHP station is introduced, supplying all the heat and electrical requirements of the factory.

Factory heat requirement = 200 TJ p.a.

Factory electricity requirement = 50 TJ p.a.

$\eta_{HO} = 75\%$ on an annual basis
$\eta_{EO} = 28\%$ on an annual basis

In this example, the nation does not need the additional electrical capacity, and η_{EO} is therefore the annual marginal efficiency of the electricity supply system, with a fuel burn of 50% coal, 25% HFO, and 25% natural gas.

Therefore Fuel consumption (HO) $= \dfrac{200}{0.75} = 267$ TJ p.a.

Fuel consumption (EO) $= \dfrac{50}{0.28} = 179$ TJ p.a.

If CHP combined efficiency $= 70\%$ on an annual basis, then

fuel consumption (CHP) $= \dfrac{200 + 50}{0.70} = 357$ TJ p.a.

Therefore fuel saving due to CHP $= 267 + 179 - 357 = 89$ TJ p.a.
In terms of different grades of fuel,

natural gas saving $= \dfrac{179}{4} = 45$ TJ p.a.

HFO saving $= 267 + \dfrac{179}{4} = 312$ TJ p.a.

Extra burn of low grade coal $= 357 - \dfrac{179}{2} = 268$ TJ p.a.

Note that a large saving of HFO consumption would have been made if the new station had been HO instead of CHP.

In the two examples above, none of the results is expressed as a percentage saving. This avoids the ambiguities arising from the choice of denominator.

Appendix C Low Temperature Schemes

C1. Introduction

Low temperature CHP schemes utilize heat from power station cooling water (CW) at, or near, normal reject temperatures. To the generation authority, the important features of such schemes are therefore:

(1) The engineering design of the generating station and its main plant is no different to that of an ordinary EO power station.
(2) There is little or no loss of electrical generation due to to the heat recovery facility.

Engineering aspects are therefore limited to matters of water transport and heat transfer at the point of use.

Most LT schemes are concerned with the growing of crops, either by enhancing the growth rate, extending the growing season or simply by saving fossil fuels which would otherwise be used for heating. Such schemes are often owned and organized as partnerships between a generation authority (supplying heat) and a product authority (who attend to crop husbandry and marketing aspects).

This appendix is limited to a short description of cooling water systems, temperature levels, and the availability of heat from power stations, together with a brief discussion of some of the commercial uses for LT heat, most of which are still in the experimental or pilot plant scales of operation.

C2. Generation Aspects

Four types of cooling water system are illustrated in Fig. 7.51. If a suitable river, sea, or lake can be utilized, the direct type (a) is generally chosen on economic grounds. If no direct CW source is available, a recirculating type of cooling system must be used, incorporating cooling towers. The evaporative type (b) is the most common, but requires a make-up of about 1% (due to evaporation) in addition to regular purging to control the salinity and solids concentration. In arid or environmentally sensitive locations, dry cooling towers are required, incorporating water/air or steam/air heat exchangers. Figure 7.51(c) and (d) illustrates two types of dry cooling tower.

The physical size of any type of cooling tower can be reduced by assisted or forced-draught fans. For base-load stations, this is generally less economic than natural draught, but may be required because of space or environmental constraints.

The CW inlet temperature to the condensers depends on the climate, the type of CW system and the cooling tower design. Temperatures of zero Celsius can be experienced in mid-winter in Northern Europe, rising to 20 °C or more in midsummer. Condensers are generally designed for a CW temperature rise of some 5 to 15 K at full load, determined by the overall station economics and possibly by cooling water system constraints.

For the LT heat recovery developer, the important parameter is the CW outlet temperature from the condensers, since this is the water which is usually the source of heat. Typical CW outlet temperatures at base-load power stations in Northern Europe are as follows:

	Midwinter °C	Midsummer °C
Direct cooled	10–15	20–25
Tower cooled (evaporative, natural draught)	10–20	30–40

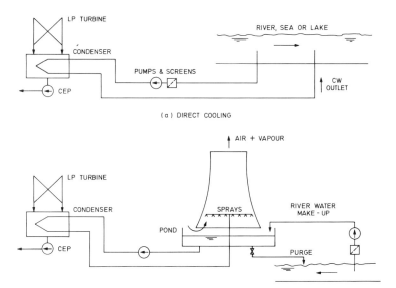

(a) DIRECT COOLING

(b) TOWER COOLING - EVAPORATIVE, NATURAL DRAUGHT

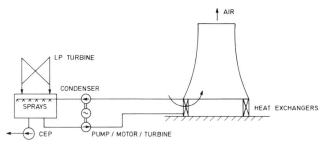

(c) TOWER COOLING - DRY, NATURAL DRAUGHT, WATER

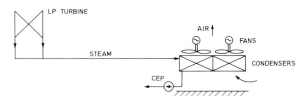

(d) TOWER COOLING - DRY, FORCED DRAUGHT, STEAM

Fig. 7.51. Types of cooling water system

A feature of steam turbine stations is therefore high CW outlet temperatures in summer and low CW outlet temperatures in winter. Ideally, however, heat consumers require the higher temperatures during the winter. It would be technically feasible for the power station to raise the CW outlet temperature artificially in the winter, but this would reduce electrical generation and incur additional costs because of the extra fuel burned at other power stations. This option is not commonly practised and, in general, the heat consumer therefore designs his system for the most onerous conditions, i.e. winter CW temperatures and maximum heat demand.

Diesel power stations have a significant cooling requirement, and the CW outlet temperature is essentially constant throughout the year, at 70 to over 100 °C depending on the engine design. Because these temperatures are significantly higher than from steam turbine stations, LT heat from diesel stations is therefore potentially more useful. Several ways of uilizing diesel LT heat have already been discussed in Sections 7.2 and 7.3.

Low temperature schemes are generally associated with base-load power stations, i.e. those intended primarily for continuous generation. Such stations give a high availability of LT heat supply without incurring out of merit operating costs. Only occasionally will there be an interruption in heat supply because of a breakdown of the whole station or of the electrical transmission system. It is important, of course, that the LT heat supply should be connected not to a single generating unit, but to several.

The expected lifetime duration of base-load operation must be considered before proceeding with a major LT scheme. Nuclear stations are generally required to operate on base-load throughout their lives, i.e. for twenty years or more. Fossil-fired stations, however, are often intended to operate on base load for a much shorter period, thereafter moving to two-shift and, eventually, peak load duty. Generating 'parks' (intended for several power stations built at different times) would be ideal for long term investment in LT schemes.

C3. Uses of LT Heat

Several uses of LT heat from diesel stations have already been discussed in Section 7.2 and 7.3, covering district heating and feed-water heating for steam raising. The uses of LT heat described below are specific to steam turbine stations (i.e. $t_s \leqslant 40$ °C) but can be applied also to diesel stations if required.

Space heating of buildings. LT heat in the range 13 to 40 °C can be used for space heating of buildings. A technically proven method is to circulate the CW to fan-assisted fin tube heat exchangers mounted on the wall or ceiling. The forced-draught feature achieves a low terminal temperature difference. The fans generate noise, and intake silencers are usually required in occupied premises. The heater units are bulky and are not suitable for small rooms, but industrial premises, stores, warehouses, and animal buildings in the vicinity of the power station can be heated in this way. There appears to be only limited application for LT space heating uses of this kind, owing to the large mass flows and consequent high costs when transporting LT heat over long distances.

Studies have been made of the economics of long distance transmission of CW to cities where heat pumps would extract heat for district heating purposes. Macadam and Jebson[6(a)] showed that, in UK economic terms, there could be a role for such schemes in the load range 50 to 200 MW(th) for transmission distances up to 30 km. However, no plans exist to build such a scheme. Large heat pump schemes already in service in several countries use heat sources other than power station CW.

Glasshouse heating. Many countries have LT glasshouse heating schemes on a pilot scale. There are several methods of heating glasshouses with CW, including soil warming, open troughs, distribution over the glasshouse roof, and water/air heat exchangers. A commercial venture in the United Kingdom, adjacent to Drax Power Station, utilizes the indirect method with a matrix of 640 fan heaters (of the type described above for the space heating of buildings). Figure 7.52 shows a photograph of this installation. It covers 8 ha of land and has a target production of 2200 tonnes of tomatoes per annum. In horticultural terms, this is a large scheme supplying about 0.8% of the UK tomato market, and saving about 4500 tonnes of fuel oil per annum.

From the viewpoint of power generation, however, the Drax tomato scheme is not large, taking 1.3% of the full load water flow which means less than 0.1% of the 3700 MW(th) maximum heat available from this 3 × 660 MW(e) station. Considering that the UK base load steam power station installed capacity is about 20 000 MW(e), there is enormous potential for further salad crop schemes of this kind.

The Drax scheme has no standby boilers. In the event of a station shutdown, the cooling tower ponds have sufficient residual heat for many hours LT supply to prevent glasshouse temperature falling disastrously. In normal operation, the average glasshouse air temperature can be maintained at about 7 K below the CW outlet temperature from the condensers.

Aquaculture. The cooling water outlet can be diverted through tanks or lagoons to provide a continuous flow of tepid water for fish farming. Compared with fish farms operating in the sea, rivers, or lakes, growth rates increase significantly, by factors of 2, 5, or more depending upon the species and conditions.

Many countries have pilot schemes, eels appearing to be the most common species. Other species reared successfully include catfish, sole, turbot, carp, trout, and oyster. The benefits appear to be essentially commercial; there are no obvious fuel savings.

Since the CW is used directly, there can be problems of water quality. Some water

Fig. 7.52. Matrix of fan heaters in glasshouse. (Reproduced by permission of the CEGB)

sources are not suitable for fish, particularly if they contain industrial pollutants. Many power stations chlorinate the CW system on a regular basis to prevent slime formation in condensers and culverts, and special provisions may be necessary to avoid over-chlorination which would kill the fish. Silt can be another problem, and regular cleaning of fish tanks/lagoons may be necessary. With large fish farms, excrement levels in the return water could become a problem requiring direct purge at some locations.

Agriculture. Open fields can be heated with CW by troughs (raising the air temperature) or by buried pipes (raising the soil temperature). Research in the United States and West Germany has concentrated upon soil warming. The Agrotherm[67] research project in West Germany has used polyethylene pipes, 55 mm bore, buried 0.75 m deep and 1 m apart, laid in a 7 ha field. Maize, potatoes, beet, soya beans have been grown successfully, with annual yields increased by 50% to 100%. In the case of soya beans, these would not normally be grown in the German climate at all.

Buried pipes are costly to install, requiring extensive land area and pipe systems (owing to the poor heat transfer coefficient). Condenser back pressure is also likely to deteriorate marginally (compared with conventional CW systems) in order to optimize CW temperatures. Such schemes are not generally regarded as being economic except in base load power stations where the CW system would otherwise need to incorporate fan-assisted dry cooling towers. This situation could exist in some parts of the world which have a scarcity of direct CW sources as well as environmental objections to large, natural draught, evaporative type cooling towers.

7.7 REFERENCES

List of references for further reading including associated topics not specifically covered in this chapter.

1. *Ad hoc* committee on combined heat and power production and district heating, CHP, World Energy Conference, London, 1978.
2. Joint *ad hoc* committee on combined heat and power and heat pumps, CHP/HP, World Energy Conference, London, 1981.
3. *10th World Energy Conference, Istanbul, 1977*:

    (a) Paper 3.3–1. Manescu, N. *et al.*, Achievements and prospects of combined heat and power production in Romania.
    (b) Paper 3.3–2. Eichner, K.-P. *et al.*, Application of heat power coupling for the supply of cities and large urban built-up areas with district heat under consideration of manoeuvrability requirements for the generation of electric energy.
    (c) Paper 3.3–3. Schulte, H., The combined generation of heat and electricity as a means of saving primary energy.
    (d) Paper 3.3–4. Kirvela, K. *et al.*, A nuclear power plant combined with the supply of district heat and process steam.
    (e) Paper 3.3–5. Almquist, P. *et al.*, Nuclear power plants for combined production of electricity.
    (f) Paper 3.3–6. Andrzejewski, S. *et al.*, The problems of incorporating nuclear heat and power plants into the electrical power system.
    (g) Paper 3.3–7. Sokolov, Ye. Ya. *et al.*, The progress in combined heat and power generation by large combined heat and power stations.
    (h) Paper 3.2–4. Grigoryants, A. N. *et al.*, Experience in construction, start-up and operation of the Bilibino nuclear heat and power plant.

4. *11th World Energy Conference, Munich, 1980:*

   (a) Volume 1B, pp. 275–90 in German language, Kubin, M., and Vlach, J., Regional heat supply systems in the CSSR.
   (b) Volume 1B, pp. 343–60 in German language, Munser, H. *et al.*, The complex district heating supply in the German Democratic Republic—an effective contribution to the rational utilization of energy resources.
   (c) Volume 1B, pp. 400–409, Bjornsson, J., Conservation of oil in the space heating sector in Iceland through substitution by indigenous energy sources.
   (d) Volume 1B, pp. 445–64, Armencoiu, N. *et al.*, Concerns regarding a rational household energy utilization in Romania.
   (e) Volume 3, pp. 455–62, Varvarsky, V. S. *et al.*, Activities on the construction of a powerful heat pump station in Moscow.
   (f) Volume 1B, pp. 510–18, Kiciman, S., Long term electricity generation in Turkey and national survey of combined heat and power generation potential.
   (g) Volume 1A, pp. 868–75, Korytnikov, V. P. *et al.*, District heating schemes with the use of nuclear sources.

5. *Proceedings of Second National Conference, Scarborough,* District Heating Association, 1977:

   (a) Paper 'h', Investigations of a combined heat and power scheme at Peterborough.
   (b) Paper 'l', Rosada, J., and Obreski, J., The role of district heating in the energy balance of Poland.
   (c) Paper 'm', Kilpinen, U., District heating in Finland.

6. *Proceedings of Third National Conference, Eastbourne,* District Heating Association, 1979:

   (a) Paper No. 8, Jebson, D. A., and Macadam, J. A., The role of large scale compressor plant in the district heating from power stations.
   (b) Paper No. 10, Lindeberg, L., The role of district heating in the energy policy of a modern industrial country.
   (c) Paper No. 11, Neuffor, H., District heating in Germany—an important energy saving technology.
   (d) Paper No. 13, Rowe, W. G. E., The Nottingham combined refuse incineration and district heating scheme–a review of its development.
   (e) Paper No. 15, Rimmen, P., Operation of a large district heating scheme at Odense, Denmark.
   (f) Paper No. 16, Postlethwaite, A. F., The feasibility of converting power stations to combined heat and power.
   (g) Paper No. 5, Jenks, P. A. D., Local heat management and operation.

7. *Proceedings of Fourth National Conference, Torquay,* District Heating Association, 1981:

   (a) Paper No. 2, Anthony, E. J., Heat load assessments.
   (b) Paper No. 3, Jebson, D. A., The effects of variations in the distribution of heat demand on the cost of district heating networks.
   (c) Paper No. 4, Postlethwaite, A. F., Selection and conversion of CHP stations.
   (d) Paper No. 5, Rowe, W. G. E., Arterial mains systems to supply UK city heat loads.

8. *Proceedings of Total Energy Conference, Brighton,* Institute of Fuel, 1971:

   (a) Paper No. 3, Gurney, J. D., and Pearson, J., The total energy installation at John Player and Sons.

(b) Paper No. 4, Weatherstone, D. E., The diesel installation at Bridgend Paper Mills.
(c) Paper No. 7, Williams, J. H., Operational experience with total energy plants in the United States.
(d) Paper No. 11, Smith, E., and Aicher, W., Heat-recovery boilers for total energy systems.
(e) Paper No. 13, Jackson, T. J., Application: gas turbines at the Beckton Sewage Treatment Works.
(f) Paper No. 15, Leijendeckers, P. H. H. *et al.,* The first 420 bed total energy hospital.
(g) Paper No. 16, Ridal, B. F., Steam and power generation for the petroleum industry.
(h) Paper No. 17, Taylor, J. R., The 120–MW total energy plant at Grenoble.

9   *Proceedings of First International Total Energy Congress, Copenhagen,* Miller Freeman Publications, Inc., 1976:

(a) pp. 36–78, Marcussen, B. T., and Hustad, K. O., Gas turbine total energy applications.
(b) pp. 79–96, Weber, D., Total energy applications for closed–cycle gas turbines.
(c) pp. 257–98, Young, C., Aldershot P. S. A. total energy power station with district heating.
(d) pp. 355–80, Rowe, I. H. *et al.,* Nuclear-based district heating for a new town development.
(e) pp. 437–59, Hausz, W., Annual storage: a catalyst for conservation.
(f) pp. 460–94, Charroppin, P. *et al.,* Underground heat storage in captive aquifer.
(g) pp. 550–74, Larsson, K., Computer optimisation of district heating schemes.
(h) pp. 748–66, Mikkelsen, W. L., Development of district heating systems in Denmark.
(i) pp. 767–91, Muir, N., Conversion of existing turbines for combined power and heat production.

10.   *Proceedings of Second International Total Energy Congress, Copenhagen,* Miller Freeman Publications, Inc., 1979:

(a) pp. 2.1–2.5, Rowe, I. H., District heating in Ontario, where we stand.
(b) pp. 4.1–4.25, Guerra, C. R. *et al.,* District heating and cooling through retrofit of public utility steam-electric stations.
(c) pp. 4.27–4.37, Price, M. E., Hereford combined heat and power station.
(d) pp. 5.15–5.38, Meyer, C. F., Evaluation of thermal energy storage for a large urban hot water district heating system in the United States.
(e) pp. 6.11–6.34, Rosbach, N., Low temperature district heating, a gradual approach.
(f) pp. 6.35–6.47, Margen, P., and Roseen, R., Solar heat for small district heating systems.
(g) pp. 7.13–7.37, El Mahgary, Y. *et al.,* Economic analysis and probabilistic site evaluation of nuclear plants producing district heat and electricity.
(h) pp. 8.1–8.11, Heden, F., Operational results from first diesel energy plant in Sweden.
(i) pp. 8.13–8.29, Smit, J. A., and Welle, M., A new concept in waste heat recovery from slow—and medium—speed diesel engines.
(j) pp. 8.31–8.39, Petersen, B., Large diesel driven heat pumps for district heating.

11. Orchard, W. H. R., and Sherratt, A. F. C. (eds), *Combined Heat and Power,* George Godwin Ltd, London and John Wiley and Sons Ltd, New York and Toronto, 1980:

   (a) Chapter 2, Wright, J. K., Energy savings and economics of whole city heating by CHP.
   (b) Chapter 3, Cassels, J. M., The CHP potential in British cities and towns.
   (c) Chapter 7, Larson, L., Development plans for a low-temperature district heating system in Odense.
   (d) Chapter 8, Margen, P. *et al.,* Minneapolis/St. Paul—a case study.
   (e) Chapter 10, Robinson, P. J., Transmission and distribution networks and the consumer—the potential for development.
   (f) Chapter 11, Postlethwaite, A. F., CHP plants for whole city heating.

12. Olszewski, M. (ed.), *Utilisation of Reject Heat,* Marcel Dekker Inc, New York and Basel, 1981:

   (a) Chapter 1, Olszewski, M., Use of low-grade reject heat.
   (b) Chapter 2, Gunn, M. E., Recovery of industrial waste heat.
   (c) Chapter 3, Reed, S. A., Coupling technology for dual purpose central station power plants.
   (d) Chapter 4, Karkheck, J. P., and Powell, J. R., Use of heat from dual-purpose power stations for district heating.

13. District heating, statistics, 1977, UNICHAL, 1979.
14. Study committee 'nomenclature and statistics', state of development of district heat in the member countries of UNICHAL, report 1977, UNICHAL.
15. Report of the study committee on nuclear energy, UNICHAL Congress, 1981, Vienna.
16. O' Callaghan, P. W., *Design and Management for Energy Conservation,* Pergamon Press, 1981.
17. Rogers, G. F. C. and Mayhew, Y. R., *Engineering Thermodynamics Work and Heat Transfer,* Longman Group Ltd, 2nd edition, 1967.
18. Oliker, I., Economic feasibility of district heat supply from coal fired power plants, American Power Conference, Chicago, 1981.
19. Tourin, R. H., District heating with combined heat and electric power generation, Volume 1 of *Advances in Energy Systems and Technology,* Academic Press, Inc, 1978.
20. Muir, N., District heating in Sweden, 1973, together with two exercises on load forecasts and costs associated with the introduction of district heating in Dublin, Stal-Laval Tech. Inf. 6/73.
21. Gubser, H. R., Combined cycle for district heating power station, *Turbomachinery International,* July/August 1980.
22. Hagen-Kabel Combined cycle/cogeneration power station supplies paper mill, *Modern Power Systems,* April 1982, pp. 51–8.
23. Marsh, P., Brewery cogeneration plant gives 82 per cent efficiency, *Modern Power Systems,* April 1982, pp. 27–31.
24. Lorentz, J. J., and Mountfort, M. V., *The Feasibility of a Geothermal District Heating Scheme,* New Zealand Energy Research and Development Committee, Report No. 55, May 1980.
25. *District Heating in IEA Countries,* International Energy Agency, Paris, 1981.
26. *Lines of Development in Energy Engineering,* Joint Convention by VDI/I. Mech E. held in Dusseldorf, May 1975:

   (a) pp. 201–204, Bublitz, D., District heating and combined heat and power generation in the Berlin area.

(b) pp. 205–12, Musil, L., The possibilities and economic significance of heat linkage to the public electricity supply business, particularly in countries with water power.
27. Papers presented at the *1st International District Heating Convention, London,* 1970, District Heating Association:

(a) Paper A5, Sokolov, E. Ya., Main systems of urban district heating in the USSR.
(b) Paper D2, Kimura, H., Present situation and future problems of district heating and city environment in Japan.
(c) Paper E1, Ryman, J–E., District heating in the Stockholm area.
(d) Paper F1, USSR State Committee, Open schemes of hot water supply.
(e) Paper H1, Delestaing, M. P., and Visseq, M. A., The Paris district heating scheme.
(f) Paper H2, Schuster, G., and Dittbrenner, A., The district heat supply system of the town of Essen.
(g) Paper H3, Zoega, J., and Kristinsson, G., The Reykjavik district heating system.
(h) Paper H5, Winkens, H. P., The application of district heating to existing town centres and new town districts related to Mannheim.
(i) Paper I1, USSR State Committee, Advantages of district heating and its technical and economic efficiency.
(j) Paper I3, Sladek, V., The production and supply of district heating in Czechoslavakia and to the town of Bratislava in particular.
(k) Paper I5, Marecki, J., and Wojcicki, J., Technical and economic aspects of district heating development in Poland.
28. Drahy, J., Combined heat and power stimulates Czechoslavakia's turbine makers, *Energy International,* October 1979, pp. 37–40.
29. Dvorak, A., The district heating power plant at Randers (Denmark), *Brown Boveri Review,* 11–80.
30. Reports of the International District Heating Convention, Warsaw, 1976:

(a) Report No. IV/13, Kilpinen, U., Operation and regulation of a big hot-water D. H. network.
(b) Report No. III/7, Hunig, H. *et al.,* Successful introduction of a heat store in the connected district heating system of Karl-Marx-Stadt (in German language).
31. Jeffs, E., Czech turbines expand Helsinki's district heating, *Energy International,* April 1976, pp. 21–2.
32. Combined heat and power is core of Romanian policy, *Energy International,* February 1977, pp. 23–6.
33. Lucas, N. J. D. (ed.), *Local Energy Centres, Conference Proceedings,* Construction Industry Conference Centre Ltd, Welwyn, UK, 1977:
(a) Paper 2, Kohler, B., Local energy centres in Sweden.
34. Convention of the International District Heating Association, Hot Springs, Va, 1978:
(a) Astrand, L. E., The Fyriskraft combined heat and power plant.
35. Jeffs, E., Oil refinery supplies waste energy to Gothenberg district heating system, *Modern Power Systems,* January 1981, pp. 47–51.
36. The Sydvarme Project—district heating from the Barsebeck nuclear power plant, Sydkraft, Malmo, Sweden.
37. District heating turbine 60–200 MW, Stal-Laval publication 357E 9.73.
38. *Stockholm—A District-heated City,* Booklet by Stockholms Energiverk.

39. Wernius, I., Hasselby power and district heating station, *Teknisk Tidskrift*, 24 February 1961.
40. *Vasteras District Heating Power Station*, Booklet by Vasteras Stads Kraftvarmeverk AB, Sweden.
41. Bitterli, J. el al., Combined heat and power station, Aubrugg, Zurich, *Sulzer Technical Review*, 3/1980.
42. Barnett, J. L. et al., Cogeneration demonstration at Riegel Textile Corporation, *Modern Power Systems*, September 1981, pp. 45–9.
43. Davis, J. P., and Nydick, S. E., US chemical plant chooses slow speed diesel cogeneration system, *Modern Power Systems*, December 1981, pp. 57–62 plus pull-out drawing.
44. Strauss, S. D., District heating links with cogeneration, *Power*, August 1979, pp. 72–5.
45. Miller, A. J. et al., Use of steam-electric power plants to provide thermal energy to urban areas, National Technical Information Service, US Department of Commerce.
46. American Power Conference, Chicago, 1980:

    (a) Oliker, I., and Muhlhauser, H. J., Technical and economic aspects of coal-fired district heating power plants in USA.
    (b) Mikkelson, W., and Hammer-Sorensen, F., District heating by cogeneration through retrofit and decentralisation.
47. *Thermal Engineering*, 1974, 21:

    (a) pp. 1–10, Sokolov, E. Ya. and Belinskii, S. Ya., Fifty years of Soviet district heating.
    (b) pp. 11–16, Melent'ev, L. A., District heating in the power industry of the USSR.
    (c) pp. 39–44, Lanin, I. S., and Kazarov, S. A., Fifty years of district heating in Leningrad.
    (d) pp. 45–9, Gromov, N. K., District heating of Moscow.
48. *Thermal Engineering*, 1976, 23:

    (a) pp. 26–38, Gromov N. K., Principles of arrangements of heating networks in towns, their automation and remote control.
49. *Thermal Engineering*, 1981, 28(3):

    (a) pp. 139–43, Averbakh, Yu. A. et al., Conversion of generating units of condensing power stations into district–heating units with single-pipe transportation of heat.
50. Ershov, I. N. et al., Fifty years of district heating in Moscow, *Teploenergetika*, No.12, pp. 4–10, 1978, in Russian language.
51. Prinz, W., How we brought district heating to Flensburg, *Building Services and Environmental Engineer*, January 1980, pp. 22–3.
52. Kruger, H., Hamburg's combined heat and power system, *Electrical Review International*, **203**, 21, December 1968, pp. 65–7.
53. *Munchen-Sendling District Heating Power Station*, Publication KWU 125–101, Kraftwerk Union AG, 1971.
54. Gesamtstudie uber die Moglichkeiten der Fernwarmeversorgung aus Heizkraftwerken in der Bundesrepublik Deutschland, eine Information des Bundesministers fur Forschung und Technologie, Bonn, 1977.
55. Boyen, J. L., *Practical Heat Recovery*, John Wiley and Sons, 1975.
56. Czwiertnia, K., Power generating equipment for thermal electric power stations,

Paper No. 18 of *Poland's Technology '79*, Centre for Progress in Technology of the Polish Federation of Engineering Associations, 1979.

57. Stucheli, A., Economic application of electric boilers for steam and hot water supplies, *Sulzer Technical Review* 2/1975.
58. Flad, J., Stadtwerke Saarbrucken erweitern Heizkraftwerk, *Energie*, November 1974, **11** pp. 391–400 (in German language).
59. Flad, J., *Combined Cycle Answers Sarrebruck's Total Energy Requirements*, Publication 3–76(3M), Alsthom, Société Generale de Constructions Electriques et Mecaniques, Belfort.
60. Bublitz, D., Stadtheizung durch Kraft-Warme-Kopplung in Berlin, *Fernwarme International*, FWI 3 (1974), 4 pp. 98–106, Sonderdruck Nr. 2667 (in German language).
61. Bublitz, D., 50 years of district heating in Berlin, *Fernwarme International*, **5** (1976) pp. 173–4 (in German language).
62. Riedlinger, R. A., *About Extracting Heat from Existing Power Stations for District Heating Purposes*, Report given to the IEA–Symposium, Berlin, 1975.
63. *The Economic Production of Industrial Power and Process Heat*, Publication AP 4000–A, W. H. Allen Sons and Co. Ltd, Bedford, England.
64. Niemiaho, H., and Walker, G. N., *Vaasa Power Station, Finland*, Publication No. B55, NEI Parsons Ltd, Newcastle-upon-Tyne, 1979.
65. Kehlhofer, R., A comparison of power plants for cogeneration of heat and electricity, *Combustion*, March 1981, pp. 22–8.
66. Clifford, D. *et al.*, Repowering of Slough Estates for optimum energy conversion, *I. Mech. E. Proceedings 1980*, Vol. 194, No. 34.
67. Luckow, H., *The Agrotherm Research Project*, Waste Heat Management and Utilisation Conference, Miami Beach, 1977.
68. DeWalle, D. R., and Chapura, A. M., Jr., Soil warming for utilization and dissipation of waste heat in Pennsylvania, *Nuclear Technology*, **38**, Apr. 1978, pp. 83–91.
69. Growing need for heat recovery, *The Heating and Air Conditioning Journal*, December 1979, pp. 32–4.
70. Save energy—eat tomatoes, *News Letter No. 110*, December 1979, Central Electricity Generating Board.
71. Aston, R. J. *et al.*, Heated water farms at inland power stations, *News Letter No. 102*, December 1976, Central Electricity Generating Board.
72. Ingram, M. V., Waste heat aquaculture in the United Kingdom—a general review and particulars of Marine Farm Ltd, Hinkley Point, Bridgwater, Somerset, England, *Proceedings, Workshop, New Brunswick, March 1978*.
73. Hooper, W. C. *et al.*, *Rearing of Brook Trout and Lake Trout in thermal Effluent of a Coal Fired Generating Station*, A paper presented at the fall meeting of the Canadian Electrical Association Thermal and Nuclear Section, Regina, 1977.
74. Boyd, L. L. *et al.*, *A demonstration of Beneficial Uses of Warm Water from Condensers of Electric Generating Plants*, Report EPA–600/7–80–099, May 1980, US Environmental Protection Agency.
75. Hubert, W. A. *et al.*, *State-of-the-art Waste Heat Utilization for Agriculture and Aquaculture*, Report EA–922 prepared by Tennessee Valley Authority for Electric Power Research Institute.
76. Grauby, A., *Thermal Releases of Nuclear Power Plants, Their Utilization for Plant and Fish Production*, Technical Meeting No. D3/8, 5th International Fair and Technical Meetings of Nuclear Industries, Basel, 1978.
77. *Combined Heat and Power Generation in the United Kingdom*, Energy Paper 35, HMSO, London, 1979.

78. England, G., *CHP: from Debate to Practical Progress?* Booklet published by Central Electricity Generating Board, February 1982.
79. *CHP/DH Feasibility Programme Stage 1* Summary report and recommendations for the Department of Energy, prepared by W. S. Atkins and Partners, Epsom, UK, 1982.
80. Ryman, J. E., *Views on Large Scale District Heating from Stockholm,* Paper prepared for 1st National Conference, District Heating Association, 1975.
81. *Combined Heat and Power in the London Borough of Southwark, a Preliminary Study,* Orchard Partners, London, 1980.
82. *Helsinki Electricity Works,* Booklet by the Helsinki Electricity Works.
83. *Sulphur in the Air of Helsinki,* Booklet by the Helsinki Electricity Works.
84. *SWD Review,* No. 20, March 1982:

    (a) Kimstra, K., and v.d. Graaf, K., Linz's diesel-driven city district heating system.
    (b) Key, J., A total energy system (TES) based on diesel power.
85. *Multi-district Heating Exhibition and Symposium, Herning, Denmark,* International Energy Agency (patron), A/S Herning-Hallen (Secretariat), 1982.

    (a) Riber, T., Cooperation and organisation in connection with the Herning project.
    (b) Nielsen, K. H., Distribution network, street network and consumers' network in Herning-Ikast.
    (c) Winkens, H. P., District heat production in Mannheim.
    (d) Kristensen, K. *et al.,* Fredericia district heating supply takes surplus heat from Superfos a/s.
    (e) Christensen, L. E. *et al.,* Barriers to district heating—ownership and financing options in the United States.
86. Armencoiu, N. *et al.,* Economic limitations of the cogeneration in Romania, *Buletinul Institutului de Studii si Proiectari Energetice (ISPE),* Bucharest, Volume XXIV, Number 3–4, 1981.
87. *Gas Turbine World and Cogeneration,* 12, 5, November 1982:

    (a) Farmer, R, Den Haag CHP station is rated at 86% efficiency.
    (b) de Biasi, V., 21 MW cogen plant allows Merck to stay in San Diego.
88. Taneja, O., Instrumentation and control aspects of retrofitting utility power plants for district heating purposes, ISA transactions, Vol. 22 No. 1, 1983.
89. Kan, G. T., Retrofitting utility power plants for cogeneration and district heating, Paper No. 104, American Power Conference, 1983.
90. Assessment of European district heating technology, interim report EPRI EM-2864, Electric Power Research Institute, 1983.

Table 7.9 Cross references by subject matter

Table 7.10 Case studies

Ref.	Location	Country	Status	Consumers	Prime mover	Fuel type	Unit size MW(e) gen.
2	Tampere	Finland	*	DH	ST	peat	60
2	Dunaujvaros	Hungary	*	I+DH	BP	Ind	17
2	Avezzano	Italy	†	I	GT–CC	NG	30
2	Mannheim–Heidelberg	W. Germany	†		ST–Hp		
3(c)	Ruhr	W. Germany	*	DH	ST	HFO/coal	37
3(d)	Olkiluoto	Finland	†	I+DH	ST	nuclear	12
3(h)	Bilibino	USSR	*	DH	EC	nuclear	25
4(e)	Lyubertsi (Moscow)	USSR	†	DH	Hp		9/14
5(a),6(g)	Peterborough	UK	†	DH	BP/GT	coal/oil	2
6(d)	Nottingham	UK	*	DH	BP	refuse/coal	40
6(e),11(c)	Odense	Denmark	*	DH	EC	coal/HFO	1
8(a)	Nottingham	UK	*	I	GT	NG	2
8(b)	Bridgend	UK	*	I	DL	HFO	1
8(e)	Beckton	UK	*	I	GT	Ind	1
8(f)	Geldrop	Netherlands	*	I	GE	NG	1
8(g)	Grangemouth	UK	*	I	BP	gas/oil	17
8(g)	Baglan Bay	UK	†	I	BP	gas/oil	30
8(h)	Grenoble	France	*	I	GT	HFO	16
9(c)	Aldershot	UK	*	DH	DL	HFO	7
9(d),10(a)	Pickering	Canada	†	DH	EC	nuclear	540
10(b)	New Jersey	USA	†	DH	EC	HFO/coal	529
10(c)	Hereford	UK	†	I	DL	HFO	7
10(d),11(d)	Minneapolis/St Paul	USA	†	DH	EC	nuclear	1000
10(g)	Helsinki	Finland	*	DH	DL	HFO	12
10(h)	Skultuna	Sweden	*	DH	EC/BP	HFO	38
20	Linkoping	Sweden	*	DH	ST	HFO/refuse	29
20	Dublin	Ireland	†	DH	GT–CC	NG/oil	100
21	Utrecht	Netherlands	*	DH	GT–CC	NG/dist	220
22	Hagen–Kabel	W. Germany	*	I	DL	HFO	2
23	Dundalk	Ireland	†	I			
24	Rotorua/Taupo	New Zealand	*	DH	HO	geo	—
25,46(b),85(a),85(b)	Herning	Denmark	†	DH	ST	coal	95
26(a),60,61	Berlin	W. Germany	*	DH	ST	coal	150
27(c),38,39,80	Stockholm	Sweden	*	DH	BP/EC	coal/HFO	34/250

Ref.	City	Country	Status	Consumers	Prime mover	Fuel	Size
27(e)	Paris	France	* *	DH	BP/EC	refuse/coal/ HFO	64
27(f)	Essen	W. Germany	*	DH	HO	coal	—
27(g)	Reykjavik	Iceland	*	DH	HO	geo	—
27(h),85(c)	Mannheim	W. Germany	*	DH + I	EC	refuse/oil/IWH	475
27(j)	Bratislava	CSSR	*	DH	BP/DL		8/25
29	Randers	Denmark	*	DH	BP	coal	48
30(a),31,82,83	Helsinki	Finland	*	DH	EC/BP/GT	coal/oil	113
34(a)	Uppsala	Sweden	*	DH	BP	HFO	200
35	Gothenberg	Sweden	†	DH	HO	IWH	—
36	Barsebeck	Sweden		DH	EC	nuclear	1000
37	Orebro	Sweden	*	DH	BP	HFO	106
40	Vasteras	Sweden	*	DH	EC	HFO	250
41	Zurich	Switzerland	*	DH	BP	NG/HFO	45
42	Riegel	USA	†	I	BP	coal	4
43	New Jersey	USA	†	I	DL	HFO	23
44	New York City	USA	*	DH	ST		
46(b)	Kalundborg	Denmark	†	DH	EC		268
47(c)	Leningrad	USSR	*	DH	ST		
47(d),50	Moscow	USSR	*	DH	ST		
51	Flensburg	W. Germany	*	DH	ST	coal	30
52	Hamburg	W. Germany	*	DH + I	BP/EC	coal/HFO	211
53	Munich	W. Germany	*	DH	GT	NG	27
58,59	Saarbrucken	W. Germany	*	DH	GT–CC	NG/dist	44
64	Vaasa	Finland	*	DH	EC	coal	165
66	Slough	UK	†	I	GT–CC	HFO	38
81	Southwark	UK	*	DH	EC	coal	250
84(a)	Linz	Austria	*	DH	DL	HFO	10
84(b)	Amsterdam	Netherlands	*	I	DL	HFO	3
85(d)	Fredericia	Denmark	*	DH	HO	IWH/oil	—
87(a)	Den Haag	Netherlands	*	DH	GT–CC	NG	100
87(b)	San Diego	US	†	I	GT–CC	NG	21

Status (at the time reference was written):
† scheme studied, planned, or projected.
* * scheme with operating experience.

Consumers:
DH district heating.
I industrial.

Prime mover:
ST steam turbine (unspecified).

BP back pressure steam turbine.
EC extraction condensing steam turbine.
GT gas turbine.
CC combined cycle.
DL diesel.
Hp heat pump.
GE gas engine.
HO heat only.

Fuel type:
Ind industrial by-product fuel.
NG natural gas.
HFO heavy fuel oil.
dist distillate oil.
geo geothermal.
IWH industrial waste heat.

Table 7.11 National situations and plans

Country	References
Australia	1
Austria	14,26(b)
Belgium	3,14
Bulgaria	1
Canada	1,10(a)
Rep. of China	1
Czechoslavakia	4(a),27(j)
Denmark	1,14
East Germany	1,3(b),4(b)
Finland	1 3(d),5(c),14
France	1,14
Hungary	1
Italy	1
Japan	1,27(b)
Rep. of Korea	1
Mexico	1
Netherlands	1,14
New Zealand	1
Poland	1,3(f),5(b),27(k),56
Romania	3(a),4(d),32,86
Spain	1
Sweden	1,3(e),6(b),14,20,33(a)
Switzerland	1,14
Thailand	1
Turkey	4(f)
United Kingdom	1,7(c),11(a),11(b),77,78,79
United States	8(c),18,19,44,45,85(e)
USSR	1,3(g),4(g),27(i),47(a),47(b)
West Germany	1,3(c),6(c),14,54,85(c)
Zambia	1

Energy—Present and Future Options, Volume 2
Edited by D. Merrick
© 1984 John Wiley & Sons Ltd

B. S. WESTON
National Coal Board,
Coal House, Lyon Road,
Harrow, Middlesex
and
K. R. SHAW
Arthur Andersen and Co.,
1 Surrey Street,
London WC2.

8

Energy Systems

8.1 INTRODUCTION

The energy production and supply industries have grown from simple local enterprises into the large, complex concerns of today. Transport costs for all forms of energy have fallen steadily, so that large individual production units now compete both with each other and with those of alternative energy sources. These interactions make a systems approach necessary for a full appreciation of the behaviour of present energy markets and the economics of energy supplies.

The nature of these interactions is illustrated in Section 8.2 for the case of electricity generation. However, this example on the one hand is simplified by the existence of only one product, and on the other made more complex because of the limited possibilities for storage. The effects of product quality and storage are therefore considered in Section 8.3, where other energy carriers are discussed. Finally progress with the contribution of large computer models to describe and predict the overall behaviour of the energy system is reviewed in Section 8.4.

8.2 ELECTRICITY

8.2.1 Introduction

Electricity supply networks are complex and expensive systems. Because of the high cost of both construction and operation of generating plant, large centralized supply systems have developed so as best to exploit the economies of scale. The limited possibilities for storing electricity at an acceptable cost have led to the development of procedures for matching supply to requirements, taking into account the variations in the demand for electricity and the various types of generating plant designs.

Investment in the existing electricity supply systems is substantial. It has been estimated that the electricity industry accounts for 20% of the fixed capital formation in the USA[1]. Decisions about generating plant choice are therefore of considerable importance. In this section, some simple economic models for assessing and optimizing generating options are examined.

Electricity supply appears in one respect to be a simple industry; it usually produces only one product—electricity. Combined heat and power is an exception to this, although it represents only a small fraction of power generation capacity. However, on closer inspection, the contention that electricity is a single product needs to be qualified because of the variations of demand from minute to minute, day to day, and seasonally. As storage is impractical or expensive, plant that is required for meeting peak demands will be idle much of the remaining time. It is therefore important to distinguish between electricity produced at times of peak demand, and that produced when demand is low, because of the repercussions of demand variability on costs. This in turn may be reflected in pricing policy. Figure 8.1 shows typical diurnal demand variations for the CEGB.

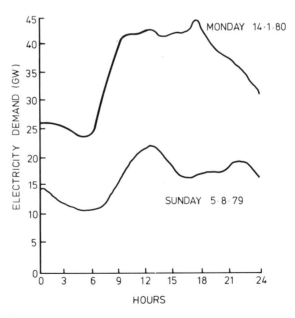

Fig. 8.1. CEGB electricity demand. *Source:* CEGB Annual Report and Accounts, 1979–1980

The duty of a plant in the system may be characterized by its average annual load factor, defined by:

load factor = electricity generated/electricity that would be generated if the plant were operated continuously at the (maximum) rated capacity.

The number of hours for which operation is possible is limited by breakdowns and the requirement for maintenance. An annual availability may therefore also be defined as follows:

availability = hours available for operation/8760

Plants which are operated at a high load factor, near to their availability, are termed base load plants. Other plants, whose duty is to assist in meeting the variations in demand and which operate at lower load factors are referred to as mid-merit or peaking plants.

A further complication is that there are a number of types of generating plant available. Some involve high construction costs, but modest running costs (for example, nuclear power stations), whereas others have low capital costs and high running costs (for example, gas turbines). A simple illustration of why different plant types are suited to particular uses is given in Fig. 8.2. This shows that plant A produces electricity at a lower average cost if the load

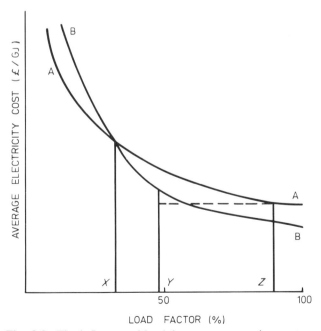

Fig. 8.2. The influence of load factor on generating costs

factor is below X, but that plant B is preferred above this value. X is known as the break-even load factor. Plant A has the lower running costs, and plant B the higher capital cost.

If the availability, Y, of plant types A and B is the same, it can be seen that type B would be preferred on the basis of cost. However, if only type B has this availability and type A has a higher availability (Z), type A would be preferred as, at their respective availabilities, the average cost of generation from the type A plant is lower.

This is a simple model which ignores differences in the duties of the plants, although if their roles are similar, it is a useful approach. However, a more detailed analysis requires the characteristics of the system to be taken into account.

From the above graphical model, the importance of discussions about plant availabilities may be appreciated. For example, the Science Policy Research Unit of Sussex University has suggested that the superior availability of CANDU nuclear reactors over PWRs may make the former attractive despite their higher capital costs. Similar considerations may apply to the comparison between fossil-fuelled plant and PWRs.

8.2.2 Supply and Demand

Meeting the demand for electricity involves operating plant at a variety of load factors.

The operating pattern of an individual plant is not that of the demand pattern, but the operations of a number of power stations are co-ordinated to supply the demand. The difference between the demand and the operating pattern of generating plant is illustrated in Fig. 8.3. In this representation of demand, only the proportion of time that the demand is greater than a specified level is given. The total quantity of electricity generated corresponds to the areas under the curves, which are equal because supply equals demand. Because the availability of plant is always less than 100%, the system capacity is greater than the maximum demand. A large proportion of the supply can be produced by power stations operating at base load. In the UK, it is estimated that about 60% of power is generated by base load plant. However, there remains a substantial supply at lower load factors. These lower load factor duties are often met by ageing plant that was originally designed for base load, although some plant may be purpose-built for low load factor duties.

It is interesting to note that the proportion of the supply which can be generated by base load plant increases as the availability decreases. This has

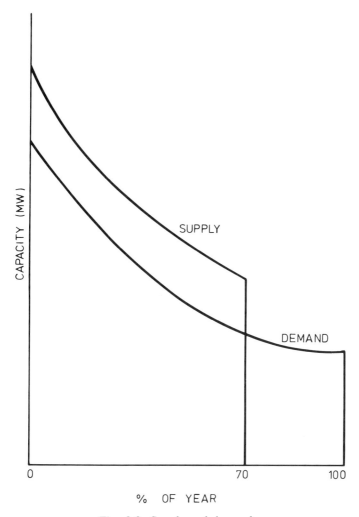

Fig. 8.3. Supply and demand

a surprising corollary; as the availability of a plant type approaches the load factor below which another technology is preferred, the proportion of overall generation provided by the plant type increases, until suddenly it is not required at all!

8.2.3 Pricing

As was noted earlier, for much of the day (even in the season of peak demand) some plant is available for generation but the electricity is not required. The installed capacity is sufficient to meet peak demands, so that an increase in demand at off-peak times can be accommodated without additional plant. Only the costs associated with running the plant for the extra period are incurred. These costs may be little more than the fuel cost. Such cost differences for demands occurring at different times may be reflected in pricing structures. In the UK, the most publicized of these is the off-peak domestic tariff. This is aimed at encouraging customers to carry out space heating and water heating overnight (when electricity demand is low), using storage heaters and highly insulated hot-water tanks. Although the actual costs òf providing off-peak electricity vary from hour to hour and day to day, the tariff system simplifies this to two costs.

8.2.4 Operating Plant

New generating plant is installed as warranted by demand growth and the retirement of old plant. However, once plant is built, the construction cost is not relevant to decisions about when it is operated, and therefore its annual average load factor. Only the running costs relative to those of other plants influence these decisions.

The generating plant in the system can be listed in order of operating cost. This list is called the merit order, and is the main consideration determining the duties of the various plants. In order to minimize the total annual cost of operating the system, those plants with the lowest running costs are operated in preference to (and therefore at a higher load factor than) those plants with higher running costs.

8.2.5 Choosing Plant

The position in the merit order and how this will change with time have to be taken into account when choosing new plant. If the life histories of the possible options are similar, it may be reasonable to compare the options without reference to the system of which they would be a part. For example, a discounted cash flow analysis of the costs of the options may be sufficient. Where the duties of the plants being compared are different, or become so in time, then their values for a given specific utility system must be determined.

One of the most important comparisons today in new generating plant investment is that between fossil-fuelled and nuclear plant. Nuclear plant often takes longer to build, introducing a substantial burden equivalent to an increase in capital costs. Once completed, however, a nuclear power station is likely to be required to operate on base load throughout its lifetime. By contrast, a fossil-fuelled plant may be expected to be relegated to lower load factor duties towards the end of its lifetime by the introduction of new low running cost plant (fossil or nuclear). A simple view of this is that the average cost of the electricity produced by a fossil plant is higher than would otherwise be expected. This view is correct, but can be misleading since, as a result of the relegation, the costs of the generating system *as a whole* are lower than they would have been without this change.

8.2.6 Uncertainty

Because power plant has a long life, a proper assessment of the cost implications of an investment option involves considering a long time period—usually at least thirty years. Over this period of time, it is necessary to estimate the demand, the technical performance of current and future plant, and fuel prices. Forecasting these factors is difficult, as evidenced by the marked failure of previous forecasts. Energy forecasting and modelling are discussed in more detail in Section 8.4, but it may be noted that uncertainty about the future affects decision making for energy utilization options in general.

A particularly important source of uncertainty in power plant ordering is the nature of the load–duration curve in the future. One approach to coping with this uncertainty would be to make a number of estimates of future demand and then to plan for the highest. This would probably ensure that the demand could be met, but may lead to expensive plant being under-utilized. Alternatively, the most likely case could be planned for, but power cuts or rationing may then be necessary if a different case occurs. Because of the importance of electricity in industrial and process applications, considerable costs could result from shortages in supply. Therefore, provision against under-estimation of demand may be seen as an appropriate insurance. But what sort of plant should the provision include?

An approach to answering this question is to test various plant selections against possible demands, and to choose the options with the lowest expected cost. The logic of this can be illustrated by using demand duration curves to represent the different forecasts of demand, see Figs. 8.4, 8.5 and 8.6. If these are weighted by their probabilities, they can be re-formed as an expected demand duration curve, shown in Fig. 8.7. A plant building plan to satisfy this

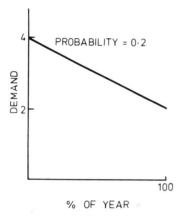

Fig. 8.4. High demand growth

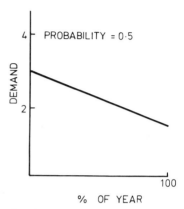

Fig. 8.5. Medium demand growth

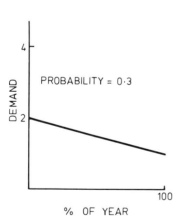

Fig. 8.6. Low demand growth

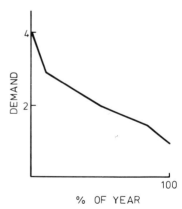

Fig. 8.7. The expected demand duration curve

would satisfy the peak demand even in the high case, but would have lower expected costs than a plan that best satisfied only the high case. This is because the improbability of the high case is reflected in the weighting, so that the insurance role of the plant is achieved by plant with a low capital cost.

8.3 OTHER ENERGY CARRIERS

The principal factor determining energy usage appears to be the level of economic activity (for example, measured by GDP per capita). Where industrial activity is low, sun, wind, and biomass can satisfy energy requirements. As industrialization proceeds, there is a tendency for life styles to change, so that energy intensive activities are introduced. For example, tumble driers are used to dry clothing rather than the wind.

In the case of the United States of America, high personal mobility, the widespread use of air conditioning, large scale industrial activity, and other factors combine to result in the highest national per capita energy consumption. The per capita energy consumptions of various countries are compared in Table 8.1.

The main energy carriers are electricity, coal (including lignite), oil, and natural gas. With the exception of electricity, these are also major items of international trade and together satisfy most of the energy demands in developed countries.

Although the electricity supply system has many features in common with that of other energy carriers, some important differences also exist. These are

Table 8.1 Per capita energy consumption (1978)

Country	GJ/capita
Brazil	23
Ethiopia	1
India	5
Italy	95
Japan	112
Mexico	41
Peru	19
Spain	70
Uganda	1
UK	152
USA	333
USSR	161
Zambia	14

Source: World Energy Supplies 1973–1978, United Nations, New York, 1979.

discussed in the present section, with particular reference to coal, oil, and natural gas.

8.3.1 Storage and Transport

Much of the complexity of electricity supplies arises because of the close coupling required between supply and demand (that is, variations in demand must be met by simultaneous variations in generation). For fossil fuels, relatively cheap storage allows supply and demand to be decoupled.

Seasonal storage of coal and oil allows mines and oilfields to produce at a steady rate, resulting in a high utilization of production capacity. To a lesser extent, natural gas is stored seasonally (in aquifers or depleted oil or gas fields). The costs of storage are not, however, negligible; coal and heating oil can be bought at lower prices in the summer than in the winter, to encourage consumers to undertake some storage themselves, and to reflect the holding costs of suppliers.

Demand variations over a day or a week are also met by storage. For coal and oil, this is normally storage at the users' premises, whereas for natural gas, gasholders, LNG plant, and the inherent storage capacity of the supply pipeline system all make contributions (particularly to meeting daily variations).

Transportation of fossil fuels is also generally less costly than for electricity. The oil industry has developed a global supply network, with tankers carrying up to 0.5 Mt, connecting the oilfields of the Middle East, Africa, and South America, with the markets of Europe, the Far East, and the USA. The international coal trade has also grown considerably over recent years, and has been the subject of a number of studies (for example, work by IEA Coal Research, Economic Assessment Service, London). Railways and lorries carry coal by land, and barges and colliers by water. Liquefied natural gas is shipped by sea, and pipelines carry natural gas increasingly large distances by land, from Mexico to the USA and soon from Siberia to Western Europe.

Industrialization began near deposits of iron ore and coal. Transport by wagon was too expensive to permit the movement of fuel over long distances. Even with the advent of canals and then the railways, industry generally remained sited near coal, rather than coal being brought to industry. The ease and low cost of transporting oil changed the established pattern of industrialization little, oil being brought to industry. The advantages of a developed industrial infrastructure and skilled workforce offset the relatively small transport costs involved.

If coal replaces oil as the principal traded fuel, it remains to be seen whether the higher costs of coal transport are sufficient to cause industry to relocate to be sited near coal reserves, or whether coal will be brought to the existing industrial areas.

8.3.2 Products

Although electricity can be sub-divided into peak and off-peak demands of various types, only in combined heat and power are different products made simultaneously. In the coal and oil industries, however, this is a common occurrence.

Crude oil is the raw material for a plethora of final products. It is distilled into various fractions, which in turn are transformed to other products by processes such as cracking and reforming. The products are diverse, ranging from power station fuel to petrol. Many of them are made to meet closely defined specifications, like petrol and derv. Different crude oils are suited to different types of refining processes, and give a different product slate. In the UK, for example, North Sea oil is exported while heavier Middle East crude oils are imported. As with oil, a wide range of coal types exist. Some coals are suitable for coke manufacture for the iron and steel industry, while others do not possess the characteristics that are necessary for the production of a satisfactory coke. Coal having a wide range of particle sizes and a relatively high ash content is used for power stations, whereas clean lump coal is used in small industrial boilers.

The production of a number of different products influences the economics of the supply system. The best mix of products depends not only on the type of feedstock available, but also on the processes which have been developed to transform this input and, most importantly, on the state of the markets for these products.

When several products are made together, the costs of manufacture may be reflected in the prices of the products in a number of ways. The relative prices of two products depend on the market requirements for these products, which in turn are determined by competition with other fuels and the balance of supply and demand. Naphtha is priced more highly than residual oil, because the market for the former is strong (it is a petrochemical feedstock and an important component and precursor of petrol), while that of the latter is weak (because of competition with coal). The market for naphtha is sufficiently strong that heavier products are upgraded to increase production. Costs are important to the extent that if, by blending procedures or other conversion processes, a 'product slate' of greater market value can be achieved without a proportionate increase in costs, the supplier modifies his operations to increase the supply of some products (e.g. naphtha) at the expense of others.

Ideally, the suppliers may be expected to recover their costs, thereby setting the average price of products. The differentials between the prices of products are set by the marginal cost of transformation, that is the additional cost of converting one product into another (taking into account the revenue lost from not having the product that is used as the feedstock in the transformation). Adjustments are made until no gain can be made by shifting output between the products. In reality, at least in oil refining, the ideal is not achieved. Some products are in surplus while others are in shortage, and adjustments to production are unable to keep pace with shifts in the demands for the various products. For example, the effects of the recession and oil price increases have resulted in a decline in the demand for fuel oil, while the demand for petrol has grown. This has been reflected in the relative prices of petrol and fuel oil, providing an incentive for the introduction of more upgrading (cracking) facilities than have been used previously in European oil refineries.

In the coal industry, the production from a single mine may supply a number of markets. In the UK, a deep mine frequently supplies coal for power generation, several grades for the industrial markets, and possibly some domestic coal. The output from the mine is processed by a coal preparation plant, which sizes the coal, removes some mineral matter, and blends the various streams to produce a range of final products of various size ranges and ash contents. More of the lower ash and higher market value grades could be produced, but this would involve washing more coal (at greater expense for the coal preparation plant), washing more severely (and therefore discarding more combustible material as waste with the mineral matter), or by reducing the quality (and therefore the value) of the power station fuel. Such changes in coal preparation practice may be required in response to a shift in the pattern of coal demand. In the UK, for example, the industrial market is expected to grow while eventually the power station market may possibly decline.

8.3.3 Interactions

There are few uses of energy for which there is no alternative means of meeting the same demand using a different fuel. For example, in steam raising (both for industrial processes and power generation) each of the fossil fuels may be used with the appropriate combustion equipment. At the other extreme, the use of oil in transport and petrochemicals cannot readily be replaced by coal used directly, although, even here, substitute liquid fuels may be made from coal given a sufficient economic incentive.

Because fuels may be substituted for one another, either directly or by inter-conversions, the choice of fuel for a particular final demand is determined by price. The demand for a fuel is therefore affected not only by its own price, but also by those of the other fuels which could meet the same demand.

However, price changes do not always result in an immediate and corresponding change in demand. Although the CEGB was able to reduce its consumption of oil almost immediately in response to oil price rises (by changing the relative positions of coal- and oil-fired stations in the merit order), substitution for the use of fuel oil in industry is slow because of the high cost of boiler plant and its long lifetime.

The rate of change in fuel usage can also vary from industry to industry. This may mean that relatively costly substitutions may be occurring while apparently less expensive ones remain to be made. For example, it is possible that, as natural gas supplies become depleted, a substitute for natural gas may be manufactured from coal and be used in industrial boilers, even although from a global viewpoint it may be less expensive to install new boilers to burn coal. This situation can arise because the different energy carriers require different amounts of investment at the point of final consumption. Since consumers tend to require shorter payback times for investment than the energy utilities, energy carriers that need a relatively low investment at the point of consumption (such as gas and electricity) may appear attractive. However, other energy carriers could offer a lower overall cost on a uniform rate of return.

8.4 MODELS

8.4.1 Why Models are Needed

Models are a means of coping with complexity. The purpose of a model is to provide a basis for choosing from the range of possible actions, when it is too difficult to assess the implications of the actions intuitively.

The planning and operation of the energy supply industries has long been the subject of modelling studies. Lead times in development of new supply sources, transportation and conversion processes are sufficiently long that the state of the demands which they are developed to serve can only be known approximately; demand forecasting models can assist here. The efficient matching of available supplies to demands is also difficult, and requires large linear programming models (for example, the operation of oil refineries is optimized to meet the prevailing market requirements in this way).

Much of the work on modelling energy systems has been directed towards the construction of computer models of various types. Probably the most relevant to energy policy are those models which predict the investments in new plant required in order to meet future energy needs. Such models are widely used by the energy industries and by governments, as a means of assessing investment programmes and R & D.

Everyone, one way or another will be affected by the consequences of these decisions. It may be that we will face needlessly high energy costs, because of over-investment in new plant to meet demands that will never rise. Alternatively, under-estimates of demand could lead to high prices to ration limited supplies. For example, a large modern power station takes ten years to plan and build, and the task of forecasting the relevant demand levels is therefore formidable. The choice of fuel depends on the costs of the various options over the lifetime of the plant, possibly up to forty years. Similar choices are

Table 8.2 Types of energy model

Functions	Type	Subject (separately or in combination)
(1) operations	(1) simulation	(1) macro-economic
(2) planning	(2) optimization	(2) energy demand
		(3) energy supply
		(4) transportation

faced in other energy industries. The development of a new coal-mine, gasfield, or oilfield similarly takes a long time and requires a large commitment of capital.

Models used in examining energy systems may be divided into a number of categories by function, type and subject, as shown in Table 8.2.

8.4.2 Functions of Models

Energy models usually have one of two functions; to assist operational decisions, or to plan new investments. Models for operational decisions concern choices about resource usage, but in the context of the available plant being fixed. Such models are often highly detailed and specific. The scheduling of power stations, or the running of compressors for the gas supply industry require detailed modelling to allow adequate supply to be maintained while costs are kept as low as possible. Systems effects are important in such models if a stable and reliable supply is to be maintained (for example, to avoid a cascade of electricity supply failures).

However, of more general interest are the models used in investment and R & D planning, and to which the remainder of this section is devoted. Ideally, models would assess fully all factors, and identify an optimum course of action. This is, however, impractical. The art of modelling is not the inclusion of everything that is relevant, but the exclusion of everything that is not critical. The reason is simple; the larger and more complex the model, the more difficult and expensive it is to understand, construct, maintain, and use. If a simple model is adequate, its simplicity is a virtue.

8.4.3 Types of Model

Planning models, as with other models, can be divided into two types:

(1) Simulation. These attempt to predict the probable outcome of events, given the existing market imperfections, institutional constraints, etc. This approach is often used to investigate options so that a course of action that is robust to probable outcomes can be selected.

(2) Optimization. These predict the outcome of events in an idealized, 'fully optimized' world. They are widely used by governments (and others) to explore the impact of national policy decisions, and to identify means of approaching optimum conditions more closely.

Short and medium term planning models are usually simulation models, often called 'what if...?' models. The model explores how *in practice* developments would occur. In a simulation, the choices made by consumers of energy supplies are typically treated by studying past responses to fuel prices and modelling these reactions by price and cross-price elasticities. The correspondence of these to the actual costs experienced by consumers under the different supply options is not important. Rational and irrational preferences are treated together, the resultant action being the prime concern.

The optimization models used in long-term energy planning are linear programmes (LPs). These represent the energy system as a set of requirements, and a set of potential ways of meeting these requirements. The model matches supplies to requirements using the rationale of a perfect market economy. Where there are limitations on the availability of a particular resource (e.g. oil), it is used where it has the most value. The model computes a price for the resource that rations the available supply to the demands. Other demands are satisfied with alternative resources (e.g. coal or natural gas).

Optimizations may be performed for a series of successive 'snapshot' years, or by integrating these into an overall description of the development of the energy system with time. These approaches are called time marching and time phased models, respectively. The former method may be done more cheaply, and probably reflects more accurately historic decision-making processes, but is likely to indicate inappropriate strategies if rapid changes in price or availabilities occur. The latter approach produces an optimal solution, but is more expensive to compute, and may be difficult to understand fully.

To describe in detail an energy system, for example a national energy economy, a large amount of information is required. Even when this information is available, it is often not all incorporated in an optimization model. The reason for this is that the solution of large LP models becomes increasingly expensive in computing resources and more difficult to interpret with size. This difficulty is normally resolved either by limiting the subject matter of the model, for example by looking at only one sector of demand, or by aggregation. Small detailed models have the disadvantage that they cannot readily be combined to form an overall picture. However, aggregation can make the model less realistic, and the solution liable to 'flipping' (that is, the phenomenon of small changes in costs or prices causing sweeping changes in the modelled activity, such as an entire sector like industrial heating changing from one fuel to another).

8.4.4 Subjects of Models

Three sorts of model are important in assessing energy options; macro-economic models, energy demand models, and energy supply models.

Macro-economic models set the context of energy developments. From a basis in economic theory and a description of the national (or regional, or global) economy, projections about fuel prices, interest rates, wage settlement rates, world trade, etc., the model derives conclusions about economic growth. This may be disaggregated to sectoral growth. An example is the EURECA macro-economic energy model developed for the EEC.

Energy demand models make projections of the quantities of useful energy that the various sectors of industry, commerce, and the domestic market will require. The output of a macro-economic model may provide the background of economic growth used to compute these requirements. Projections about fuel prices are also required. Simulation models are preferred for this function.

Energy supply models match the supplies of energy to the energy demands, using appropriate combustion or utilization processes. Both simulation and optimization models are used for this function. Information on the prices of fuels, cost of equipment, the available processes and their efficiencies, limits on the quantities of fuels and possible rates of substitution are all inputs to such models. The output is the pattern of energy supplies that will meet the projected demands.

The EEC, for example, has an energy supply modelling capacity called EFOM12C, that has both simulation and optimization options. The International Energy Agency also have an energy supply LP (MARKAL).

Some supply models are concerned with energy trade and are basically transportation models. The Economic Assessment Service of IEA Coal Research are developing a coal trade model to examine the future pattern of coal flows.

8.4.5 Uncertainty

Energy demands and international energy prices will depend on the levels of industrial activity, technical developments, and the choices that people and companies make. It is not practical to model all activities that influence the events of interest, and therefore a model must take some factors as given. These factors are know as 'exogenous'. Typically these exogenous factors are economic growth and international energy prices. Unfortunately, these factors are both important and subject to major uncertainties. For example, current estimates of oil prices at the end of the century range from present levels to more than a doubling. International coal and gas prices are subject to similar uncertainties. The future impact of nuclear power arguably depends as much on public opinion as technical or economic merit.

Given such a wide range of possible scenarios, no one strategy of investment and R & D is likely to be suitable for every eventuality. One method of choosing a strategy would be to compute the costs of meeting energy requirements (or failing to, in some cases) under a particular strategy, for all reasonable scenarios. These costs could then be compared for each strategy and the best one selected. Operational Research offers a number of means of selecting the 'best' strategy. A simple method is to choose the strategy with the lowest expected costs (where expected means the cost under each outcome weighted by an estimate of its probability). Although uncertainty is recognized as an important component of energy planning, often it is difficult to assess the probabilities of the various scenarios of interest, so that the above methods cannot properly be applied.

An alternative approach is to choose the action which is least susceptible to an undesirable outcome. For example, utilities may adopt a policy of not being dependent on a single fuel. From a national viewpoint, it is also important that policy should be robust to uncertainties in supplies of a particular fuel type.

For electricity, its premium uses are such that the costs of shortage to society may greatly outweigh the costs of an equal degree of over-capacity. Lucas[2] has shown that the social costs which may arise from the inadequate provision of energy supplies should have an important influence on policy. The contribution of developing indigenous energy supplies, adopting diverse and relatively reliable sources of supply, and conservation to reduce the vulnerability of energy supplies to uncertainty has a significant value to the economy.

Models can explore the consequences of our knowledge and beliefs about the future. They may direct our attention to inconsistencies in our ideas, and omissions or defects in our plans. But their results are, at best, only as good as the information put into them.

8.5 REFERENCES

1. Navarro, P. Our stake in the electricity utility's dilemma. *Harvard Business Review,* May–June 1982.
2. Lucas, N., and Papaconstantinou, D. Energy planning under uncertainty. *Energy Policy,* June 1982.

Index